Selected Chapters from

Anton: Elementary Linear Algebra, 8th Edition

A Wiley Canada Custom Publication for

University of Calgary

MATH 211

Anton: Elementary Linear Algebra, 8th Edition (ISBN: 9780471170556)

Cover Photo Credit: ©Joao Paulo/The Image Bank

Marketing Manager:Anne-Marie Seymour
Custom Coordinator: Sara Tinteri

Printed and bound in the United States of America

John Wiley & Sons Canada, Ltd
6045 Freemont Blvd.
Mississauga, Ontario
L5R 4J3
Visit our website at: www.wiley.ca

Elementary Linear Algebra

Elementary Linear Algebra

Eighth Edition

HOWARD ANTON

Drexel University

John Wiley & Sons, Inc.
New York Chichester Weinheim Brisbane Toronto Singapore

MATHEMATICS EDITOR Barbara Holland
PRODUCTION EDITOR Ken Santor
ASSISTANT EDITOR Roseann Zappia
DESIGNER Dawn L. Stanley
PHOTO EDITOR Hilary Newman
ILLUSTRATION COORDINATOR Techsetters/Sigmund Malinowski
COVER PHOTO ©Joao Paulo/The Image Bank

This book was set in Times New Roman PS by Techsetters, Inc. and printed and bound by Von Hoffman Press, Inc. The cover was printed by Lehigh Press, Inc.

This book is printed on acid-free paper.♾

The paper in this book was manufactured by a mill whose forest management programs include sustained yield harvesting of its timberlands. Sustained yield harvesting principles ensure that the numbers of trees cut each year does not exceed the amount of new growth.

Library of Congress Cataloging-in-Publication Data:
Anton, Howard.
Elementary linear algebra/Howard Anton.—8th ed.
p. cm.
Includes index.
ISBN 0-471-17055-0 (cl.: alk. paper)
1. Algebras, Linear. I. Title.

QA184.A57 2000b
512′.5 21—dc21 99-044926

Printed in the United States of America

10 9 8 7 6 5 4 3 2 1

1

Systems of Linear Equations and Matrices

Chapter Contents

INTRODUCTION: Information in science and mathematics is often organized into rows and columns to form rectangular arrays, called "matrices" (plural of "matrix"). Matrices are often tables of numerical data that arise from physical observations, but they also occur in various mathematical contexts. For example, we shall see in this chapter that to solve a system of equations such as

$$\begin{aligned} 5x + y &= 3 \\ 2x - y &= 4 \end{aligned}$$

all of the information required for the solution is embodied in the matrix

$$\begin{bmatrix} 5 & 1 & 3 \\ 2 & -1 & 4 \end{bmatrix}$$

and that the solution can be obtained by performing appropriate operations on this matrix. This is particularly important in developing computer programs to solve systems of linear equations because computers are well suited for manipulating arrays of numerical information. However, matrices are not simply a notational tool for solving systems of equations; they can be viewed as mathematical objects in their own right, and there is a rich and important theory associated with them that has a wide variety of applications. In this chapter we will begin the study of matrices.

1.1 INTRODUCTION TO SYSTEMS OF LINEAR EQUATIONS

Systems of linear algebraic equations and their solutions constitute one of the major topics studied in the course known as "linear algebra." In this first section we shall introduce some basic terminology and discuss a method for solving such systems.

Linear Equations Any straight line in the xy-plane can be represented algebraically by an equation of the form

$$a_1x + a_2y = b$$

where a_1, a_2, and b are real constants and a_1 and a_2 are not both zero. An equation of this form is called a linear equation in the variables x and y. More generally, we define a ***linear equation*** in the n variables $x_1, x_2, \ldots, x_n$ to be one that can be expressed in the form

$$a_1x_1 + a_2x_2 + \cdots + a_nx_n = b$$

where $a_1, a_2, \ldots, a_n$, and b are real constants. The variables in a linear equation are sometimes called ***unknowns***.

EXAMPLE 1 Linear Equations

The equations

$$x + 3y = 7, \quad y = \tfrac{1}{2}x + 3z + 1, \quad \text{and} \quad x_1 - 2x_2 - 3x_3 + x_4 = 7$$

are linear. Observe that a linear equation does not involve any products or roots of variables. All variables occur only to the first power and do not appear as arguments for trigonometric, logarithmic, or exponential functions. The equations

$$x + 3\sqrt{y} = 5, \quad 3x + 2y - z + xz = 4, \quad \text{and} \quad y = \sin x$$

are *not* linear. ◆

A ***solution*** of a linear equation $a_1x_1 + a_2x_2 + \cdots + a_nx_n = b$ is a sequence of n numbers $s_1, s_2, \ldots, s_n$ such that the equation is satisfied when we substitute $x_1 = s_1$, $x_2 = s_2, \ldots, x_n = s_n$. The set of all solutions of the equation is called its ***solution set*** or sometimes the ***general solution*** of the equation.

EXAMPLE 2 Finding a Solution Set

Find the solution set of (a) $4x - 2y = 1$, and (b) $x_1 - 4x_2 + 7x_3 = 5$.

Solution (a). To find solutions of (a), we can assign an arbitrary value to x and solve for y, or choose an arbitrary value for y and solve for x. If we follow the first approach and assign x an arbitrary value t, we obtain

$$x = t, \qquad y = 2t - \tfrac{1}{2}$$

These formulas describe the solution set in terms of an arbitrary number t, called a ***parameter***. Particular numerical solutions can be obtained by substituting specific values for t. For example, $t = 3$ yields the solution $x = 3$, $y = \frac{11}{2}$; and $t = -\frac{1}{2}$ yields the solution $x = -\frac{1}{2}$, $y = -\frac{3}{2}$.

If we follow the second approach and assign y the arbitrary value t, we obtain

$$x = \tfrac{1}{2}t + \tfrac{1}{4}, \qquad y = t$$

Although these formulas are different from those obtained above, they yield the same solution set as t varies over all possible real numbers. For example, the previous formulas gave the solution $x = 3$, $y = \frac{11}{2}$ when $t = 3$, while the formulas immediately above yield that solution when $t = \frac{11}{2}$.

Solution (b). To find the solution set of (b) we can assign arbitrary values to any two variables and solve for the third variable. In particular, if we assign arbitrary values s and t to x_2 and x_3, respectively, and solve for x_1, we obtain

$$x_1 = 5 + 4s - 7t, \qquad x_2 = s, \qquad x_3 = t$$ ◆

Linear Systems A finite set of linear equations in the variables $x_1, x_2, \ldots, x_n$ is called a ***system of linear equations*** or a ***linear system***. A sequence of numbers $s_1, s_2, \ldots, s_n$ is called a ***solution*** of the system if $x_1 = s_1, x_2 = s_2, \ldots, x_n = s_n$ is a solution of every equation in the system. For example, the system

$$\begin{aligned} 4x_1 - x_2 + 3x_3 &= -1 \\ 3x_1 + x_2 + 9x_3 &= -4 \end{aligned}$$

has the solution $x_1 = 1$, $x_2 = 2$, $x_3 = -1$ since these values satisfy both equations. However, $x_1 = 1$, $x_2 = 8$, $x_3 = 1$ is not a solution since these values satisfy only the first of the two equations in the system.

Not all systems of linear equations have solutions. For example, if we multiply the second equation of the system

$$\begin{aligned} x + y &= 4 \\ 2x + 2y &= 6 \end{aligned}$$

by $\frac{1}{2}$, it becomes evident that there are no solutions since the resulting equivalent system

$$\begin{aligned} x + y &= 4 \\ x + y &= 3 \end{aligned}$$

has contradictory equations.

A system of equations that has no solutions is said to be ***inconsistent***; if there is at least one solution of the system, it is called ***consistent***. To illustrate the possibilities that can occur in solving systems of linear equations, consider a general system of two linear equations in the unknowns x and y:

$$\begin{aligned} a_1x + b_1y &= c_1 \quad (a_1, b_1 \text{ not both zero}) \\ a_2x + b_2y &= c_2 \quad (a_2, b_2 \text{ not both zero}) \end{aligned}$$

The graphs of these equations are lines; call them l_1 and l_2. Since a point (x, y) lies on a line if and only if the numbers x and y satisfy the equation of the line, the solutions of the system of equations correspond to points of intersection of l_1 and l_2. There are three possibilities illustrated in Figure 1.1.1:

- The lines l_1 and l_2 may be parallel, in which case there is no intersection and consequently no solution to the system.

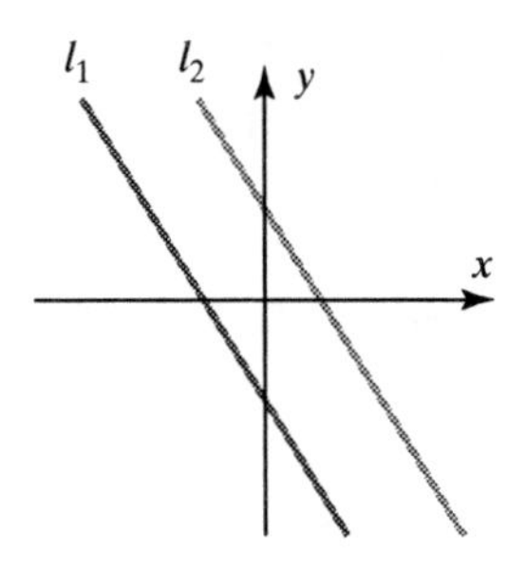

(a) No solution

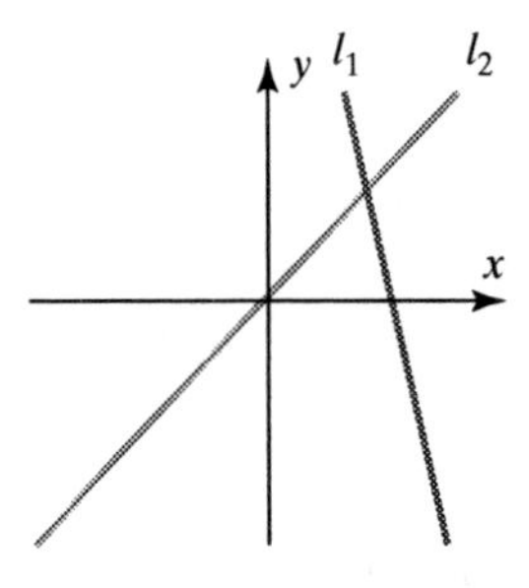

(b) One solution

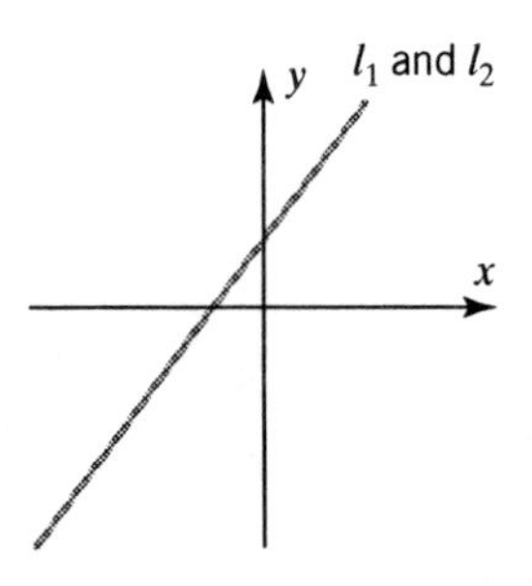

(c) Infinitely many solutions

Figure 1.1.1

- The lines l_1 and l_2 may intersect at only one point, in which case the system has exactly one solution.
- The lines l_1 and l_2 may coincide, in which case there are infinitely many points of intersection and consequently infinitely many solutions to the system.

Although we have considered only two equations with two unknowns here, we will show later that the same three possibilities hold for arbitrary linear systems:

Every system of linear equations has either no solutions, exactly one solution, or infinitely many solutions.

An arbitrary system of m linear equations in n unknowns can be written as

$$\begin{aligned} a_{11}x_1 &+ a_{12}x_2 + \cdots + a_{1n}x_n = b_1 \\ a_{21}x_1 &+ a_{22}x_2 + \cdots + a_{2n}x_n = b_2 \\ &\vdots \\ a_{m1}x_1 &+ a_{m2}x_2 + \cdots + a_{mn}x_n = b_m \end{aligned}$$

where $x_1, x_2, \ldots, x_n$ are the unknowns and the subscripted a's and b's denote constants. For example, a general system of three linear equations in four unknowns can be written as

$$\begin{aligned} a_{11}x_1 + a_{12}x_2 + a_{13}x_3 + a_{14}x_4 &= b_1 \\ a_{21}x_1 + a_{22}x_2 + a_{23}x_3 + a_{24}x_4 &= b_2 \\ a_{31}x_1 + a_{32}x_2 + a_{33}x_3 + a_{34}x_4 &= b_3 \end{aligned}$$

The double subscripting on the coefficients of the unknowns is a useful device that is used to specify the location of the coefficient in the system. The first subscript on the coefficient a_{ij} indicates the equation in which the coefficient occurs, and the second subscript indicates which unknown it multiplies. Thus, a_{12} is in the first equation and multiplies unknown x_2.

Augmented Matrices If we mentally keep track of the location of the +'s, the x's, and the =’s, a system of m linear equations in n unknowns can be abbreviated by writing only the rectangular array of numbers:

$$\begin{bmatrix} a_{11} & a_{12} & \cdots & a_{1n} & b_1 \\ a_{21} & a_{22} & \cdots & a_{2n} & b_2 \\ \vdots & \vdots & & \vdots & \vdots \\ a_{m1} & a_{m2} & \cdots & a_{mn} & b_m \end{bmatrix}$$

This is called the ***augmented matrix*** for the system. (The term *matrix* is used in mathematics to denote a rectangular array of numbers. Matrices arise in many contexts, which we will consider in more detail in later sections.) For example, the augmented matrix for the system of equations

$$\begin{aligned} x_1 + x_2 + 2x_3 &= 9 \\ 2x_1 + 4x_2 - 3x_3 &= 1 \\ 3x_1 + 6x_2 - 5x_3 &= 0 \end{aligned}$$

is

$$\begin{bmatrix} 1 & 1 & 2 & 9 \\ 2 & 4 & -3 & 1 \\ 3 & 6 & -5 & 0 \end{bmatrix}$$

REMARK. When constructing an augmented matrix, the unknowns must be written in the same order in each equation and the constants must be on the right.

The basic method for solving a system of linear equations is to replace the given system by a new system that has the same solution set but which is easier to solve. This new system is generally obtained in a series of steps by applying the following three types of operations to eliminate unknowns systematically.

1. Multiply an equation through by a nonzero constant.
2. Interchange two equations.
3. Add a multiple of one equation to another.

Since the rows (horizontal lines) of an augmented matrix correspond to the equations in the associated system, these three operations correspond to the following operations on the rows of the augmented matrix.

1. Multiply a row through by a nonzero constant.
2. Interchange two rows.
3. Add a multiple of one row to another row.

Elementary Row Operations These are called ***elementary row operations***. The following example illustrates how these operations can be used to solve systems of linear equations. Since a systematic procedure for finding solutions will be derived in the next section, it is not necessary to worry about how the steps in this example were selected. The main effort at this time should be devoted to understanding the computations and the discussion.

EXAMPLE 3 Using Elementary Row Operations

In the left column below we solve a system of linear equations by operating on the equations in the system, and in the right column we solve the same system by operating on the rows of the augmented matrix.

$$\begin{aligned} x + y + 2z &= 9 \\ 2x + 4y - 3z &= 1 \\ 3x + 6y - 5z &= 0 \end{aligned} \qquad \begin{bmatrix} 1 & 1 & 2 & 9 \\ 2 & 4 & -3 & 1 \\ 3 & 6 & -5 & 0 \end{bmatrix}$$

Add -2 times the first equation to the second to obtain

Add -2 times the first row to the second to obtain

$$\begin{aligned} x + y + 2z &= 9 \\ 2y - 7z &= -17 \\ 3x + 6y - 5z &= 0 \end{aligned} \qquad \begin{bmatrix} 1 & 1 & 2 & 9 \\ 0 & 2 & -7 & -17 \\ 3 & 6 & -5 & 0 \end{bmatrix}$$

Add -3 times the first equation to the third to obtain

Add -3 times the first row to the third to obtain

$$\begin{aligned} x + y + 2z &= 9 \\ 2y - 7z &= -17 \\ 3y - 11z &= -27 \end{aligned} \qquad \begin{bmatrix} 1 & 1 & 2 & 9 \\ 0 & 2 & -7 & -17 \\ 0 & 3 & -11 & -27 \end{bmatrix}$$

Multiply the second equation by $\frac{1}{2}$ to obtain

$$\begin{aligned} x + y + 2z &= 9 \\ y - \tfrac{7}{2}z &= -\tfrac{17}{2} \\ 3y - 11z &= -27 \end{aligned}$$

Multiply the second row by $\frac{1}{2}$ to obtain

$$\begin{bmatrix} 1 & 1 & 2 & 9 \\ 0 & 1 & -\frac{7}{2} & -\frac{17}{2} \\ 0 & 3 & -11 & -27 \end{bmatrix}$$

Add -3 times the second equation to the third to obtain

$$\begin{aligned} x + y + 2z &= 9 \\ y - \tfrac{7}{2}z &= -\tfrac{17}{2} \\ -\tfrac{1}{2}z &= -\tfrac{3}{2} \end{aligned}$$

Add -3 times the second row to the third to obtain

$$\begin{bmatrix} 1 & 1 & 2 & 9 \\ 0 & 1 & -\frac{7}{2} & -\frac{17}{2} \\ 0 & 0 & -\frac{1}{2} & -\frac{3}{2} \end{bmatrix}$$

Multiply the third equation by -2 to obtain

$$\begin{aligned} x + y + 2z &= 9 \\ y - \tfrac{7}{2}z &= -\tfrac{17}{2} \\ z &= 3 \end{aligned}$$

Multiply the third row by -2 to obtain

$$\begin{bmatrix} 1 & 1 & 2 & 9 \\ 0 & 1 & -\frac{7}{2} & -\frac{17}{2} \\ 0 & 0 & 1 & 3 \end{bmatrix}$$

Add -1 times the second equation to the first to obtain

$$\begin{aligned} x \qquad + \tfrac{11}{2}z &= \tfrac{35}{2} \\ y - \tfrac{7}{2}z &= -\tfrac{17}{2} \\ z &= 3 \end{aligned}$$

Add -1 times the second row to the first to obtain

$$\begin{bmatrix} 1 & 0 & \frac{11}{2} & \frac{35}{2} \\ 0 & 1 & -\frac{7}{2} & -\frac{17}{2} \\ 0 & 0 & 1 & 3 \end{bmatrix}$$

Add $-\frac{11}{2}$ times the third equation to the first and $\frac{7}{2}$ times the third equation to the second to obtain

$$\begin{aligned} x \qquad &= 1 \\ y \quad &= 2 \\ z &= 3 \end{aligned}$$

Add $-\frac{11}{2}$ times the third row to the first and $\frac{7}{2}$ times the third row to the second to obtain

$$\begin{bmatrix} 1 & 0 & 0 & 1 \\ 0 & 1 & 0 & 2 \\ 0 & 0 & 1 & 3 \end{bmatrix}$$

The solution $x = 1$, $y = 2$, $z = 3$ is now evident. ◆

Exercise Set 1.1

1. Which of the following are linear equations in x_1, x_2, and x_3?

(a) $x_1 + 5x_2 - \sqrt{2}x_3 = 1$ (b) $x_1 + 3x_2 + x_1x_3 = 2$ (c) $x_1 = -7x_2 + 3x_3$
(d) $x_1^{-2} + x_2 + 8x_3 = 5$ (e) $x_1^{3/5} - 2x_2 + x_3 = 4$ (f) $\pi x_1 - \sqrt{2}x_2 + \frac{1}{3}x_3 = 7^{1/3}$

2. Given that k is a constant, which of the following are linear equations?

(a) $x_1 - x_2 + x_3 = \sin k$ (b) $kx_1 - \dfrac{1}{k}x_2 = 9$ (c) $2^k x_1 + 7x_2 - x_3 = 0$

3. Find the solution set of each of the following linear equations.

(a) $7x - 5y = 3$ (b) $3x_1 - 5x_2 + 4x_3 = 7$
(c) $-8x_1 + 2x_2 - 5x_3 + 6x_4 = 1$ (d) $3v - 8w + 2x - y + 4z = 0$

4. Find the augmented matrix for each of the following systems of linear equations.

(a) $\begin{aligned} 3x_1 - 2x_2 &= -1 \\ 4x_1 + 5x_2 &= 3 \\ 7x_1 + 3x_2 &= 2 \end{aligned}$ (b) $\begin{aligned} 2x_1 \qquad + 2x_3 &= 1 \\ 3x_1 - x_2 + 4x_3 &= 7 \\ 6x_1 + x_2 - x_3 &= 0 \end{aligned}$ (c) $\begin{aligned} x_1 + 2x_2 \qquad - x_4 + x_5 &= 1 \\ 3x_2 + x_3 \qquad - x_5 &= 2 \\ x_3 + 7x_4 \qquad &= 1 \end{aligned}$ (d) $\begin{aligned} x_1 \qquad &= 1 \\ x_2 \quad &= 2 \\ x_3 &= 3 \end{aligned}$

5. Find a system of linear equations corresponding to the augmented matrix.

(a) $\begin{bmatrix} 2 & 0 & 0 \\ 3 & -4 & 0 \\ 0 & 1 & 1 \end{bmatrix}$ (b) $\begin{bmatrix} 3 & 0 & -2 & 5 \\ 7 & 1 & 4 & -3 \\ 0 & -2 & 1 & 7 \end{bmatrix}$

(c) $\begin{bmatrix} 7 & 2 & 1 & -3 & 5 \\ 1 & 2 & 4 & 0 & 1 \end{bmatrix}$ (d) $\begin{bmatrix} 1 & 0 & 0 & 0 & 7 \\ 0 & 1 & 0 & 0 & -2 \\ 0 & 0 & 1 & 0 & 3 \\ 0 & 0 & 0 & 1 & 4 \end{bmatrix}$

6. (a) Find a linear equation in the variables x and y that has the general solution $x = 5 + 2t$, $y = t$.

(b) Show that $x = t$, $y = \frac{1}{2}t - \frac{5}{2}$ is also the general solution of the equation in part (a).

7. The curve $y = ax^2 + bx + c$ shown in the accompanying figure passes through the points (x_1, y_1), (x_2, y_2), and (x_3, y_3). Show that the coefficients a, b, and c are a solution of the system of linear equations whose augmented matrix is

$$\begin{bmatrix} x_1^2 & x_1 & 1 & y_1 \\ x_2^2 & x_2 & 1 & y_2 \\ x_3^2 & x_3 & 1 & y_3 \end{bmatrix}$$

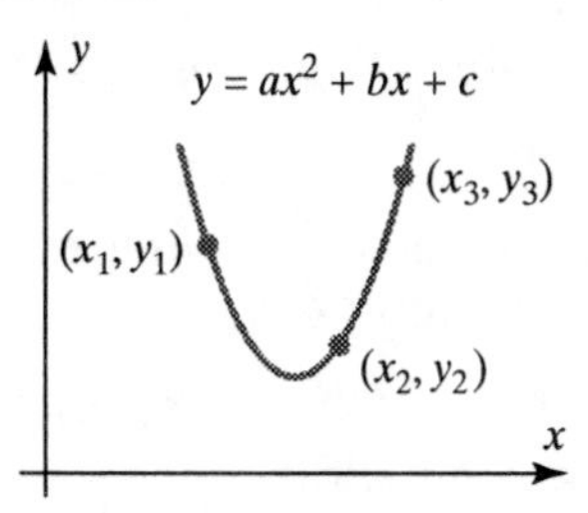

Figure Ex-7

8. Consider the system of equations

$$\begin{aligned} x + y + 2z &= a \\ x \quad\;\; + z &= b \\ 2x + y + 3z &= c \end{aligned}$$

Show that for this system to be consistent, the constants a, b, and c must satisfy $c = a + b$.

9. Show that if the linear equations $x_1 + kx_2 = c$ and $x_1 + lx_2 = d$ have the same solution set, then the equations are identical.

Discussion and Discovery

10. For which value(s) of the constant k does the system

$$\begin{aligned} x - \;\, y &= 3 \\ 2x - 2y &= k \end{aligned}$$

have no solutions? Exactly one solution? Infinitely many solutions? Explain your reasoning.

11. Consider the system of equations

$$\begin{aligned} ax + by &= k \\ cx + dy &= l \\ ex + fy &= m \end{aligned}$$

What can you say about the relative positions of the lines $ax + by = k$, $cx + dy = l$, and $ex + fy = m$ when

(a) the system has no solutions;

(b) the system has exactly one solution;

(c) the system has infinitely many solutions?

12. If the system of equations in Exercise 11 is consistent, explain why at least one equation can be discarded from the system without altering the solution set.

13. If $k = l = m = 0$ in Exercise 11, explain why the system must be consistent. What can be said about the point of intersection of the three lines if the system has exactly one solution?

1.2 GAUSSIAN ELIMINATION

In this section we shall develop a systematic procedure for solving systems of linear equations. The procedure is based on the idea of reducing the augmented matrix of a system to another augmented matrix that is simple enough that the solution of the system can be found by inspection.

Echelon Forms In Example 3 of the last section, we solved a linear system in the unknowns x, y, and z by reducing the augmented matrix to the form

$$\begin{bmatrix} 1 & 0 & 0 & 1 \\ 0 & 1 & 0 & 2 \\ 0 & 0 & 1 & 3 \end{bmatrix}$$

from which the solution $x = 1$, $y = 2$, $z = 3$ became evident. This is an example of a matrix that is in ***reduced row-echelon form***. To be of this form a matrix must have the following properties:

1. If a row does not consist entirely of zeros, then the first nonzero number in the row is a 1. We call this a ***leading* 1**.
2. If there are any rows that consist entirely of zeros, then they are grouped together at the bottom of the matrix.
3. In any two successive rows that do not consist entirely of zeros, the leading 1 in the lower row occurs farther to the right than the leading 1 in the higher row.
4. Each column that contains a leading 1 has zeros everywhere else.

A matrix that has the first three properties is said to be in ***row-echelon form***. (Thus, a matrix in reduced row-echelon form is of necessity in row-echelon form, but not conversely.)

EXAMPLE 1 Row-Echelon and Reduced Row-Echelon Form

The following matrices are in reduced row-echelon form.

$$\begin{bmatrix} 1 & 0 & 0 & 4 \\ 0 & 1 & 0 & 7 \\ 0 & 0 & 1 & -1 \end{bmatrix}, \quad \begin{bmatrix} 1 & 0 & 0 \\ 0 & 1 & 0 \\ 0 & 0 & 1 \end{bmatrix}, \quad \begin{bmatrix} 0 & 1 & -2 & 0 & 1 \\ 0 & 0 & 0 & 1 & 3 \\ 0 & 0 & 0 & 0 & 0 \\ 0 & 0 & 0 & 0 & 0 \end{bmatrix}, \quad \begin{bmatrix} 0 & 0 \\ 0 & 0 \end{bmatrix}$$

The following matrices are in row-echelon form.

$$\begin{bmatrix} 1 & 4 & -3 & 7 \\ 0 & 1 & 6 & 2 \\ 0 & 0 & 1 & 5 \end{bmatrix}, \quad \begin{bmatrix} 1 & 1 & 0 \\ 0 & 1 & 0 \\ 0 & 0 & 0 \end{bmatrix}, \quad \begin{bmatrix} 0 & 1 & 2 & 6 & 0 \\ 0 & 0 & 1 & -1 & 0 \\ 0 & 0 & 0 & 0 & 1 \end{bmatrix}$$

We leave it for you to confirm that each of the matrices in this example satisfies all of the requirements for its stated form. ♦

EXAMPLE 2 More on Row-Echelon and Reduced Row-Echelon Form

As the last example illustrates, a matrix in row-echelon form has zeros below each leading 1, whereas a matrix in reduced row-echelon form has zeros below *and above* each leading 1. Thus, with any real numbers substituted for the $*$'s, all matrices of the following types are in row-echelon form:

$$\begin{bmatrix} 1 & * & * & * \\ 0 & 1 & * & * \\ 0 & 0 & 1 & * \\ 0 & 0 & 0 & 1 \end{bmatrix}, \quad \begin{bmatrix} 1 & * & * & * \\ 0 & 1 & * & * \\ 0 & 0 & 1 & * \\ 0 & 0 & 0 & 0 \end{bmatrix},$$

$$\begin{bmatrix} 1 & * & * & * \\ 0 & 1 & * & * \\ 0 & 0 & 0 & 0 \\ 0 & 0 & 0 & 0 \end{bmatrix}, \quad \begin{bmatrix} 0 & 1 & * & * & * & * & * & * & * & * \\ 0 & 0 & 0 & 1 & * & * & * & * & * & * \\ 0 & 0 & 0 & 0 & 1 & * & * & * & * & * \\ 0 & 0 & 0 & 0 & 0 & 1 & * & * & * & * \\ 0 & 0 & 0 & 0 & 0 & 0 & 0 & 0 & 1 & * \end{bmatrix}$$

Moreover, all matrices of the following types are in reduced row-echelon form:

$$\begin{bmatrix} 1 & 0 & 0 & 0 \\ 0 & 1 & 0 & 0 \\ 0 & 0 & 1 & 0 \\ 0 & 0 & 0 & 1 \end{bmatrix}, \quad \begin{bmatrix} 1 & 0 & 0 & * \\ 0 & 1 & 0 & * \\ 0 & 0 & 1 & * \\ 0 & 0 & 0 & 0 \end{bmatrix},$$

$$\begin{bmatrix} 1 & 0 & * & * \\ 0 & 1 & * & * \\ 0 & 0 & 0 & 0 \\ 0 & 0 & 0 & 0 \end{bmatrix}, \quad \begin{bmatrix} 0 & 1 & * & 0 & 0 & 0 & * & * & 0 & * \\ 0 & 0 & 0 & 1 & 0 & 0 & * & * & 0 & * \\ 0 & 0 & 0 & 0 & 1 & 0 & * & * & 0 & * \\ 0 & 0 & 0 & 0 & 0 & 1 & * & * & 0 & * \\ 0 & 0 & 0 & 0 & 0 & 0 & 0 & 0 & 1 & * \end{bmatrix}$$

◆

If, by a sequence of elementary row operations, the augmented matrix for a system of linear equations is put in reduced row-echelon form, then the solution set of the system will be evident by inspection or after a few simple steps. The next example illustrates this situation.

EXAMPLE 3 Solutions of Four Linear Systems

Suppose that the augmented matrix for a system of linear equations has been reduced by row operations to the given reduced row-echelon form. Solve the system.

(a) $\begin{bmatrix} 1 & 0 & 0 & 5 \\ 0 & 1 & 0 & -2 \\ 0 & 0 & 1 & 4 \end{bmatrix}$ (b) $\begin{bmatrix} 1 & 0 & 0 & 4 & -1 \\ 0 & 1 & 0 & 2 & 6 \\ 0 & 0 & 1 & 3 & 2 \end{bmatrix}$

(c) $\begin{bmatrix} 1 & 6 & 0 & 0 & 4 & -2 \\ 0 & 0 & 1 & 0 & 3 & 1 \\ 0 & 0 & 0 & 1 & 5 & 2 \\ 0 & 0 & 0 & 0 & 0 & 0 \end{bmatrix}$ (d) $\begin{bmatrix} 1 & 0 & 0 & 0 \\ 0 & 1 & 2 & 0 \\ 0 & 0 & 0 & 1 \end{bmatrix}$

Solution (a). The corresponding system of equations is

$$\begin{aligned} x_1 \qquad\qquad &= 5 \\ x_2 \qquad &= -2 \\ x_3 &= 4 \end{aligned}$$

By inspection, $x_1 = 5$, $x_2 = -2$, $x_3 = 4$.

Solution (b). The corresponding system of equations is

$$\begin{aligned} x_1 \quad\quad\quad + 4x_4 &= -1 \\ x_2 \quad\quad + 2x_4 &= 6 \\ x_3 + 3x_4 &= 2 \end{aligned}$$

Since x_1, x_2, and x_3 correspond to leading 1's in the augmented matrix, we call them ***leading variables***. The nonleading variables (in this case x_4) are called ***free variables***. Solving for the leading variables in terms of the free variable gives

$$\begin{aligned} x_1 &= -1 - 4x_4 \\ x_2 &= 6 - 2x_4 \\ x_3 &= 2 - 3x_4 \end{aligned}$$

From this form of the equations we see that the free variable x_4 can be assigned an arbitrary value, say t, which then determines the values of the leading variables x_1, x_2, and x_3. Thus there are infinitely many solutions, and the general solution is given by the formulas

$$x_1 = -1 - 4t, \quad x_2 = 6 - 2t, \quad x_3 = 2 - 3t, \quad x_4 = t$$

Solution (c). The row of zeros leads to the equation $0x_1 + 0x_2 + 0x_3 + 0x_4 + 0x_5 = 0$, which places no restrictions on the solutions (why?). Thus, we can omit this equation and write the corresponding system as

$$\begin{aligned} x_1 + 6x_2 \quad\quad\quad + 4x_5 &= -2 \\ x_3 \quad\quad + 3x_5 &= 1 \\ x_4 + 5x_5 &= 2 \end{aligned}$$

Here the leading variables are x_1, x_3, and x_4, and the free variables are x_2 and x_5. Solving for the leading variables in terms of the free variables gives

$$\begin{aligned} x_1 &= -2 - 6x_2 - 4x_5 \\ x_3 &= 1 - 3x_5 \\ x_4 &= 2 - 5x_5 \end{aligned}$$

Since x_5 can be assigned an arbitrary value, t, and x_2 can be assigned an arbitrary value, s, there are infinitely many solutions. The general solution is given by the formulas

$$x_1 = -2 - 6s - 4t, \quad x_2 = s, \quad x_3 = 1 - 3t, \quad x_4 = 2 - 5t, \quad x_5 = t$$

Solution (d). The last equation in the corresponding system of equations is

$$0x_1 + 0x_2 + 0x_3 = 1$$

Since this equation cannot be satisfied, there is no solution to the system. ♦

Elimination Methods We have just seen how easy it is to solve a system of linear equations once its augmented matrix is in reduced row-echelon form. Now we shall give a step-by-step ***elimination*** procedure that can be used to reduce any matrix to reduced row-echelon form. As we state each step in the procedure, we shall illustrate the idea by reducing the following matrix to reduced row-echelon form.

$$\begin{bmatrix} 0 & 0 & -2 & 0 & 7 & 12 \\ 2 & 4 & -10 & 6 & 12 & 28 \\ 2 & 4 & -5 & 6 & -5 & -1 \end{bmatrix}$$

Step 1. Locate the leftmost column that does not consist entirely of zeros.

$$\begin{bmatrix} 0 & 0 & -2 & 0 & 7 & 12 \\ 2 & 4 & -10 & 6 & 12 & 28 \\ 2 & 4 & -5 & 6 & -5 & -1 \end{bmatrix}$$

Leftmost nonzero column (first column)

Step 2. Interchange the top row with another row, if necessary, to bring a nonzero entry to the top of the column found in Step 1.

$$\begin{bmatrix} 2 & 4 & -10 & 6 & 12 & 28 \\ 0 & 0 & -2 & 0 & 7 & 12 \\ 2 & 4 & -5 & 6 & -5 & -1 \end{bmatrix}$$

← The first and second rows in the preceding matrix were interchanged.

Step 3. If the entry that is now at the top of the column found in Step 1 is a, multiply the first row by $1/a$ in order to introduce a leading 1.

$$\begin{bmatrix} 1 & 2 & -5 & 3 & 6 & 14 \\ 0 & 0 & -2 & 0 & 7 & 12 \\ 2 & 4 & -5 & 6 & -5 & -1 \end{bmatrix}$$

← The first row of the preceding matrix was multiplied by $\frac{1}{2}$.

Step 4. Add suitable multiples of the top row to the rows below so that all entries below the leading 1 become zeros.

$$\begin{bmatrix} 1 & 2 & -5 & 3 & 6 & 14 \\ 0 & 0 & -2 & 0 & 7 & 12 \\ 0 & 0 & 5 & 0 & -17 & -29 \end{bmatrix}$$

← -2 times the first row of the preceding matrix was added to the third row.

Step 5. Now cover the top row in the matrix and begin again with Step 1 applied to the submatrix that remains. Continue in this way until the *entire* matrix is in row-echelon form.

$$\begin{bmatrix} 1 & 2 & -5 & 3 & 6 & 14 \\ 0 & 0 & -2 & 0 & 7 & 12 \\ 0 & 0 & 5 & 0 & -17 & -29 \end{bmatrix}$$

Leftmost nonzero column in the submatrix (third column)

$$\begin{bmatrix} 1 & 2 & -5 & 3 & 6 & 14 \\ 0 & 0 & 1 & 0 & -\frac{7}{2} & -6 \\ 0 & 0 & 5 & 0 & -17 & -29 \end{bmatrix}$$

← The first row in the submatrix was multiplied by $-\frac{1}{2}$ to introduce a leading 1.

$$\begin{bmatrix} 1 & 2 & -5 & 3 & 6 & 14 \\ 0 & 0 & 1 & 0 & -\frac{7}{2} & -6 \\ 0 & 0 & 0 & 0 & \frac{1}{2} & 1 \end{bmatrix}$$

← -5 times the first row of the submatrix was added to the second row of the submatrix to introduce a zero below the leading 1.

$$\begin{bmatrix} 1 & 2 & -5 & 3 & 6 & 14 \\ 0 & 0 & 1 & 0 & -\frac{7}{2} & -6 \\ 0 & 0 & 0 & 0 & \frac{1}{2} & 1 \end{bmatrix}$$

← The top row in the submatrix was covered, and we returned again to Step 1.

Leftmost nonzero column in the new submatrix (fifth column)

$$\begin{bmatrix} 1 & 2 & -5 & 3 & 6 & 14 \\ 0 & 0 & 1 & 0 & -\frac{7}{2} & -6 \\ 0 & 0 & 0 & 0 & 1 & 2 \end{bmatrix}$$

⟵ The first (and only) row in the new submatrix was multiplied by 2 to introduce a leading 1.

The *entire* matrix is now in row-echelon form. To find the reduced row-echelon form we need the following additional step.

Step 6. Beginning with the last nonzero row and working upward, add suitable multiples of each row to the rows above to introduce zeros above the leading 1's.

$$\begin{bmatrix} 1 & 2 & -5 & 3 & 6 & 14 \\ 0 & 0 & 1 & 0 & 0 & 1 \\ 0 & 0 & 0 & 0 & 1 & 2 \end{bmatrix}$$

⟵ $\frac{7}{2}$ times the third row of the preceding matrix was added to the second row.

$$\begin{bmatrix} 1 & 2 & -5 & 3 & 0 & 2 \\ 0 & 0 & 1 & 0 & 0 & 1 \\ 0 & 0 & 0 & 0 & 1 & 2 \end{bmatrix}$$

⟵ -6 times the third row was added to the first row.

$$\begin{bmatrix} 1 & 2 & 0 & 3 & 0 & 7 \\ 0 & 0 & 1 & 0 & 0 & 1 \\ 0 & 0 & 0 & 0 & 1 & 2 \end{bmatrix}$$

⟵ 5 times the second row was added to the first row.

The last matrix is in reduced row-echelon form.

If we use only the first five steps, the above procedure produces a row-echelon form and is called ***Gaussian elimination***. Carrying the procedure through to the sixth step and producing a matrix in reduced row-echelon form is called ***Gauss–Jordan elimination***.

REMARK. It can be shown that *every matrix has a unique reduced row-echelon form*; that is, one will arrive at the same reduced row-echelon form for a given matrix no matter how the row operations are varied. (A proof of this result can be found in the article "The Reduced Row Echelon Form of a Matrix Is Unique: A Simple Proof," by Thomas Yuster, *Mathematics Magazine*, Vol. 57, No. 2, 1984, pp. 93–94.) In contrast, *a row-echelon form of a given matrix is not unique*: different sequences of row operations can produce different row-echelon forms.

EXAMPLE 4 Gauss–Jordan Elimination

Solve by Gauss–Jordan elimination.

$$\begin{aligned} x_1 + 3x_2 - 2x_3 \phantom{{}-2x_4} + 2x_5 \phantom{{}-3x_6} &= 0 \\ 2x_1 + 6x_2 - 5x_3 - 2x_4 + 4x_5 - 3x_6 &= -1 \\ 5x_3 + 10x_4 \phantom{{}+4x_5} + 15x_6 &= 5 \\ 2x_1 + 6x_2 \phantom{{}-5x_3} + 8x_4 + 4x_5 + 18x_6 &= 6 \end{aligned}$$

Solution.

The augmented matrix for the system is

$$\begin{bmatrix} 1 & 3 & -2 & 0 & 2 & 0 & 0 \\ 2 & 6 & -5 & -2 & 4 & -3 & -1 \\ 0 & 0 & 5 & 10 & 0 & 15 & 5 \\ 2 & 6 & 0 & 8 & 4 & 18 & 6 \end{bmatrix}$$

Karl Friedrich Gauss

Wilhelm Jordan

Karl Friedrich Gauss (1777–1855) was a German mathematician and scientist. Sometimes called the "prince of mathematicians," Gauss ranks with Isaac Newton and Archimedes as one of the three greatest mathematicians who ever lived. In the entire history of mathematics there may never have been a child so precocious as Gauss—by his own account he worked out the rudiments of arithmetic before he could talk. One day, before he was even three years old, his genius became apparent to his parents in a very dramatic way. His father was preparing the weekly payroll for the laborers under his charge while the boy watched quietly from a corner. At the end of the long and tedious calculation, Gauss informed his father that there was an error in the result and stated the answer, which he had worked out in his head. To the astonishment of his parents, a check of the computations showed Gauss to be correct!

In his doctoral dissertation Gauss gave the first complete proof of the fundamental theorem of algebra, which states that every polynomial equation has as many solutions as its degree. At age 19 he solved a problem that baffled Euclid, inscribing a regular polygon of seventeen sides in a circle using straightedge and compass; and in 1801, at age 24, he published his first masterpiece, *Disquisitiones Arithmeticae*, considered by many to be one of the most brilliant achievements in mathematics. In that paper Gauss systematized the study of number theory (properties of the integers) and formulated the basic concepts that form the foundation of the subject.

Among his myriad achievements, Gauss discovered the Gaussian or "bell-shaped" curve that is fundamental in probability, gave the first geometric interpretation of complex numbers and established their fundamental role in mathematics, developed methods of characterizing surfaces intrinsically by means of the curves that they contain, developed the theory of conformal (angle-preserving) maps, and discovered non-Euclidean geometry 30 years before the ideas were published by others. In physics he made major contributions to the theory of lenses and capillary action, and with Wilhelm Weber he did fundamental work in electromagnetism. Gauss invented the heliotrope, bifilar magnetometer, and an electrotelegraph.

Gauss was deeply religious and aristocratic in demeanor. He mastered foreign languages with ease, read extensively, and enjoyed mineralogy and botany as hobbies. He disliked teaching and was usually cool and discouraging to other mathematicians, possibly because he had already anticipated their work. It has been said that if Gauss had published all of his discoveries, the current state of mathematics would be advanced by 50 years. He was without a doubt the greatest mathematician of the modern era.

Wilhelm Jordan (1842–1899) was a German engineer who specialized in geodesy. His contribution to solving linear systems appeared in his popular book, *Handbuch der Vermessungskunde* (*Handbook of Geodesy*), in 1888.

Adding -2 times the first row to the second and fourth rows gives

$$\begin{bmatrix} 1 & 3 & -2 & 0 & 2 & 0 & 0 \\ 0 & 0 & -1 & -2 & 0 & -3 & -1 \\ 0 & 0 & 5 & 10 & 0 & 15 & 5 \\ 0 & 0 & 4 & 8 & 0 & 18 & 6 \end{bmatrix}$$

Multiplying the second row by -1 and then adding -5 times the new second row to the third row and -4 times the new second row to the fourth row gives

$$\begin{bmatrix} 1 & 3 & -2 & 0 & 2 & 0 & 0 \\ 0 & 0 & 1 & 2 & 0 & 3 & 1 \\ 0 & 0 & 0 & 0 & 0 & 0 & 0 \\ 0 & 0 & 0 & 0 & 0 & 6 & 2 \end{bmatrix}$$

Interchanging the third and fourth rows and then multiplying the third row of the resulting matrix by $\frac{1}{6}$ gives the row-echelon form

$$\begin{bmatrix} 1 & 3 & -2 & 0 & 2 & 0 & 0 \\ 0 & 0 & 1 & 2 & 0 & 3 & 1 \\ 0 & 0 & 0 & 0 & 0 & 1 & \frac{1}{3} \\ 0 & 0 & 0 & 0 & 0 & 0 & 0 \end{bmatrix}$$

Adding -3 times the third row to the second row and then adding 2 times the second row of the resulting matrix to the first row yields the reduced row-echelon form

$$\begin{bmatrix} 1 & 3 & 0 & 4 & 2 & 0 & 0 \\ 0 & 0 & 1 & 2 & 0 & 0 & 0 \\ 0 & 0 & 0 & 0 & 0 & 1 & \frac{1}{3} \\ 0 & 0 & 0 & 0 & 0 & 0 & 0 \end{bmatrix}$$

The corresponding system of equations is

$$\begin{aligned} x_1 + 3x_2 \quad + 4x_4 + 2x_5 \quad &= 0 \\ x_3 + 2x_4 \quad &= 0 \\ x_6 &= \tfrac{1}{3} \end{aligned}$$

(We have discarded the last equation, $0x_1 + 0x_2 + 0x_3 + 0x_4 + 0x_5 + 0x_6 = 0$, since it will be satisfied automatically by the solutions of the remaining equations.) Solving for the leading variables, we obtain

$$\begin{aligned} x_1 &= -3x_2 - 4x_4 - 2x_5 \\ x_3 &= -2x_4 \\ x_6 &= \tfrac{1}{3} \end{aligned}$$

If we assign the free variables x_2, x_4, and x_5 arbitrary values r, s, and t, respectively, the general solution is given by the formulas

$$x_1 = -3r - 4s - 2t, \quad x_2 = r, \quad x_3 = -2s, \quad x_4 = s, \quad x_5 = t, \quad x_6 = \tfrac{1}{3}$$ ♦

Back-Substitution It is sometimes preferable to solve a system of linear equations by using Gaussian elimination to bring the augmented matrix into row-echelon form without continuing all the way to the reduced row-echelon form. When this is done, the corresponding system of equations can be solved by a technique called ***back-substitution***. The next example illustates the idea.

EXAMPLE 5 Example 4 Solved by Back-Substitution

From the computations in Example 4, a row-echelon form of the augmented matrix is

$$\begin{bmatrix} 1 & 3 & -2 & 0 & 2 & 0 & 0 \\ 0 & 0 & 1 & 2 & 0 & 3 & 1 \\ 0 & 0 & 0 & 0 & 0 & 1 & \frac{1}{3} \\ 0 & 0 & 0 & 0 & 0 & 0 & 0 \end{bmatrix}$$

To solve the corresponding system of equations

$$\begin{aligned}
x_1 + 3x_2 - 2x_3 \qquad + 2x_5 \qquad &= 0 \\
x_3 + 2x_4 \qquad + 3x_6 &= 1 \\
x_6 &= \tfrac{1}{3}
\end{aligned}$$

we proceed as follows:

Step 1. Solve the equations for the leading variables.

$$\begin{aligned}
x_1 &= -3x_2 + 2x_3 - 2x_5 \\
x_3 &= 1 - 2x_4 - 3x_6 \\
x_6 &= \tfrac{1}{3}
\end{aligned}$$

Step 2. Beginning with the bottom equation and working upward, successively substitute each equation into all the equations above it.

Substituting $x_6 = \frac{1}{3}$ into the second equation yields

$$\begin{aligned}
x_1 &= -3x_2 + 2x_3 - 2x_5 \\
x_3 &= -2x_4 \\
x_6 &= \tfrac{1}{3}
\end{aligned}$$

Substituting $x_3 = -2x_4$ into the first equation yields

$$\begin{aligned}
x_1 &= -3x_2 - 4x_4 - 2x_5 \\
x_3 &= -2x_4 \\
x_6 &= \tfrac{1}{3}
\end{aligned}$$

Step 3. Assign arbitrary values to the free variables, if any.

If we assign x_2, x_4, and x_5 the arbitrary values r, s, and t, respectively, the general solution is given by the formulas

$$x_1 = -3r - 4s - 2t, \quad x_2 = r, \quad x_3 = -2s, \quad x_4 = s, \quad x_5 = t, \quad x_6 = \tfrac{1}{3}$$

This agrees with the solution obtained in Example 4. ◆

REMARK. The arbitrary values that are assigned to the free variables are often called ***parameters***. Although we shall generally use the letters $r, s, t, \ldots$ for the parameters, any letters that do not conflict with the variable names may be used.

EXAMPLE 6 Gaussian Elimination

Solve

$$\begin{aligned}
x + \ y + 2z &= 9 \\
2x + 4y - 3z &= 1 \\
3x + 6y - 5z &= 0
\end{aligned}$$

by Gaussian elimination and back-substitution.

Solution.

This is the system in Example 3 of Section 1.1. In that example we converted the augmented matrix

$$\begin{bmatrix} 1 & 1 & 2 & 9 \\ 2 & 4 & -3 & 1 \\ 3 & 6 & -5 & 0 \end{bmatrix}$$

to the row-echelon form

$$\begin{bmatrix} 1 & 1 & 2 & 9 \\ 0 & 1 & -\frac{7}{2} & -\frac{17}{2} \\ 0 & 0 & 1 & 3 \end{bmatrix}$$

The system corresponding to this matrix is

$$\begin{aligned} x + y + 2z &= 9 \\ y - \tfrac{7}{2}z &= -\tfrac{17}{2} \\ z &= 3 \end{aligned}$$

Solving for the leading variables yields

$$\begin{aligned} x &= 9 - y - 2z \\ y &= -\tfrac{17}{2} + \tfrac{7}{2}z \\ z &= 3 \end{aligned}$$

Substituting the bottom equation into those above yields

$$\begin{aligned} x &= 3 - y \\ y &= 2 \\ z &= 3 \end{aligned}$$

and substituting the second equation into the top yields $x = 1, y = 2, z = 3$. This agrees with the result found by Gauss–Jordan elimination in Example 3 of Section 1.1. ♦

Homogeneous Linear Systems A system of linear equations is said to be ***homogeneous*** if the constant terms are all zero; that is, the system has the form

$$\begin{array}{ccccccc} a_{11}x_1 & + & a_{12}x_2 & + \cdots + & a_{1n}x_n & = 0 \\ a_{21}x_1 & + & a_{22}x_2 & + \cdots + & a_{2n}x_n & = 0 \\ \vdots & & \vdots & & \vdots & \vdots \\ a_{m1}x_1 & + & a_{m2}x_2 & + \cdots + & a_{mn}x_n & = 0 \end{array}$$

Every homogeneous system of linear equations is consistent, since all such systems have $x_1 = 0, x_2 = 0, \ldots, x_n = 0$ as a solution. This solution is called the ***trivial solution***; if there are other solutions, they are called ***nontrivial solutions***.

Because a homogeneous linear system always has the trivial solution, there are only two possibilities for its solutions:

- The system has only the trivial solution.
- The system has infinitely many solutions in addition to the trivial solution.

In the special case of a homogeneous linear system of two equations in two unknowns, say

$$a_1x + b_1y = 0 \quad (a_1, b_1 \text{ not both zero})$$
$$a_2x + b_2y = 0 \quad (a_2, b_2 \text{ not both zero})$$

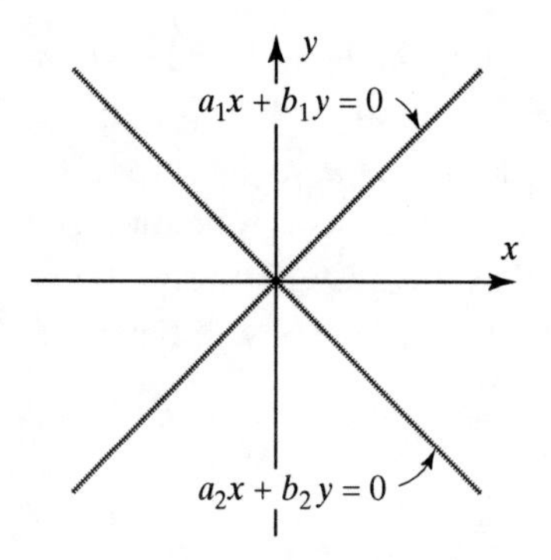

(a) Only the trivial solution

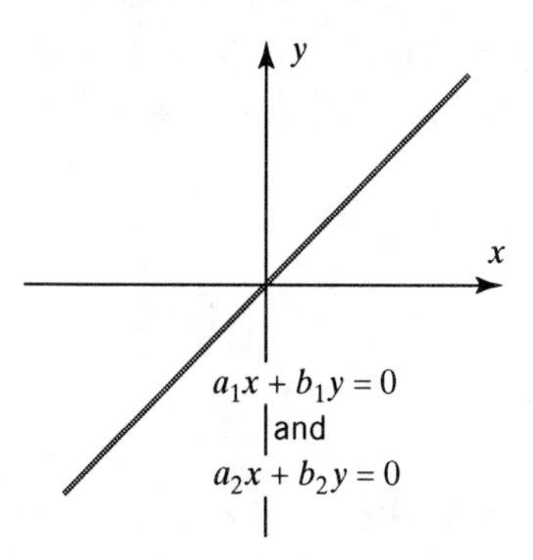

(b) Infinitely many solutions

Figure 1.2.1

the graphs of the equations are lines through the origin, and the trivial solution corresponds to the point of intersection at the origin (Figure 1.2.1).

There is one case in which a homogeneous system is assured of having nontrivial solutions, namely, whenever the system involves more unknowns than equations. To see why, consider the following example of four equations in five unknowns.

EXAMPLE 7 Gauss–Jordan Elimination

Solve the following homogeneous system of linear equations by using Gauss–Jordan elimination.

$$\begin{aligned} 2x_1 + 2x_2 - x_3 + x_5 &= 0 \\ -x_1 - x_2 + 2x_3 - 3x_4 + x_5 &= 0 \\ x_1 + x_2 - 2x_3 - x_5 &= 0 \\ x_3 + x_4 + x_5 &= 0 \end{aligned} \tag{1}$$

Solution.

The augmented matrix for the system is

$$\begin{bmatrix} 2 & 2 & -1 & 0 & 1 & 0 \\ -1 & -1 & 2 & -3 & 1 & 0 \\ 1 & 1 & -2 & 0 & -1 & 0 \\ 0 & 0 & 1 & 1 & 1 & 0 \end{bmatrix}$$

Reducing this matrix to reduced row-echelon form, we obtain

$$\begin{bmatrix} 1 & 1 & 0 & 0 & 1 & 0 \\ 0 & 0 & 1 & 0 & 1 & 0 \\ 0 & 0 & 0 & 1 & 0 & 0 \\ 0 & 0 & 0 & 0 & 0 & 0 \end{bmatrix}$$

The corresponding system of equations is

$$\begin{aligned} x_1 + x_2 + x_5 &= 0 \\ x_3 + x_5 &= 0 \\ x_4 &= 0 \end{aligned} \tag{2}$$

Solving for the leading variables yields

$$\begin{aligned} x_1 &= -x_2 - x_5 \\ x_3 &= -x_5 \\ x_4 &= 0 \end{aligned}$$

Thus, the general solution is

$$x_1 = -s - t, \quad x_2 = s, \quad x_3 = -t, \quad x_4 = 0, \quad x_5 = t$$

Note that the trivial solution is obtained when $s = t = 0$. ◆

Example 7 illustrates two important points about solving homogeneous systems of linear equations. First, none of the three elementary row operations alters the final column of zeros in the augmented matrix, so that the system of equations corresponding to the reduced row-echelon form of the augmented matrix must also be a homogeneous system [see system (2)]. Second, depending on whether the reduced row-echelon form of the augmented matrix has any zero rows, the number of equations in the reduced system is the same as or less than the number of equations in the original system [compare systems (1) and (2)]. Thus, if the given homogeneous system has m equations in n unknowns with $m < n$, and if there are r nonzero rows in the reduced row-echelon form of the augmented matrix, we will have $r < n$. It follows that the system of equations corresponding to the reduced row-echelon form of the augmented matrix will have the form

$$\begin{array}{lllll} \cdots x_{k_1} & & & + \Sigma(\,) = 0 \\ & \cdots x_{k_2} & & + \Sigma(\,) = 0 \\ & & \ddots & \vdots \\ & & & x_{k_r} + \Sigma(\,) = 0 \end{array} \tag{3}$$

where $x_{k_1}, x_{k_2}, \ldots, x_{k_r}$ are the leading variables and $\Sigma(\,)$ denotes sums (possibly all different) that involve the $n - r$ free variables [compare system (3) with system (2) above]. Solving for the leading variables gives

$$\begin{aligned} x_{k_1} &= -\Sigma(\,) \\ x_{k_2} &= -\Sigma(\,) \\ &\vdots \\ x_{k_r} &= -\Sigma(\,) \end{aligned}$$

As in Example 7, we can assign arbitrary values to the free variables on the right-hand side and thus obtain infinitely many solutions to the system.

In summary, we have the following important theorem.

Theorem 1.2.1

A homogeneous system of linear equations with more unknowns than equations has infinitely many solutions.

REMARK. Note that Theorem 1.2.1 applies only to homogeneous systems. A nonhomogeneous system with more unknowns than equations need not be consistent (Exercise 28); however, if the system is consistent, it will have infinitely many solutions. This will be proved later.

Computer Solution of Linear Systems

In applications it is not uncommon to encounter large linear systems that must be solved by computer. Most computer algorithms for solving such systems are based on Gaussian elimination or Gauss–Jordan elimination, but the basic procedures are often modified to deal with such issues as

- Reducing roundoff errors
- Minimizing the use of computer memory space
- Solving the system with maximum speed

Some of these matters will be considered in Chapter 9. For hand computations fractions are an annoyance that often cannot be avoided. However, in some cases it is possible to avoid them by varying the elementary row operations in the right way. Thus, once the methods of Gaussian elimination and Gauss–Jordan elimination have been mastered, the reader may wish to vary the steps in specific problems to avoid fractions (see Exercise 18).

REMARK. Since Gauss–Jordan elimination avoids the use of back-substitution, it would seem that this method would be the more efficient of the two methods we have considered.

It can be argued that this statement is true when solving small systems by hand since Gauss–Jordan elimination actually involves less writing. However, for large systems of equations, it has been shown that the Gauss–Jordan elimination method requires about 50% more operations than Gaussian elimination. This is an important consideration when working on computers.

Exercise Set 1.2

1. Which of the following 3×3 matrices are in reduced row-echelon form?

(a) $\begin{bmatrix} 1 & 0 & 0 \\ 0 & 1 & 0 \\ 0 & 0 & 1 \end{bmatrix}$ (b) $\begin{bmatrix} 1 & 0 & 0 \\ 0 & 1 & 0 \\ 0 & 0 & 0 \end{bmatrix}$ (c) $\begin{bmatrix} 0 & 1 & 0 \\ 0 & 0 & 1 \\ 0 & 0 & 0 \end{bmatrix}$ (d) $\begin{bmatrix} 1 & 0 & 0 \\ 0 & 0 & 1 \\ 0 & 0 & 0 \end{bmatrix}$ (e) $\begin{bmatrix} 1 & 0 & 0 \\ 0 & 0 & 0 \\ 0 & 0 & 1 \end{bmatrix}$

(f) $\begin{bmatrix} 0 & 1 & 0 \\ 1 & 0 & 0 \\ 0 & 0 & 0 \end{bmatrix}$ (g) $\begin{bmatrix} 1 & 1 & 0 \\ 0 & 1 & 0 \\ 0 & 0 & 0 \end{bmatrix}$ (h) $\begin{bmatrix} 1 & 0 & 2 \\ 0 & 1 & 3 \\ 0 & 0 & 0 \end{bmatrix}$ (i) $\begin{bmatrix} 0 & 0 & 1 \\ 0 & 0 & 0 \\ 0 & 0 & 0 \end{bmatrix}$ (j) $\begin{bmatrix} 0 & 0 & 0 \\ 0 & 0 & 0 \\ 0 & 0 & 0 \end{bmatrix}$

2. Which of the following 3×3 matrices are in row-echelon form?

(a) $\begin{bmatrix} 1 & 0 & 0 \\ 0 & 1 & 0 \\ 0 & 0 & 1 \end{bmatrix}$ (b) $\begin{bmatrix} 1 & 2 & 0 \\ 0 & 1 & 0 \\ 0 & 0 & 0 \end{bmatrix}$ (c) $\begin{bmatrix} 1 & 0 & 0 \\ 0 & 1 & 0 \\ 0 & 2 & 0 \end{bmatrix}$ (d) $\begin{bmatrix} 1 & 3 & 4 \\ 0 & 0 & 1 \\ 0 & 0 & 0 \end{bmatrix}$

(e) $\begin{bmatrix} 1 & 5 & -3 \\ 0 & 1 & 1 \\ 0 & 0 & 0 \end{bmatrix}$ (f) $\begin{bmatrix} 1 & 2 & 3 \\ 0 & 0 & 0 \\ 0 & 0 & 1 \end{bmatrix}$

3. In each part determine whether the matrix is in row-echelon form, reduced row-echelon form, both, or neither.

(a) $\begin{bmatrix} 1 & 2 & 0 & 3 & 0 \\ 0 & 0 & 1 & 1 & 0 \\ 0 & 0 & 0 & 0 & 1 \\ 0 & 0 & 0 & 0 & 0 \end{bmatrix}$ (b) $\begin{bmatrix} 1 & 0 & 0 & 5 \\ 0 & 0 & 1 & 3 \\ 0 & 1 & 0 & 4 \end{bmatrix}$ (c) $\begin{bmatrix} 1 & 0 & 3 & 1 \\ 0 & 1 & 2 & 4 \end{bmatrix}$

(d) $\begin{bmatrix} 1 & -7 & 5 & 5 \\ 0 & 1 & 3 & 2 \end{bmatrix}$ (e) $\begin{bmatrix} 1 & 3 & 0 & 2 & 0 \\ 1 & 0 & 2 & 2 & 0 \\ 0 & 0 & 0 & 0 & 1 \\ 0 & 0 & 0 & 0 & 0 \end{bmatrix}$ (f) $\begin{bmatrix} 0 & 0 \\ 0 & 0 \\ 0 & 0 \end{bmatrix}$

4. In each part suppose that the augmented matrix for a system of linear equations has been reduced by row operations to the given reduced row-echelon form. Solve the system.

(a) $\begin{bmatrix} 1 & 0 & 0 & -3 \\ 0 & 1 & 0 & 0 \\ 0 & 0 & 1 & 7 \end{bmatrix}$ (b) $\begin{bmatrix} 1 & 0 & 0 & -7 & 8 \\ 0 & 1 & 0 & 3 & 2 \\ 0 & 0 & 1 & 1 & -5 \end{bmatrix}$

(c) $\begin{bmatrix} 1 & -6 & 0 & 0 & 3 & -2 \\ 0 & 0 & 1 & 0 & 4 & 7 \\ 0 & 0 & 0 & 1 & 5 & 8 \\ 0 & 0 & 0 & 0 & 0 & 0 \end{bmatrix}$ (d) $\begin{bmatrix} 1 & -3 & 0 & 0 \\ 0 & 0 & 1 & 0 \\ 0 & 0 & 0 & 1 \end{bmatrix}$

5. In each part suppose that the augmented matrix for a system of linear equations has been reduced by row operations to the given row-echelon form. Solve the system.

(a) $\begin{bmatrix} 1 & -3 & 4 & 7 \\ 0 & 1 & 2 & 2 \\ 0 & 0 & 1 & 5 \end{bmatrix}$ (b) $\begin{bmatrix} 1 & 0 & 8 & -5 & 6 \\ 0 & 1 & 4 & -9 & 3 \\ 0 & 0 & 1 & 1 & 2 \end{bmatrix}$

(c) $\begin{bmatrix} 1 & 7 & -2 & 0 & -8 & -3 \\ 0 & 0 & 1 & 1 & 6 & 5 \\ 0 & 0 & 0 & 1 & 3 & 9 \\ 0 & 0 & 0 & 0 & 0 & 0 \end{bmatrix}$ (d) $\begin{bmatrix} 1 & -3 & 7 & 1 \\ 0 & 1 & 4 & 0 \\ 0 & 0 & 0 & 1 \end{bmatrix}$

6. Solve each of the following systems by Gauss–Jordan elimination.

(a) $\begin{aligned} x_1 + x_2 + 2x_3 &= 8 \\ -x_1 - 2x_2 + 3x_3 &= 1 \\ 3x_1 - 7x_2 + 4x_3 &= 10 \end{aligned}$ (b) $\begin{aligned} 2x_1 + 2x_2 + 2x_3 &= 0 \\ -2x_1 + 5x_2 + 2x_3 &= 1 \\ 8x_1 + x_2 + 4x_3 &= -1 \end{aligned}$

(c) $\begin{aligned} x - y + 2z - w &= -1 \\ 2x + y - 2z - 2w &= -2 \\ -x + 2y - 4z + w &= 1 \\ 3x \qquad\qquad - 3w &= -3 \end{aligned}$ (d) $\begin{aligned} -2b + 3c &= 1 \\ 3a + 6b - 3c &= -2 \\ 6a + 6b + 3c &= 5 \end{aligned}$

7. Solve each of the systems in Exercise 6 by Gaussian elimination.

8. Solve each of the following systems by Gauss–Jordan elimination.

(a) $\begin{aligned} 2x_1 - 3x_2 &= -2 \\ 2x_1 + x_2 &= 1 \\ 3x_1 + 2x_2 &= 1 \end{aligned}$ (b) $\begin{aligned} 3x_1 + 2x_2 - x_3 &= -15 \\ 5x_1 + 3x_2 + 2x_3 &= 0 \\ 3x_1 + x_2 + 3x_3 &= 11 \\ -6x_1 - 4x_2 + 2x_3 &= 30 \end{aligned}$

(c) $\begin{aligned} 4x_1 - 8x_2 &= 12 \\ 3x_1 - 6x_2 &= 9 \\ -2x_1 + 4x_2 &= -6 \end{aligned}$ (d) $\begin{aligned} 10y - 4z + w &= 1 \\ x + 4y - z + w &= 2 \\ 3x + 2y + z + 2w &= 5 \\ -2x - 8y + 2z - 2w &= -4 \\ x - 6y + 3z \qquad &= 1 \end{aligned}$

9. Solve each of the systems in Exercise 8 by Gaussian elimination.

10. Solve each of the following systems by Gauss–Jordan elimination.

(a) $\begin{aligned} 5x_1 - 2x_2 + 6x_3 &= 0 \\ -2x_1 + x_2 + 3x_3 &= 1 \end{aligned}$ (b) $\begin{aligned} x_1 - 2x_2 + x_3 - 4x_4 &= 1 \\ x_1 + 3x_2 + 7x_3 + 2x_4 &= 2 \\ x_1 - 12x_2 - 11x_3 - 16x_4 &= 5 \end{aligned}$ (c) $\begin{aligned} w + 2x - y &= 4 \\ x - y &= 3 \\ w + 3x - 2y &= 7 \\ 2u + 4v + w + 7x \qquad &= 7 \end{aligned}$

11. Solve each of the systems in Exercise 10 by Gaussian elimination.

12. Without using pencil and paper, determine which of the following homogeneous systems have nontrivial solutions.

(a) $\begin{aligned} 2x_1 - 3x_2 + 4x_3 - x_4 &= 0 \\ 7x_1 + x_2 - 8x_3 + 9x_4 &= 0 \\ 2x_1 + 8x_2 + x_3 - x_4 &= 0 \end{aligned}$ (b) $\begin{aligned} x_1 + 3x_2 - x_3 &= 0 \\ x_2 - 8x_3 &= 0 \\ 4x_3 &= 0 \end{aligned}$

(c) $\begin{aligned} a_{11}x_1 + a_{12}x_2 + a_{13}x_3 &= 0 \\ a_{21}x_1 + a_{22}x_2 + a_{23}x_3 &= 0 \end{aligned}$ (d) $\begin{aligned} 3x_1 - 2x_2 &= 0 \\ 6x_1 - 4x_2 &= 0 \end{aligned}$

13. Solve the following homogeneous systems of linear equations by any method.

(a) $\begin{aligned} 2x_1 + x_2 + 3x_3 &= 0 \\ x_1 + 2x_2 \qquad &= 0 \\ x_2 + x_3 &= 0 \end{aligned}$ (b) $\begin{aligned} 3x_1 + x_2 + x_3 + x_4 &= 0 \\ 5x_1 - x_2 + x_3 - x_4 &= 0 \end{aligned}$ (c) $\begin{aligned} 2x + 2y + 4z &= 0 \\ w \qquad - y - 3z &= 0 \\ 2w + 3x + y + z &= 0 \\ -2w + x + 3y - 2z &= 0 \end{aligned}$

14. Solve the following homogeneous systems of linear equations by any method.

(a)
$$\begin{aligned} 2x - y - 3z &= 0 \\ -x + 2y - 3z &= 0 \\ x + y + 4z &= 0 \end{aligned}$$

(b)
$$\begin{aligned} v + 3w - 2x &= 0 \\ 2u + v - 4w + 3x &= 0 \\ 2u + 3v + 2w - x &= 0 \\ -4u - 3v + 5w - 4x &= 0 \end{aligned}$$

(c)
$$\begin{aligned} x_1 + 3x_2 + x_4 &= 0 \\ x_1 + 4x_2 + 2x_3 &= 0 \\ -2x_2 - 2x_3 - x_4 &= 0 \\ 2x_1 - 4x_2 + x_3 + x_4 &= 0 \\ x_1 - 2x_2 - x_3 + x_4 &= 0 \end{aligned}$$

15. Solve the following systems by any method.

(a)
$$\begin{aligned} 2I_1 - I_2 + 3I_3 + 4I_4 &= 9 \\ I_1 - 2I_3 + 7I_4 &= 11 \\ 3I_1 - 3I_2 + I_3 + 5I_4 &= 8 \\ 2I_1 + I_2 + 4I_3 + 4I_4 &= 10 \end{aligned}$$

(b)
$$\begin{aligned} Z_3 + Z_4 + Z_5 &= 0 \\ -Z_1 - Z_2 + 2Z_3 - 3Z_4 + Z_5 &= 0 \\ Z_1 + Z_2 - 2Z_3 - Z_5 &= 0 \\ 2Z_1 + 2Z_2 - Z_3 + Z_5 &= 0 \end{aligned}$$

16. Solve the following systems, where a, b, and c are constants.

(a)
$$\begin{aligned} 2x + y &= a \\ 3x + 6y &= b \end{aligned}$$

(b)
$$\begin{aligned} x_1 + x_2 + x_3 &= a \\ 2x_1 + 2x_3 &= b \\ 3x_2 + 3x_3 &= c \end{aligned}$$

17. For which values of a will the following system have no solutions? Exactly one solution? Infinitely many solutions?

$$\begin{aligned} x + 2y - 3z &= 4 \\ 3x - y + 5z &= 2 \\ 4x + y + (a^2 - 14)z &= a + 2 \end{aligned}$$

18. Reduce

$$\begin{bmatrix} 2 & 1 & 3 \\ 0 & -2 & -29 \\ 3 & 4 & 5 \end{bmatrix}$$

to reduced row-echelon form without introducing any fractions.

19. Find two different row-echelon forms of

$$\begin{bmatrix} 1 & 3 \\ 2 & 7 \end{bmatrix}$$

20. Solve the following system of nonlinear equations for the unknown angles α, β, and γ, where $0 \le \alpha \le 2\pi$, $0 \le \beta \le 2\pi$, and $0 \le \gamma < \pi$.

$$\begin{aligned} 2\sin\alpha - \cos\beta + 3\tan\gamma &= 3 \\ 4\sin\alpha + 2\cos\beta - 2\tan\gamma &= 2 \\ 6\sin\alpha - 3\cos\beta + \tan\gamma &= 9 \end{aligned}$$

21. Show that the following nonlinear system has 18 solutions if $0 \le \alpha \le 2\pi$, $0 \le \beta \le 2\pi$, and $0 \le \gamma < 2\pi$.

$$\begin{aligned} \sin\alpha + 2\cos\beta + 3\tan\gamma &= 0 \\ 2\sin\alpha + 5\cos\beta + 3\tan\gamma &= 0 \\ -\sin\alpha - 5\cos\beta + 5\tan\gamma &= 0 \end{aligned}$$

22. For which value(s) of λ does the system of equations

$$\begin{aligned} (\lambda - 3)x + y &= 0 \\ x + (\lambda - 3)y &= 0 \end{aligned}$$

have nontrivial solutions?

23. Solve the system

$$\begin{aligned} 2x_1 - \ x_2 \qquad &= \lambda x_1 \\ 2x_1 - \ x_2 + x_3 &= \lambda x_2 \\ -2x_1 + 2x_2 + x_3 &= \lambda x_3 \end{aligned}$$

for x_1, x_2, and x_3 in the two cases $\lambda = 1, \lambda = 2$.

24. Solve the following system for x, y, and z.

$$\begin{aligned} \frac{1}{x} + \frac{2}{y} - \frac{4}{z} &= 1 \\ \frac{2}{x} + \frac{3}{y} + \frac{8}{z} &= 0 \\ -\frac{1}{x} + \frac{9}{y} + \frac{10}{z} &= 5 \end{aligned}$$

25. Find the coefficients a, b, c, and d so that the curve shown in the accompanying figure is the graph of the equation $y = ax^3 + bx^2 + cx + d$.

26. Find coefficients a, b, c, and d so that the curve shown in the accompanying figure is given by the equation $ax^2 + ay^2 + bx + cy + d = 0$.

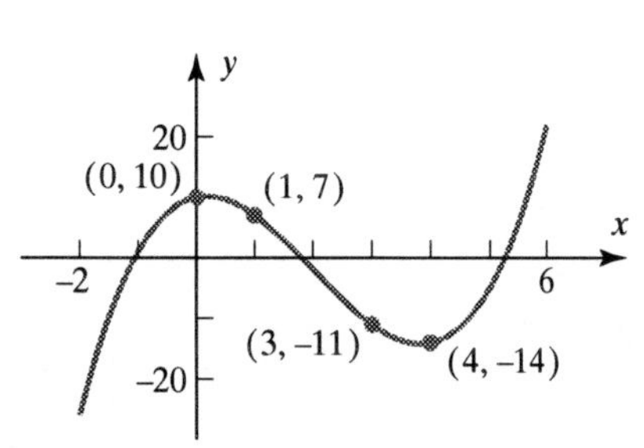

Figure Ex-25

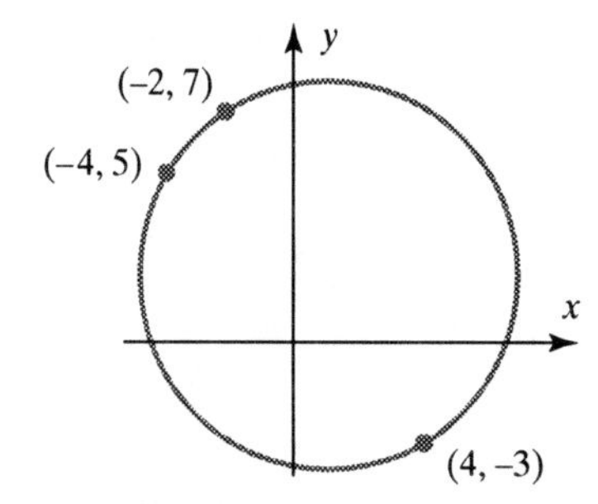

Figure Ex-26

27. (a) Show that if $ad - bc \neq 0$, then the reduced row-echelon form of

$$\begin{bmatrix} a & b \\ c & d \end{bmatrix} \text{ is } \begin{bmatrix} 1 & 0 \\ 0 & 1 \end{bmatrix}$$

(b) Use part (a) to show that the system

$$\begin{aligned} ax + by &= k \\ cx + dy &= l \end{aligned}$$

has exactly one solution when $ad - bc \neq 0$.

28. Find an inconsistent linear system that has more unknowns than equations.

Discussion and Discovery

29. Discuss the possible reduced row-echelon forms of

$$\begin{bmatrix} a & b & c \\ d & e & f \\ g & h & i \end{bmatrix}$$

30. Consider the system of equations

$$\begin{aligned} ax + by &= 0 \\ cx + dy &= 0 \\ ex + fy &= 0 \end{aligned}$$

Discuss the relative positions of the lines $ax + by = 0$, $cx + dy = 0$, and $ex + fy = 0$ when (a) the system has only the trivial solution, and (b) the system has nontrivial solutions.

31. Indicate whether the statement is always true or sometimes false. Justify your answer by giving a logical argument or a counterexample.

(a) If a matrix is reduced to reduced row-echelon form by two different sequences of elementary row operations, the resulting matrices will be different.

(b) If a matrix is reduced to row-echelon form by two different sequences of elementary row operations, the resulting matrices might be different.

(c) If the reduced row-echelon form of the augmented matrix for a linear system has a row of zeros, then the system must have infinitely many solutions.

(d) If three lines in the xy-plane are sides of a triangle, then the system of equations formed from their equations has three solutions, one corresponding to each vertex.

32. Indicate whether the statement is always true or sometimes false. Justify your answer by giving a logical argument or a counterexample.

(a) A linear system of three equations in five unknowns must be consistent.

(b) A linear system of five equations in three unknowns cannot be consistent.

(c) If a linear system of n equations in n unknowns has n leading 1's in the reduced row-echelon form of its augmented matrix, then the system has exactly one solution.

(d) If a linear system of n equations in n unknowns has two equations that are multiples of one another, then the system is inconsistent.

1.3 MATRICES AND MATRIX OPERATIONS

Rectangular arrays of real numbers arise in many contexts other than as augmented matrices for systems of linear equations. In this section we begin our study of matrix theory by giving some of the fundamental definitions of the subject. We shall see how matrices can be combined through the arithmetic operations of addition, subtraction, and multiplication.

Matrix Notation and Terminology In Section 1.2 we used rectangular arrays of numbers, called *augmented matrices*, to abbreviate systems of linear equations. However, rectangular arrays of numbers occur in other contexts as well. For example, the following rectangular array with three rows and seven columns might describe the number of hours that a student spent studying three subjects during a certain week:

	Mon.	Tues.	Wed.	Thurs.	Fri.	Sat.	Sun.
Math	2	3	2	4	1	4	2
History	0	3	1	4	3	2	2
Language	4	1	3	1	0	0	2

If we suppress the headings, then we are left with the following rectangular array of numbers with three rows and seven columns called a "matrix":

$$\begin{bmatrix} 2 & 3 & 2 & 4 & 1 & 4 & 2 \\ 0 & 3 & 1 & 4 & 3 & 2 & 2 \\ 4 & 1 & 3 & 1 & 0 & 0 & 2 \end{bmatrix}$$

More generally, we make the following definition.

Definition

A ***matrix*** is a rectangular array of numbers. The numbers in the array are called the ***entries*** in the matrix.

EXAMPLE 1 Examples of Matrices

Some examples of matrices are

$$\begin{bmatrix} 1 & 2 \\ 3 & 0 \\ -1 & 4 \end{bmatrix}, \quad [2 \quad 1 \quad 0 \quad -3], \quad \begin{bmatrix} e & \pi & -\sqrt{2} \\ 0 & \frac{1}{2} & 1 \\ 0 & 0 & 0 \end{bmatrix}, \quad \begin{bmatrix} 1 \\ 3 \end{bmatrix}, \quad [4] \quad \blacklozenge$$

The ***size*** of a matrix is described in terms of the number of rows (horizontal lines) and columns (vertical lines) it contains. For example, the first matrix in Example 1 has three rows and two columns, so its size is 3 by 2 (written 3×2). In a size description, the first number always denotes the number of rows and the second denotes the number of columns. The remaining matrices in Example 1 have sizes 1×4, 3×3, 2×1, and 1×1, respectively. A matrix with only one column is called a ***column matrix*** (or a ***column vector***), and a matrix with only one row is called a ***row matrix*** (or a ***row vector***). Thus, in Example 1 the 2×1 matrix is a column matrix, the 1×4 matrix is a row matrix, and the 1×1 matrix is both a row matrix and a column matrix. (The term *vector* has another meaning that we will discuss in subsequent chapters.)

REMARK. It is common practice to omit the brackets on a 1×1 matrix. Thus, we might write 4 rather than [4]. Although this makes it impossible to tell whether 4 denotes the number "four" or the 1×1 matrix whose entry is "four," this rarely causes problems, since it is usually possible to tell which is meant from the context in which the symbol appears.

We shall use capital letters to denote matrices and lowercase letters to denote numerical quantities; thus, we might write

$$A = \begin{bmatrix} 2 & 1 & 7 \\ 3 & 4 & 2 \end{bmatrix} \quad \text{or} \quad C = \begin{bmatrix} a & b & c \\ d & e & f \end{bmatrix}$$

When discussing matrices, it is common to refer to numerical quantities as ***scalars***. Unless stated otherwise, *scalars will be real numbers*; complex scalars will be considered in Chapter 10.

The entry that occurs in row i and column j of a matrix A will be denoted by a_{ij}. Thus, a general 3×4 matrix might be written as

$$A = \begin{bmatrix} a_{11} & a_{12} & a_{13} & a_{14} \\ a_{21} & a_{22} & a_{23} & a_{24} \\ a_{31} & a_{32} & a_{33} & a_{34} \end{bmatrix}$$

and a general $m \times n$ matrix as

$$A = \begin{bmatrix} a_{11} & a_{12} & \cdots & a_{1n} \\ a_{21} & a_{22} & \cdots & a_{2n} \\ \vdots & \vdots & & \vdots \\ a_{m1} & a_{m2} & \cdots & a_{mn} \end{bmatrix} \tag{1}$$

When compactness of notation is desired, the preceding matrix can be written as

$$[a_{ij}]_{m\times n} \quad \text{or} \quad [a_{ij}]$$

the first notation being used when it is important in the discussion to know the size and the second when the size need not be emphasized. Usually, we shall match the letter denoting a matrix with the letter denoting its entries; thus, for a matrix B we would generally use b_{ij} for the entry in row i and column j and for a matrix C we would use the notation c_{ij}.

The entry in row i and column j of a matrix A is also commonly denoted by the symbol $(A)_{ij}$. Thus, for matrix (1) above, we have

$$(A)_{ij} = a_{ij}$$

and for the matrix

$$A = \begin{bmatrix} 2 & -3 \\ 7 & 0 \end{bmatrix}$$

we have $(A)_{11} = 2$, $(A)_{12} = -3$, $(A)_{21} = 7$, and $(A)_{22} = 0$.

Row and column matrices are of special importance, and it is common practice to denote them by boldface lowercase letters rather than capital letters. For such matrices double subscripting of the entries is unnecessary. Thus, a general $1 \times n$ row matrix $\mathbf{a}$ and a general $m \times 1$ column matrix $\mathbf{b}$ would be written as

$$\mathbf{a} = [a_1 \quad a_2 \quad \cdots \quad a_n] \quad \text{and} \quad \mathbf{b} = \begin{bmatrix} b_1 \\ b_2 \\ \vdots \\ b_m \end{bmatrix}$$

A matrix A with n rows and n columns is called a ***square matrix of order n***, and the shaded entries $a_{11}, a_{22}, \ldots, a_{nn}$ in (2) are said to be on the ***main diagonal*** of A.

$$\begin{bmatrix} a_{11} & a_{12} & \cdots & a_{1n} \\ a_{21} & a_{22} & \cdots & a_{2n} \\ \vdots & \vdots & & \vdots \\ a_{n1} & a_{n2} & \cdots & a_{nn} \end{bmatrix} \tag{2}$$

Operations on Matrices So far, we have used matrices to abbreviate the work in solving systems of linear equations. For other applications, however, it is desirable to develop an "arithmetic of matrices" in which matrices can be added, subtracted, and multiplied in a useful way. The remainder of this section will be devoted to developing this arithmetic.

Definition

Two matrices are defined to be ***equal*** if they have the same size and their corresponding entries are equal.

In matrix notation, if $A = [a_{ij}]$ and $B = [b_{ij}]$ have the same size, then $A = B$ if and only if $(A)_{ij} = (B)_{ij}$, or equivalently, $a_{ij} = b_{ij}$ for all i and j.

EXAMPLE 2 Equality of Matrices

Consider the matrices

$$A = \begin{bmatrix} 2 & 1 \\ 3 & x \end{bmatrix}, \quad B = \begin{bmatrix} 2 & 1 \\ 3 & 5 \end{bmatrix}, \quad C = \begin{bmatrix} 2 & 1 & 0 \\ 3 & 4 & 0 \end{bmatrix}$$

If $x = 5$, then $A = B$, but for all other values of x the matrices A and B are not equal, since not all of their corresponding entries are equal. There is no value of x for which $A = C$ since A and C have different sizes. ♦

Definition

If A and B are matrices of the same size, then the ***sum*** $A + B$ is the matrix obtained by adding the entries of B to the corresponding entries of A, and the ***difference*** $A - B$ is the matrix obtained by subtracting the entries of B from the corresponding entries of A. Matrices of different sizes cannot be added or subtracted.

In matrix notation, if $A = [a_{ij}]$ and $B = [b_{ij}]$ have the same size, then

$$(A + B)_{ij} = (A)_{ij} + (B)_{ij} = a_{ij} + b_{ij} \quad \text{and} \quad (A - B)_{ij} = (A)_{ij} - (B)_{ij} = a_{ij} - b_{ij}$$

EXAMPLE 3 Addition and Subtraction

Consider the matrices

$$A = \begin{bmatrix} 2 & 1 & 0 & 3 \\ -1 & 0 & 2 & 4 \\ 4 & -2 & 7 & 0 \end{bmatrix}, \quad B = \begin{bmatrix} -4 & 3 & 5 & 1 \\ 2 & 2 & 0 & -1 \\ 3 & 2 & -4 & 5 \end{bmatrix}, \quad C = \begin{bmatrix} 1 & 1 \\ 2 & 2 \end{bmatrix}$$

Then

$$A + B = \begin{bmatrix} -2 & 4 & 5 & 4 \\ 1 & 2 & 2 & 3 \\ 7 & 0 & 3 & 5 \end{bmatrix} \quad \text{and} \quad A - B = \begin{bmatrix} 6 & -2 & -5 & 2 \\ -3 & -2 & 2 & 5 \\ 1 & -4 & 11 & -5 \end{bmatrix}$$

The expressions $A + C$, $B + C$, $A - C$, and $B - C$ are undefined. ♦

Definition

If A is any matrix and c is any scalar, then the ***product*** cA is the matrix obtained by multiplying each entry of the matrix A by c. The matrix cA is said to be a ***scalar multiple*** of A.

In matrix notation, if $A = [a_{ij}]$, then

$$(cA)_{ij} = c(A)_{ij} = ca_{ij}$$

EXAMPLE 4 Scalar Multiples

For the matrices

$$A = \begin{bmatrix} 2 & 3 & 4 \\ 1 & 3 & 1 \end{bmatrix}, \quad B = \begin{bmatrix} 0 & 2 & 7 \\ -1 & 3 & -5 \end{bmatrix}, \quad C = \begin{bmatrix} 9 & -6 & 3 \\ 3 & 0 & 12 \end{bmatrix}$$

we have

$$2A = \begin{bmatrix} 4 & 6 & 8 \\ 2 & 6 & 2 \end{bmatrix}, \quad (-1)B = \begin{bmatrix} 0 & -2 & -7 \\ 1 & -3 & 5 \end{bmatrix}, \quad \tfrac{1}{3}C = \begin{bmatrix} 3 & -2 & 1 \\ 1 & 0 & 4 \end{bmatrix}$$

It is common practice to denote $(-1)B$ by $-B$. ♦

If $A_1, A_2, \ldots, A_n$ are matrices of the same size and $c_1, c_2, \ldots, c_n$ are scalars, then an expression of the form

$$c_1A_1 + c_2A_2 + \cdots + c_nA_n$$

is called a ***linear combination*** of $A_1, A_2, \ldots, A_n$ with ***coefficients*** $c_1, c_2, \ldots, c_n$. For example, if A, B, and C are the matrices in Example 4, then

$$\begin{aligned} 2A - B + \tfrac{1}{3}C &= 2A + (-1)B + \tfrac{1}{3}C \\ &= \begin{bmatrix} 4 & 6 & 8 \\ 2 & 6 & 2 \end{bmatrix} + \begin{bmatrix} 0 & -2 & -7 \\ 1 & -3 & 5 \end{bmatrix} + \begin{bmatrix} 3 & -2 & 1 \\ 1 & 0 & 4 \end{bmatrix} = \begin{bmatrix} 7 & 2 & 2 \\ 4 & 3 & 11 \end{bmatrix} \end{aligned}$$

is the linear combination of A, B, and C with scalar coefficients 2, -1, and $\frac{1}{3}$.

Thus far we have defined multiplication of a matrix by a scalar but not the multiplication of two matrices. Since matrices are added by adding corresponding entries and subtracted by subtracting corresponding entries, it would seem natural to define multiplication of matrices by multiplying corresponding entries. However, it turns out that such a definition would not be very useful for most problems. Experience has led mathematicians to the following more useful definition of matrix multiplication.

Definition

If A is an $m \times r$ matrix and B is an $r \times n$ matrix, then the ***product*** AB is the $m \times n$ matrix whose entries are determined as follows. To find the entry in row i and column j of AB, single out row i from the matrix A and column j from the matrix B. Multiply the corresponding entries from the row and column together and then add up the resulting products.

EXAMPLE 5 Multiplying Matrices

Consider the matrices

$$A = \begin{bmatrix} 1 & 2 & 4 \\ 2 & 6 & 0 \end{bmatrix}, \quad B = \begin{bmatrix} 4 & 1 & 4 & 3 \\ 0 & -1 & 3 & 1 \\ 2 & 7 & 5 & 2 \end{bmatrix}$$

Since A is a 2×3 matrix and B is a 3×4 matrix, the product AB is a 2×4 matrix. To determine, for example, the entry in row 2 and column 3 of AB, we single out row 2 from A and column 3 from B. Then, as illustrated below, we multiply corresponding entries together and add up these products.

$$\begin{bmatrix} 1 & 2 & 4 \\ 2 & 6 & 0 \end{bmatrix} \begin{bmatrix} 4 & 1 & 4 & 3 \\ 0 & -1 & 3 & 1 \\ 2 & 7 & 5 & 2 \end{bmatrix} = \begin{bmatrix} \square & \square & \square & \square \\ \square & \square & 26 & \square \end{bmatrix}$$

$$(2 \cdot 4) + (6 \cdot 3) + (0 \cdot 5) = 26$$

The entry in row 1 and column 4 of AB is computed as follows.

$$\begin{bmatrix} 1 & 2 & 4 \\ 2 & 6 & 0 \end{bmatrix} \begin{bmatrix} 4 & 1 & 4 & 3 \\ 0 & -1 & 3 & 1 \\ 2 & 7 & 5 & 2 \end{bmatrix} = \begin{bmatrix} \square & \square & \square & 13 \\ \square & \square & \square & \square \end{bmatrix}$$

$$(1 \cdot 3) + (2 \cdot 1) + (4 \cdot 2) = 13$$

The computations for the remaining products are

$$\begin{aligned} (1 \cdot 4) + (2 \cdot 0) + (4 \cdot 2) &= 12 \\ (1 \cdot 1) - (2 \cdot 1) + (4 \cdot 7) &= 27 \\ (1 \cdot 4) + (2 \cdot 3) + (4 \cdot 5) &= 30 \\ (2 \cdot 4) + (6 \cdot 0) + (0 \cdot 2) &= 8 \\ (2 \cdot 1) - (6 \cdot 1) + (0 \cdot 7) &= -4 \\ (2 \cdot 3) + (6 \cdot 1) + (0 \cdot 2) &= 12 \end{aligned} \qquad AB = \begin{bmatrix} 12 & 27 & 30 & 13 \\ 8 & -4 & 26 & 12 \end{bmatrix}$$

♦

The definition of matrix multiplication requires that the number of columns of the first factor A be the same as the number of rows of the second factor B in order to form the product AB. If this condition is not satisfied, the product is undefined. A convenient way to determine whether a product of two matrices is defined is to write down the size of the first factor and, to the right of it, write down the size of the second factor. If, as in (3), the inside numbers are the same, then the product is defined. The outside numbers then give the size of the product.

$$\begin{array}{ccccc} A & & B & & AB \\ m \times r & & r \times n & = & m \times n \end{array} \tag{3}$$

Inside

Outside

EXAMPLE 6 Determining Whether a Product Is Defined

Suppose that A, B, and C are matrices with the following sizes:

$$\begin{array}{ccc} A & B & C \\ 3 \times 4 & 4 \times 7 & 7 \times 3 \end{array}$$

Then by (3), AB is defined and is a 3×7 matrix; BC is defined and is a 4×3 matrix; and CA is defined and is a 7×4 matrix. The products AC, CB, and BA are all undefined. ♦

In general, if $A = [a_{ij}]$ is an $m \times r$ matrix and $B = [b_{ij}]$ is an $r \times n$ matrix, then as illustrated by the shading in (4),

$$AB = \begin{bmatrix} a_{11} & a_{12} & \cdots & a_{1r} \\ a_{21} & a_{22} & \cdots & a_{2r} \\ \vdots & \vdots & & \vdots \\ a_{i1} & a_{i2} & \cdots & a_{ir} \\ \vdots & \vdots & & \vdots \\ a_{m1} & a_{m2} & \cdots & a_{mr} \end{bmatrix} \begin{bmatrix} b_{11} & b_{12} & \cdots & b_{1j} & \cdots & b_{1n} \\ b_{21} & b_{22} & \cdots & b_{2j} & \cdots & b_{2n} \\ \vdots & \vdots & & \vdots & & \vdots \\ b_{r1} & b_{r2} & \cdots & b_{rj} & \cdots & b_{rn} \end{bmatrix} \tag{4}$$

the entry $(AB)_{ij}$ in row i and column j of AB is given by

$$(AB)_{ij} = a_{i1}b_{1j} + a_{i2}b_{2j} + a_{i3}b_{3j} + \cdots + a_{ir}b_{rj} \tag{5}$$

Partitioned Matrices A matrix can be subdivided or ***partitioned*** into smaller matrices by inserting horizontal and vertical rules between selected rows and columns. For example, below are three possible partitions of a general 3×4 matrix A—the first is a partition of A into four ***submatrices*** A_{11}, A_{12}, A_{21}, and A_{22}; the second is a partition of A into its row matrices $\mathbf{r}_1$, $\mathbf{r}_2$, and $\mathbf{r}_3$; and the third is a partition of A into its column matrices $\mathbf{c}_1$, $\mathbf{c}_2$, $\mathbf{c}_3$, and $\mathbf{c}_4$:

$$A = \left[\begin{array}{ccc|c} a_{11} & a_{12} & a_{13} & a_{14} \\ a_{21} & a_{22} & a_{23} & a_{24} \\ \hline a_{31} & a_{32} & a_{33} & a_{34} \end{array}\right] = \begin{bmatrix} A_{11} & A_{12} \\ A_{21} & A_{22} \end{bmatrix}$$

$$A = \left[\begin{array}{cccc} a_{11} & a_{12} & a_{13} & a_{14} \\ \hline a_{21} & a_{22} & a_{23} & a_{24} \\ \hline a_{31} & a_{32} & a_{33} & a_{34} \end{array}\right] = \begin{bmatrix} \mathbf{r}_1 \\ \mathbf{r}_2 \\ \mathbf{r}_3 \end{bmatrix}$$

$$A = \left[\begin{array}{c|c|c|c} a_{11} & a_{12} & a_{13} & a_{14} \\ a_{21} & a_{22} & a_{23} & a_{24} \\ a_{31} & a_{32} & a_{33} & a_{34} \end{array}\right] = [\mathbf{c}_1 \quad \mathbf{c}_2 \quad \mathbf{c}_3 \quad \mathbf{c}_4]$$

Matrix Multiplication by Columns and by Rows Sometimes it may be desirable to find a particular row or column of a matrix product AB without computing the entire product. The following results, whose proofs are left as exercises, are useful for that purpose:

$$j\text{th column matrix of } AB = A[j\text{th column matrix of } B] \tag{6}$$

$$i\text{th row matrix of } AB = [i\text{th row matrix of } A]B \tag{7}$$

EXAMPLE 7 Example 5 Revisited

If A and B are the matrices in Example 5, then from (6) the second column matrix of AB can be obtained by the computation

$$\begin{bmatrix} 1 & 2 & 4 \\ 2 & 6 & 0 \end{bmatrix} \begin{bmatrix} 1 \\ -1 \\ 7 \end{bmatrix} = \begin{bmatrix} 27 \\ -4 \end{bmatrix}$$

Second column of B ↑ ↑ Second column of AB

and from (7) the first row matrix of AB can be obtained by the computation

$$\underbrace{[1 \quad 2 \quad 4]}_{\text{First row of } A}\begin{bmatrix} 4 & 1 & 4 & 3 \\ 0 & -1 & 3 & 1 \\ 2 & 7 & 5 & 2 \end{bmatrix} = \underbrace{[12 \quad 27 \quad 30 \quad 13]}_{\text{First row of } AB}$$

♦

If $\mathbf{a}_1, \mathbf{a}_2, \ldots, \mathbf{a}_m$ denote the row matrices of A and $\mathbf{b}_1, \mathbf{b}_2, \ldots, \mathbf{b}_n$ denote the column matrices of B, then it follows from Formulas (6) and (7) that

$$AB = A[\mathbf{b}_1 \quad \mathbf{b}_2 \quad \cdots \quad \mathbf{b}_n] = [A\mathbf{b}_1 \quad A\mathbf{b}_2 \quad \cdots \quad A\mathbf{b}_n] \tag{8}$$

(AB computed column by column)

$$AB = \begin{bmatrix} \mathbf{a}_1 \\ \mathbf{a}_2 \\ \vdots \\ \mathbf{a}_m \end{bmatrix} B = \begin{bmatrix} \mathbf{a}_1 B \\ \mathbf{a}_2 B \\ \vdots \\ \mathbf{a}_m B \end{bmatrix} \tag{9}$$

(AB computed row by row)

REMARK. Formulas (8) and (9) are special cases of a more general procedure for multiplying partitioned matrices (see Exercises 15–17).

Matrix Products as Linear Combinations Row and column matrices provide an alternative way of thinking about matrix multiplication. For example, suppose that

$$A = \begin{bmatrix} a_{11} & a_{12} & \cdots & a_{1n} \\ a_{21} & a_{22} & \cdots & a_{2n} \\ \vdots & \vdots & & \vdots \\ a_{m1} & a_{m2} & \cdots & a_{mn} \end{bmatrix} \quad \text{and} \quad \mathbf{x} = \begin{bmatrix} x_1 \\ x_2 \\ \vdots \\ x_n \end{bmatrix}$$

Then

$$A\mathbf{x} = \begin{bmatrix} a_{11}x_1 + a_{12}x_2 + \cdots + a_{1n}x_n \\ a_{21}x_1 + a_{22}x_2 + \cdots + a_{2n}x_n \\ \vdots \\ a_{m1}x_1 + a_{m2}x_2 + \cdots + a_{mn}x_n \end{bmatrix} = x_1\begin{bmatrix} a_{11} \\ a_{21} \\ \vdots \\ a_{m1} \end{bmatrix} + x_2\begin{bmatrix} a_{12} \\ a_{22} \\ \vdots \\ a_{m2} \end{bmatrix} + \cdots + x_n\begin{bmatrix} a_{1n} \\ a_{2n} \\ \vdots \\ a_{mn} \end{bmatrix} \tag{10}$$

In words, (10) tells us that *the product $A\mathbf{x}$ of a matrix A with a column matrix $\mathbf{x}$ is a linear combination of the column matrices of A with the coefficients coming from the matrix $\mathbf{x}$.* In the exercises we ask the reader to show that *the product $\mathbf{y}A$ of a $1 \times m$ matrix $\mathbf{y}$ with an $m \times n$ matrix A is a linear combination of the row matrices of A with scalar coefficients coming from $\mathbf{y}$.*

EXAMPLE 8 Linear Combinations

The matrix product

$$\begin{bmatrix} -1 & 3 & 2 \\ 1 & 2 & -3 \\ 2 & 1 & -2 \end{bmatrix} \begin{bmatrix} 2 \\ -1 \\ 3 \end{bmatrix} = \begin{bmatrix} 1 \\ -9 \\ -3 \end{bmatrix}$$

can be written as the linear combination of column matrices

$$2\begin{bmatrix} -1 \\ 1 \\ 2 \end{bmatrix} - 1\begin{bmatrix} 3 \\ 2 \\ 1 \end{bmatrix} + 3\begin{bmatrix} 2 \\ -3 \\ -2 \end{bmatrix} = \begin{bmatrix} 1 \\ -9 \\ -3 \end{bmatrix}$$

The matrix product

$$[1 \quad -9 \quad -3]\begin{bmatrix} -1 & 3 & 2 \\ 1 & 2 & -3 \\ 2 & 1 & -2 \end{bmatrix} = [-16 \quad -18 \quad 35]$$

can be written as the linear combination of row matrices

$$1[-1 \quad 3 \quad 2] - 9[1 \quad 2 \quad -3] - 3[2 \quad 1 \quad -2] = [-16 \quad -18 \quad 35]$$ ◆

It follows from (8) and (10) that *the jth column matrix of a product AB is a linear combination of the column matrices of A with the coefficients coming from the jth column of B.*

EXAMPLE 9 Columns of a Product AB as Linear Combinations

We showed in Example 5 that

$$AB = \begin{bmatrix} 1 & 2 & 4 \\ 2 & 6 & 0 \end{bmatrix} \begin{bmatrix} 4 & 1 & 4 & 3 \\ 0 & -1 & 3 & 1 \\ 2 & 7 & 5 & 2 \end{bmatrix} = \begin{bmatrix} 12 & 27 & 30 & 13 \\ 8 & -4 & 26 & 12 \end{bmatrix}$$

The column matrices of AB can be expressed as linear combinations of the column matrices of A as follows:

$$\begin{bmatrix} 12 \\ 8 \end{bmatrix} = 4\begin{bmatrix} 1 \\ 2 \end{bmatrix} + 0\begin{bmatrix} 2 \\ 6 \end{bmatrix} + 2\begin{bmatrix} 4 \\ 0 \end{bmatrix}$$

$$\begin{bmatrix} 27 \\ -4 \end{bmatrix} = \begin{bmatrix} 1 \\ 2 \end{bmatrix} - \begin{bmatrix} 2 \\ 6 \end{bmatrix} + 7\begin{bmatrix} 4 \\ 0 \end{bmatrix}$$

$$\begin{bmatrix} 30 \\ 26 \end{bmatrix} = 4\begin{bmatrix} 1 \\ 2 \end{bmatrix} + 3\begin{bmatrix} 2 \\ 6 \end{bmatrix} + 5\begin{bmatrix} 4 \\ 0 \end{bmatrix}$$

$$\begin{bmatrix} 13 \\ 12 \end{bmatrix} = 3\begin{bmatrix} 1 \\ 2 \end{bmatrix} + \begin{bmatrix} 2 \\ 6 \end{bmatrix} + 2\begin{bmatrix} 4 \\ 0 \end{bmatrix}$$ ◆

Matrix Form of a Linear System Matrix multiplication has an important application to systems of linear equations. Consider any system of m linear equations in n unknowns.

$$\begin{matrix} a_{11}x_1 & + & a_{12}x_2 & + \cdots + & a_{1n}x_n & = & b_1 \\ a_{21}x_1 & + & a_{22}x_2 & + \cdots + & a_{2n}x_n & = & b_2 \\ \vdots & & \vdots & & \vdots & & \vdots \\ a_{m1}x_1 & + & a_{m2}x_2 & + \cdots + & a_{mn}x_n & = & b_m \end{matrix}$$

Since two matrices are equal if and only if their corresponding entries are equal, we can replace the m equations in this system by the single matrix equation

$$\begin{bmatrix} a_{11}x_1 + a_{12}x_2 + \cdots + a_{1n}x_n \\ a_{21}x_1 + a_{22}x_2 + \cdots + a_{2n}x_n \\ \vdots \quad\quad \vdots \quad\quad\quad \vdots \\ a_{m1}x_1 + a_{m2}x_2 + \cdots + a_{mn}x_n \end{bmatrix} = \begin{bmatrix} b_1 \\ b_2 \\ \vdots \\ b_m \end{bmatrix}$$

The $m \times 1$ matrix on the left side of this equation can be written as a product to give

$$\begin{bmatrix} a_{11} & a_{12} & \cdots & a_{1n} \\ a_{21} & a_{22} & \cdots & a_{2n} \\ \vdots & \vdots & & \vdots \\ a_{m1} & a_{m2} & \cdots & a_{mn} \end{bmatrix} \begin{bmatrix} x_1 \\ x_2 \\ \vdots \\ x_n \end{bmatrix} = \begin{bmatrix} b_1 \\ b_2 \\ \vdots \\ b_m \end{bmatrix}$$

If we designate these matrices by A, $\mathbf{x}$, and $\mathbf{b}$, respectively, the original system of m equations in n unknowns has been replaced by the single matrix equation

$$A\mathbf{x} = \mathbf{b}$$

The matrix A in this equation is called the ***coefficient matrix*** of the system. The augmented matrix for the system is obtained by adjoining $\mathbf{b}$ to A as the last column; thus the augmented matrix is

$$[A \mid \mathbf{b}] = \left[\begin{array}{cccc|c} a_{11} & a_{12} & \cdots & a_{1n} & b_1 \\ a_{21} & a_{22} & \cdots & a_{2n} & b_2 \\ \vdots & \vdots & & \vdots & \vdots \\ a_{m1} & a_{m2} & \cdots & a_{mn} & b_m \end{array}\right]$$

Transpose of a Matrix We conclude this section by defining two matrix operations that have no analogs in the real numbers.

Definition

If A is any $m \times n$ matrix, then the ***transpose of A***, denoted by A^T, is defined to be the $n \times m$ matrix that results from interchanging the rows and columns of A; that is, the first column of A^T is the first row of A, the second column of A^T is the second row of A, and so forth.

EXAMPLE 10 Some Transposes

The following are some examples of matrices and their transposes.

$$A = \begin{bmatrix} a_{11} & a_{12} & a_{13} & a_{14} \\ a_{21} & a_{22} & a_{23} & a_{24} \\ a_{31} & a_{32} & a_{33} & a_{34} \end{bmatrix}, \quad B = \begin{bmatrix} 2 & 3 \\ 1 & 4 \\ 5 & 6 \end{bmatrix}, \quad C = [1 \ \ 3 \ \ 5], \quad D = [4]$$

$$A^T = \begin{bmatrix} a_{11} & a_{21} & a_{31} \\ a_{12} & a_{22} & a_{32} \\ a_{13} & a_{23} & a_{33} \\ a_{14} & a_{24} & a_{34} \end{bmatrix}, \quad B^T = \begin{bmatrix} 2 & 1 & 5 \\ 3 & 4 & 6 \end{bmatrix}, \quad C^T = \begin{bmatrix} 1 \\ 3 \\ 5 \end{bmatrix}, \quad D^T = [4]$$ ◆

Observe that not only are the columns of A^T the rows of A, but the rows of A^T are columns of A. Thus, the entry in row i and column j of A^T is the entry in row j and column i of A; that is,

$$(A^T)_{ij} = (A)_{ji} \tag{11}$$

Note the reversal of the subscripts.

In the special case where A is a square matrix, the transpose of A can be obtained by interchanging entries that are symmetrically positioned about the main diagonal. In (12) it is shown that A^T can also be obtained by "reflecting" A about its main diagonal.

$$A = \begin{bmatrix} 1 & -2 & 4 \\ 3 & 7 & 0 \\ -5 & 8 & 6 \end{bmatrix} \rightarrow \begin{bmatrix} 1 & -2 & 4 \\ 3 & 7 & 0 \\ -5 & 8 & 6 \end{bmatrix} \rightarrow A^T = \begin{bmatrix} 1 & 3 & -5 \\ -2 & 7 & 8 \\ 4 & 0 & 6 \end{bmatrix} \tag{12}$$

Interchange entries that are symmetrically positioned about the main diagonal.

Definition

If A is a square matrix, then the ***trace of A***, denoted by $\text{tr}(A)$, is defined to be the sum of the entries on the main diagonal of A. The trace of A is undefined if A is not a square matrix.

EXAMPLE 11 Trace of a Matrix

The following are examples of matrices and their traces.

$$A = \begin{bmatrix} a_{11} & a_{12} & a_{13} \\ a_{21} & a_{22} & a_{23} \\ a_{31} & a_{32} & a_{33} \end{bmatrix}, \quad B = \begin{bmatrix} -1 & 2 & 7 & 0 \\ 3 & 5 & -8 & 4 \\ 1 & 2 & 7 & -3 \\ 4 & -2 & 1 & 0 \end{bmatrix}$$

$$\text{tr}(A) = a_{11} + a_{22} + a_{33} \qquad \text{tr}(B) = -1 + 5 + 7 + 0 = 11$$

◆

Exercise Set 1.3

1. Suppose that A, B, C, D, and E are matrices with the following sizes:

A	B	C	D	E
(4×5)	(4×5)	(5×2)	(4×2)	(5×4)

Determine which of the following matrix expressions are defined. For those which are defined, give the size of the resulting matrix.

(a) BA (b) $AC + D$ (c) $AE + B$ (d) $AB + B$
(e) $E(A + B)$ (f) $E(AC)$ (g) E^TA (h) $(A^T + E)D$

2. Solve the following matrix equation for a, b, c, and d.

$$\begin{bmatrix} a-b & b+c \\ 3d+c & 2a-4d \end{bmatrix} = \begin{bmatrix} 8 & 1 \\ 7 & 6 \end{bmatrix}$$

3. Consider the matrices

$$A = \begin{bmatrix} 3 & 0 \\ -1 & 2 \\ 1 & 1 \end{bmatrix}, \quad B = \begin{bmatrix} 4 & -1 \\ 0 & 2 \end{bmatrix}, \quad C = \begin{bmatrix} 1 & 4 & 2 \\ 3 & 1 & 5 \end{bmatrix}, \quad D = \begin{bmatrix} 1 & 5 & 2 \\ -1 & 0 & 1 \\ 3 & 2 & 4 \end{bmatrix}, \quad E = \begin{bmatrix} 6 & 1 & 3 \\ -1 & 1 & 2 \\ 4 & 1 & 3 \end{bmatrix}$$

Compute the following (where possible).

(a) $D+E$ (b) $D-E$ (c) $5A$ (d) $-7C$
(e) $2B-C$ (f) $4E-2D$ (g) $-3(D+2E)$ (h) $A-A$
(i) $\text{tr}(D)$ (j) $\text{tr}(D-3E)$ (k) $4\ \text{tr}(7B)$ (l) $\text{tr}(A)$

4. Using the matrices in Exercise 3, compute the following (where possible).

(a) $2A^T+C$ (b) D^T-E^T (c) $(D-E)^T$ (d) B^T+5C^T
(e) $\frac{1}{2}C^T-\frac{1}{4}A$ (f) $B-B^T$ (g) $2E^T-3D^T$ (h) $(2E^T-3D^T)^T$

5. Using the matrices in Exercise 3, compute the following (where possible).

(a) AB (b) BA (c) $(3E)D$ (d) $(AB)C$
(e) $A(BC)$ (f) CC^T (g) $(DA)^T$ (h) $(C^TB)A^T$
(i) $\text{tr}(DD^T)$ (j) $\text{tr}(4E^T-D)$ (k) $\text{tr}(C^TA^T+2E^T)$

6. Using the matrices in Exercise 3, compute the following (where possible).

(a) $(2D^T-E)A$ (b) $(4B)C+2B$ (c) $(-AC)^T+5D^T$
(d) $(BA^T-2C)^T$ (e) $B^T(CC^T-A^TA)$ (f) $D^TE^T-(ED)^T$

7. Let

$$A = \begin{bmatrix} 3 & -2 & 7 \\ 6 & 5 & 4 \\ 0 & 4 & 9 \end{bmatrix} \quad \text{and} \quad B = \begin{bmatrix} 6 & -2 & 4 \\ 0 & 1 & 3 \\ 7 & 7 & 5 \end{bmatrix}$$

Use the method of Example 7 to find

(a) the first row of AB (b) the third row of AB (c) the second column of AB
(d) the first column of BA (e) the third row of AA (f) the third column of AA

8. Let A and B be the matrices in Exercise 7.

(a) Express each column matrix of AB as a linear combination of the column matrices of A.
(b) Express each column matrix of BA as a linear combination of the column matrices of B.

9. Let

$$\mathbf{y} = [y_1 \quad y_2 \quad \cdots \quad y_m] \quad \text{and} \quad A = \begin{bmatrix} a_{11} & a_{12} & \cdots & a_{1n} \\ a_{21} & a_{22} & \cdots & a_{2n} \\ \vdots & \vdots & & \vdots \\ a_{m1} & a_{m2} & \cdots & a_{mn} \end{bmatrix}$$

Show that the product $\mathbf{y}A$ can be expressed as a linear combination of the row matrices of A with the scalar coefficients coming from $\mathbf{y}$.

10. Let A and B be the matrices in Exercise 7.

(a) Use the result in Exercise 9 to express each row matrix of AB as a linear combination of the row matrices of B.
(b) Use the result in Exercise 9 to express each row matrix of BA as a linear combination of the row matrices of A.

11. Let C, D, and E be the matrices in Exercise 3. Using as few computations as possible, determine the entry in row 2 and column 3 of $C(DE)$.

12. (a) Show that if AB and BA are both defined, then AB and BA are square matrices.
(b) Show that if A is an $m \times n$ matrix and $A(BA)$ is defined, then B is an $n \times m$ matrix.

13. In each part find matrices A, $\mathbf{x}$, and $\mathbf{b}$ that express the given system of linear equations as a single matrix equation $A\mathbf{x} = \mathbf{b}$.

(a)
$$\begin{aligned} 2x_1 - 3x_2 + 5x_3 &= 7 \\ 9x_1 - x_2 + x_3 &= -1 \\ x_1 + 5x_2 + 4x_3 &= 0 \end{aligned}$$

(b)
$$\begin{aligned} 4x_1 \qquad\quad - 3x_3 + x_4 &= 1 \\ 5x_1 + x_2 \qquad\quad - 8x_4 &= 3 \\ 2x_1 - 5x_2 + 9x_3 - x_4 &= 0 \\ 3x_2 - x_3 + 7x_4 &= 2 \end{aligned}$$

14. In each part, express the matrix equation as a system of linear equations.

(a) $\begin{bmatrix} 3 & -1 & 2 \\ 4 & 3 & 7 \\ -2 & 1 & 5 \end{bmatrix}\begin{bmatrix} x_1 \\ x_2 \\ x_3 \end{bmatrix} = \begin{bmatrix} 2 \\ -1 \\ 4 \end{bmatrix}$ (b) $\begin{bmatrix} 3 & -2 & 0 & 1 \\ 5 & 0 & 2 & -2 \\ 3 & 1 & 4 & 7 \\ -2 & 5 & 1 & 6 \end{bmatrix}\begin{bmatrix} w \\ x \\ y \\ z \end{bmatrix} = \begin{bmatrix} 0 \\ 0 \\ 0 \\ 0 \end{bmatrix}$

15. If A and B are partitioned into submatrices, for example,

$$A = \left[\begin{array}{c|c} A_{11} & A_{12} \\ \hline A_{21} & A_{22} \end{array}\right] \quad \text{and} \quad B = \left[\begin{array}{c|c} B_{11} & B_{12} \\ \hline B_{21} & B_{22} \end{array}\right]$$

then AB can be expressed as

$$AB = \left[\begin{array}{c|c} A_{11}B_{11} + A_{12}B_{21} & A_{11}B_{12} + A_{12}B_{22} \\ \hline A_{21}B_{11} + A_{22}B_{21} & A_{21}B_{12} + A_{22}B_{22} \end{array}\right]$$

provided the sizes of the submatrices of A and B are such that the indicated operations can be performed. This method of multiplying partitioned matrices is called ***block multiplication***. In each part compute the product by block multiplication. Check your results by multiplying directly.

(a) $A = \left[\begin{array}{cc|cc} -1 & 2 & 1 & 5 \\ 0 & -3 & 4 & 2 \\ \hline 1 & 5 & 6 & 1 \end{array}\right], \quad B = \left[\begin{array}{cc|c} 2 & 1 & 4 \\ -3 & 5 & 2 \\ \hline 7 & -1 & 5 \\ 0 & 3 & -3 \end{array}\right]$

(b) $A = \left[\begin{array}{ccc|c} -1 & 2 & 1 & 5 \\ 0 & -3 & 4 & 2 \\ 1 & 5 & 6 & 1 \end{array}\right], \quad B = \left[\begin{array}{cc|c} 2 & 1 & 4 \\ -3 & 5 & 2 \\ 7 & -1 & 5 \\ \hline 0 & 3 & -3 \end{array}\right]$

16. Adapt the method of Exercise 15 to compute the following products by block multiplication.

(a) $\left[\begin{array}{ccc|c} 3 & -1 & 0 & -3 \\ 2 & 1 & 4 & 5 \end{array}\right]\left[\begin{array}{ccc} 2 & -4 & 1 \\ 3 & 0 & 2 \\ 1 & -3 & 5 \\ \hline 2 & 1 & 4 \end{array}\right]$ (b) $\left[\begin{array}{cc} 2 & -5 \\ 1 & 3 \\ 0 & 5 \\ \hline 1 & 4 \end{array}\right]\begin{bmatrix} 2 & -1 & 3 & -4 \\ 0 & 1 & 5 & 7 \end{bmatrix}$

(c) $\left[\begin{array}{ccc|cc} 1 & 0 & 0 & 0 & 0 \\ 0 & 1 & 0 & 0 & 0 \\ 0 & 0 & 1 & 0 & 0 \\ \hline 0 & 0 & 0 & 2 & 0 \\ 0 & 0 & 0 & -1 & 2 \end{array}\right]\left[\begin{array}{cc} 3 & 3 \\ -1 & 4 \\ 1 & 5 \\ \hline 2 & -2 \\ 1 & 6 \end{array}\right]$

17. In each part determine whether block multiplication can be used to compute AB from the given partitions. If so, compute the product by block multiplication.

(a) $A = \left[\begin{array}{rrr|r} -1 & 2 & 1 & 5 \\ 0 & -3 & 4 & 2 \\ \hline 1 & 5 & 6 & 1 \end{array}\right], \quad B = \left[\begin{array}{rr|r} 2 & 1 & 4 \\ -3 & 5 & 2 \\ \hline 7 & -1 & 5 \\ 0 & 3 & -3 \end{array}\right]$

(b) $A = \left[\begin{array}{rrrr} -1 & 2 & 1 & 5 \\ 0 & -3 & 4 & 2 \\ \hline 1 & 5 & 6 & 1 \end{array}\right], \quad B = \left[\begin{array}{r|r|r} 2 & 1 & 4 \\ -3 & 5 & 2 \\ 7 & -1 & 5 \\ 0 & 3 & -3 \end{array}\right]$

18. (a) Show that if A has a row of zeros and B is any matrix for which AB is defined, then AB also has a row of zeros.

(b) Find a similar result involving a column of zeros.

19. Let A be any $m \times n$ matrix and let 0 be the $m \times n$ matrix each of whose entries is zero. Show that if $kA = 0$, then $k = 0$ or $A = 0$.

20. Let I be the $n \times n$ matrix whose entry in row i and column j is

$$\begin{cases} 1 & \text{if} \quad i = j \\ 0 & \text{if} \quad i \neq j \end{cases}$$

Show that $AI = IA = A$ for every $n \times n$ matrix A.

21. In each part find a 6×6 matrix $[a_{ij}]$ that satisfies the stated condition. Make your answers as general as possible by using letters rather than specific numbers for the nonzero entries.

(a) $a_{ij} = 0$ if $i \neq j$ (b) $a_{ij} = 0$ if $i > j$ (c) $a_{ij} = 0$ if $i < j$ (d) $a_{ij} = 0$ if $|i - j| > 1$

22. Find the 4×4 matrix $A = [a_{ij}]$ whose entries satisfy the stated condition.

(a) $a_{ij} = i + j$ (b) $a_{ij} = i^{j-1}$ (c) $a_{ij} = \begin{cases} 1 & \text{if} \quad |i - j| > 1 \\ -1 & \text{if} \quad |i - j| \leq 1 \end{cases}$

23. Prove: If A and B are $n \times n$ matrices, then $\text{tr}(A + B) = \text{tr}(A) + \text{tr}(B)$.

Discussion and Discovery

24. Describe three different methods for computing a matrix product, and illustrate the methods by computing some product AB three different ways.

25. How many 3×3 matrices A can you find such that

$$A \begin{bmatrix} x \\ y \\ z \end{bmatrix} = \begin{bmatrix} x + y \\ x - y \\ 0 \end{bmatrix}$$

for all choices of x, y, and z?

26. How many 3×3 matrices A can you find such that

$$A \begin{bmatrix} x \\ y \\ z \end{bmatrix} = \begin{bmatrix} xy \\ 0 \\ 0 \end{bmatrix}$$

for all choices of x, y, and z?

27. A matrix B is said to be a ***square root*** of a matrix A if $BB = A$.

(a) Find two square roots of $A = \begin{bmatrix} 2 & 2 \\ 2 & 2 \end{bmatrix}$.

(b) How many different square roots can you find of $A = \begin{bmatrix} 5 & 0 \\ 0 & 9 \end{bmatrix}$?

(c) Do you think that every 2×2 matrix has at least one square root? Explain your reasoning.

28. Let 0 denote a 2×2 matrix, each of whose entries is zero.

(a) Is there a 2×2 matrix A such that $A \neq 0$ and $AA = 0$? Justify your answer.

(b) Is there a 2×2 matrix A such that $A \neq 0$ and $AA = A$? Justify your answer.

29. Indicate whether the statement is always true or sometimes false. Justify your answer with a logical argument or a counterexample.

(a) The expressions $\operatorname{tr}(AA^T)$ and $\operatorname{tr}(A^TA)$ are always defined, regardless of the size of A.

(b) $\operatorname{tr}(AA^T) = \operatorname{tr}(A^TA)$ for every matrix A.

(c) If the first column of A has all zeros, then so does the first column of every product AB.

(d) If the first row of A has all zeros, then so does the first row of every product AB.

30. Indicate whether the statement is always true or sometimes false. Justify your answer with a logical argument or a counterexample.

(a) If A is a square matrix with two identical rows, then AA has two identical rows.

(b) If A is a square matrix and AA has a column of zeros, then A must have a column of zeros.

(c) If B is an $n \times n$ matrix whose entries are positive even integers, and if A is an $n \times n$ matrix whose entries are positive integers, then the entries of AB and BA are positive even integers.

(d) If the matrix sum $AB + BA$ is defined, then A and B must be square.

1.4 INVERSES; RULES OF MATRIX ARITHMETIC

In this section we shall discuss some properties of the arithmetic operations on matrices. We shall see that many of the basic rules of arithmetic for real numbers also hold for matrices but a few do not.

Properties of Matrix Operations For real numbers a and b, we always have $ab = ba$, which is called the *commutative law for multiplication.* For matrices, however, AB and BA need not be equal. Equality can fail to hold for three reasons: It can happen that the product AB is defined but BA is undefined. For example, this is the case if A is a 2×3 matrix and B is a 3×4 matrix. Also, it can happen that AB and BA are both defined but have different sizes. This is the situation if A is a 2×3 matrix and B is a 3×2 matrix. Finally, as Example 1 shows, it is possible to have $AB \neq BA$ even if both AB and BA are defined and have the same size.

EXAMPLE 1 AB and BA Need Not Be Equal

Consider the matrices

$$A = \begin{bmatrix} -1 & 0 \\ 2 & 3 \end{bmatrix}, \quad B = \begin{bmatrix} 1 & 2 \\ 3 & 0 \end{bmatrix}$$

Multiplying gives

$$AB = \begin{bmatrix} -1 & -2 \\ 11 & 4 \end{bmatrix}, \quad BA = \begin{bmatrix} 3 & 6 \\ -3 & 0 \end{bmatrix}$$

Thus, $AB \neq BA$. ◆

Although the commutative law for multiplication is not valid in matrix arithmetic, many familiar laws of arithmetic are valid for matrices. Some of the most important ones and their names are summarized in the following theorem.

Theorem 1.4.1 Properties of Matrix Arithmetic

Assuming that the sizes of the matrices are such that the indicated operations can be performed, the following rules of matrix arithmetic are valid.

(a) $A+B=B+A$ **(Commutative law for addition)**
(b) $A+(B+C)=(A+B)+C$ **(Associative law for addition)**
(c) $A(BC)=(AB)C$ **(Associative law for multiplication)**
(d) $A(B+C)=AB+AC$ **(Left distributive law)**
(e) $(B+C)A=BA+CA$ **(Right distributive law)**
(f) $A(B-C)=AB-AC$
(g) $(B-C)A=BA-CA$
(h) $a(B+C)=aB+aC$
(i) $a(B-C)=aB-aC$
(j) $(a+b)C=aC+bC$
(k) $(a-b)C=aC-bC$
(l) $a(bC)=(ab)C$
(m) $a(BC)=(aB)C=B(aC)$

To prove the equalities in this theorem we must show that the matrix on the left side has the same size as the matrix on the right side and that corresponding entries on the two sides are equal. With the exception of the associative law in part (*c*), the proofs all follow the same general pattern. We shall prove part (*d*) as an illustration. The proof of the associative law, which is more complicated, is outlined in the exercises.

Proof (*d*). We must show that $A(B+C)$ and $AB+AC$ have the same size and that corresponding entries are equal. To form $A(B+C)$, the matrices B and C must have the same size, say $m \times n$, and the matrix A must then have m columns, so its size must be of the form $r \times m$. This makes $A(B+C)$ an $r \times n$ matrix. It follows that $AB+AC$ is also an $r \times n$ matrix and, consequently, $A(B+C)$ and $AB+AC$ have the same size.

Suppose that $A=[a_{ij}]$, $B=[b_{ij}]$, and $C=[c_{ij}]$. We want to show that corresponding entries of $A(B+C)$ and $AB+AC$ are equal; that is,

$$[A(B+C)]_{ij}=[AB+AC]_{ij}$$

for all values of i and j. But from the definitions of matrix addition and matrix multiplication we have

$$\begin{aligned}[A(B+C)]_{ij} &= a_{i1}(b_{1j}+c_{1j})+a_{i2}(b_{2j}+c_{2j})+\cdots+a_{im}(b_{mj}+c_{mj}) \\ &= (a_{i1}b_{1j}+a_{i2}b_{2j}+\cdots+a_{im}b_{mj})+(a_{i1}c_{1j}+a_{i2}c_{2j}+\cdots+a_{im}c_{mj}) \\ &= [AB]_{ij}+[AC]_{ij}=[AB+AC]_{ij}\end{aligned}$$ ■

REMARK. Although the operations of matrix addition and matrix multiplication were defined for pairs of matrices, associative laws (*b*) and (*c*) enable us to denote sums and products of three matrices as $A+B+C$ and ABC without inserting any parentheses. This is justified by the fact that no matter how parentheses are inserted, the associative laws guarantee that the same end result will be obtained. In general, *given any sum or any product of matrices, pairs of parentheses can be inserted or deleted anywhere within the expression without affecting the end result.*

EXAMPLE 2 Associativity of Matrix Multiplication

As an illustration of the associative law for matrix multiplication, consider

$$A = \begin{bmatrix} 1 & 2 \\ 3 & 4 \\ 0 & 1 \end{bmatrix}, \quad B = \begin{bmatrix} 4 & 3 \\ 2 & 1 \end{bmatrix}, \quad C = \begin{bmatrix} 1 & 0 \\ 2 & 3 \end{bmatrix}$$

Then

$$AB = \begin{bmatrix} 1 & 2 \\ 3 & 4 \\ 0 & 1 \end{bmatrix} \begin{bmatrix} 4 & 3 \\ 2 & 1 \end{bmatrix} = \begin{bmatrix} 8 & 5 \\ 20 & 13 \\ 2 & 1 \end{bmatrix} \quad \text{and} \quad BC = \begin{bmatrix} 4 & 3 \\ 2 & 1 \end{bmatrix} \begin{bmatrix} 1 & 0 \\ 2 & 3 \end{bmatrix} = \begin{bmatrix} 10 & 9 \\ 4 & 3 \end{bmatrix}$$

Thus,

$$(AB)C = \begin{bmatrix} 8 & 5 \\ 20 & 13 \\ 2 & 1 \end{bmatrix} \begin{bmatrix} 1 & 0 \\ 2 & 3 \end{bmatrix} = \begin{bmatrix} 18 & 15 \\ 46 & 39 \\ 4 & 3 \end{bmatrix}$$

and

$$A(BC) = \begin{bmatrix} 1 & 2 \\ 3 & 4 \\ 0 & 1 \end{bmatrix} \begin{bmatrix} 10 & 9 \\ 4 & 3 \end{bmatrix} = \begin{bmatrix} 18 & 15 \\ 46 & 39 \\ 4 & 3 \end{bmatrix}$$

so $(AB)C = A(BC)$, as guaranteed by Theorem 1.4.1*c*. ◆

Zero Matrices

A matrix, all of whose entries are zero, such as

$$\begin{bmatrix} 0 & 0 \\ 0 & 0 \end{bmatrix}, \quad \begin{bmatrix} 0 & 0 & 0 \\ 0 & 0 & 0 \\ 0 & 0 & 0 \end{bmatrix}, \quad \begin{bmatrix} 0 & 0 & 0 & 0 \\ 0 & 0 & 0 & 0 \end{bmatrix}, \quad \begin{bmatrix} 0 \\ 0 \\ 0 \\ 0 \end{bmatrix}, \quad [0]$$

is called a ***zero matrix***. A zero matrix will be denoted by 0; if it is important to emphasize the size, we shall write $0_{m \times n}$ for the $m \times n$ zero matrix. Moreover, in keeping with our convention of using boldface symbols for matrices with one column, we will denote a zero matrix with one column by $\mathbf{0}$.

If A is any matrix and 0 is the zero matrix with the same size, it is obvious that $A + 0 = 0 + A = A$. The matrix 0 plays much the same role in these matrix equations as the number 0 plays in the numerical equations $a + 0 = 0 + a = a$.

Since we already know that some of the rules of arithmetic for real numbers do not carry over to matrix arithmetic, it would be foolhardy to assume that all the properties of the real number zero carry over to zero matrices. For example, consider the following two standard results in the arithmetic of real numbers.

- If $ab = ac$ and $a \neq 0$, then $b = c$. (This is called the *cancellation law.*)
- If $ad = 0$, then at least one of the factors on the left is 0.

As the next example shows, the corresponding results are not generally true in matrix arithmetic.

EXAMPLE 3 The Cancellation Law Does Not Hold

Consider the matrices

$$A = \begin{bmatrix} 0 & 1 \\ 0 & 2 \end{bmatrix}, \quad B = \begin{bmatrix} 1 & 1 \\ 3 & 4 \end{bmatrix}, \quad C = \begin{bmatrix} 2 & 5 \\ 3 & 4 \end{bmatrix}, \quad D = \begin{bmatrix} 3 & 7 \\ 0 & 0 \end{bmatrix}$$

You should verify that

$$AB = AC = \begin{bmatrix} 3 & 4 \\ 6 & 8 \end{bmatrix} \quad \text{and} \quad AD = \begin{bmatrix} 0 & 0 \\ 0 & 0 \end{bmatrix}$$

Thus, although $A \neq 0$, it is *incorrect* to cancel the A from both sides of the equation $AB = AC$ and write $B = C$. Also, $AD = 0$, yet $A \neq 0$ and $D \neq 0$. Thus, the cancellation law is not valid for matrix multiplication, and it is possible for a product of matrices to be zero without either factor being zero. ♦

In spite of the above example, there are a number of familiar properties of the real number 0 that *do* carry over to zero matrices. Some of the more important ones are summarized in the next theorem. The proofs are left as exercises.

Theorem 1.4.2 Properties of Zero Matrices

Assuming that the sizes of the matrices are such that the indicated operations can be performed, the following rules of matrix arithmetic are valid.

(*a*) $A + 0 = 0 + A = A$

(*b*) $A - A = 0$

(*c*) $0 - A = -A$

(*d*) $A0 = 0; \quad 0A = 0$

Identity Matrices Of special interest are square matrices with 1's on the main diagonal and 0's off the main diagonal, such as

$$\begin{bmatrix} 1 & 0 \\ 0 & 1 \end{bmatrix}, \quad \begin{bmatrix} 1 & 0 & 0 \\ 0 & 1 & 0 \\ 0 & 0 & 1 \end{bmatrix}, \quad \begin{bmatrix} 1 & 0 & 0 & 0 \\ 0 & 1 & 0 & 0 \\ 0 & 0 & 1 & 0 \\ 0 & 0 & 0 & 1 \end{bmatrix}, \quad \text{and so on.}$$

A matrix of this form is called an ***identity matrix*** and is denoted by I. If it is important to emphasize the size, we shall write I_n for the $n \times n$ identity matrix.

If A is an $m \times n$ matrix, then, as illustrated in the next example,

$$AI_n = A \quad \text{and} \quad I_m A = A$$

Thus, an identity matrix plays much the same role in matrix arithmetic as the number 1 plays in the numerical relationships $a \cdot 1 = 1 \cdot a = a$.

EXAMPLE 4 Multiplication by an Identity Matrix

Consider the matrix

$$A = \begin{bmatrix} a_{11} & a_{12} & a_{13} \\ a_{21} & a_{22} & a_{23} \end{bmatrix}$$

Then

$$I_2A = \begin{bmatrix} 1 & 0 \\ 0 & 1 \end{bmatrix} \begin{bmatrix} a_{11} & a_{12} & a_{13} \\ a_{21} & a_{22} & a_{23} \end{bmatrix} = \begin{bmatrix} a_{11} & a_{12} & a_{13} \\ a_{21} & a_{22} & a_{23} \end{bmatrix} = A$$

and

$$AI_3 = \begin{bmatrix} a_{11} & a_{12} & a_{13} \\ a_{21} & a_{22} & a_{23} \end{bmatrix} \begin{bmatrix} 1 & 0 & 0 \\ 0 & 1 & 0 \\ 0 & 0 & 1 \end{bmatrix} = \begin{bmatrix} a_{11} & a_{12} & a_{13} \\ a_{21} & a_{22} & a_{23} \end{bmatrix} = A$$ ♦

As the next theorem shows, identity matrices arise naturally in studying reduced row-echelon forms of *square* matrices.

Theorem 1.4.3

If R is the reduced row-echelon form of an $n \times n$ matrix A, then either R has a row of zeros or R is the identity matrix I_n.

Proof. Suppose that the reduced row-echelon form of A is

$$R = \begin{bmatrix} r_{11} & r_{12} & \cdots & r_{1n} \\ r_{21} & r_{22} & \cdots & r_{2n} \\ \vdots & \vdots & & \vdots \\ r_{n1} & r_{n2} & \cdots & r_{nn} \end{bmatrix}$$

Either the last row in this matrix consists entirely of zeros or it does not. If not, the matrix contains no zero rows, and consequently each of the n rows has a leading entry of 1. Since these leading 1's occur progressively further to the right as we move down the matrix, each of these 1's must occur on the main diagonal. Since the other entries in the same column as one of these 1's are zero, R must be I_n. Thus, either R has a row of zeros or $R = I_n$. ■

Definition

If A is a square matrix, and if a matrix B of the same size can be found such that $AB = BA = I$, then A is said to be ***invertible*** and B is called an ***inverse*** of A. If no such matrix B can be found, then A is said to be ***singular***.

EXAMPLE 5 Verifying the Inverse Requirements

The matrix

$$B = \begin{bmatrix} 3 & 5 \\ 1 & 2 \end{bmatrix} \quad \text{is an inverse of} \quad A = \begin{bmatrix} 2 & -5 \\ -1 & 3 \end{bmatrix}$$

since

$$AB = \begin{bmatrix} 2 & -5 \\ -1 & 3 \end{bmatrix} \begin{bmatrix} 3 & 5 \\ 1 & 2 \end{bmatrix} = \begin{bmatrix} 1 & 0 \\ 0 & 1 \end{bmatrix} = I$$

and

$$BA = \begin{bmatrix} 3 & 5 \\ 1 & 2 \end{bmatrix} \begin{bmatrix} 2 & -5 \\ -1 & 3 \end{bmatrix} = \begin{bmatrix} 1 & 0 \\ 0 & 1 \end{bmatrix} = I$$ ♦

EXAMPLE 6 A Matrix with No Inverse

The matrix

$$A = \begin{bmatrix} 1 & 4 & 0 \\ 2 & 5 & 0 \\ 3 & 6 & 0 \end{bmatrix}$$

is singular. To see why, let

$$B = \begin{bmatrix} b_{11} & b_{12} & b_{13} \\ b_{21} & b_{22} & b_{23} \\ b_{31} & b_{32} & b_{33} \end{bmatrix}$$

be any 3×3 matrix. The third column of BA is

$$\begin{bmatrix} b_{11} & b_{12} & b_{13} \\ b_{21} & b_{22} & b_{23} \\ b_{31} & b_{32} & b_{33} \end{bmatrix} \begin{bmatrix} 0 \\ 0 \\ 0 \end{bmatrix} = \begin{bmatrix} 0 \\ 0 \\ 0 \end{bmatrix}$$

Thus,

$$BA \neq I = \begin{bmatrix} 1 & 0 & 0 \\ 0 & 1 & 0 \\ 0 & 0 & 1 \end{bmatrix}$$

♦

Properties of Inverses It is reasonable to ask whether an invertible matrix can have more than one inverse. The next theorem shows that the answer is no—*an invertible matrix has exactly one inverse.*

Theorem 1.4.4

If B and C are both inverses of the matrix A, then $B = C$.

Proof. Since B is an inverse of A, we have $BA = I$. Multiplying both sides on the right by C gives $(BA)C = IC = C$. But $(BA)C = B(AC) = BI = B$, so that $C = B$. ■

As a consequence of this important result, we can now speak of "the" inverse of an invertible matrix. If A is invertible, then its inverse will be denoted by the symbol A^{-1}. Thus,

$$AA^{-1} = I \quad \text{and} \quad A^{-1}A = I$$

The inverse of A plays much the same role in matrix arithmetic that the reciprocal a^{-1} plays in the numerical relationships $aa^{-1} = 1$ and $a^{-1}a = 1$.

In the next section we shall develop a method for finding inverses of invertible matrices of any size; however, the following theorem gives conditions under which a 2×2 matrix is invertible and provides a simple formula for the inverse.

Theorem 1.4.5

The matrix

$$A = \begin{bmatrix} a & b \\ c & d \end{bmatrix}$$

is invertible if $ad - bc \neq 0$, *in which case the inverse is given by the formula*

$$A^{-1} = \frac{1}{ad - bc}\begin{bmatrix} d & -b \\ -c & a \end{bmatrix} = \begin{bmatrix} \dfrac{d}{ad - bc} & -\dfrac{b}{ad - bc} \\ -\dfrac{c}{ad - bc} & \dfrac{a}{ad - bc} \end{bmatrix}$$

Proof. We leave it for the reader to verify that $AA^{-1} = I_2$ and $A^{-1}A = I_2$. ■

Theorem 1.4.6

If A and B are invertible matrices of the same size, then AB is invertible and

$$(AB)^{-1} = B^{-1}A^{-1}$$

Proof. If we can show that $(AB)(B^{-1}A^{-1}) = (B^{-1}A^{-1})(AB) = I$, then we will have simultaneously shown that the matrix AB is invertible and that $(AB)^{-1} = B^{-1}A^{-1}$. But $(AB)(B^{-1}A^{-1}) = A(BB^{-1})A^{-1} = AIA^{-1} = AA^{-1} = I$. A similar argument shows that $(B^{-1}A^{-1})(AB) = I$. ■

Although we will not prove it, this result can be extended to include three or more factors; that is,

A product of any number of invertible matrices is invertible, and the inverse of the product is the product of the inverses in the reverse order.

EXAMPLE 7 Inverse of a Product

Consider the matrices

$$A = \begin{bmatrix} 1 & 2 \\ 1 & 3 \end{bmatrix}, \quad B = \begin{bmatrix} 3 & 2 \\ 2 & 2 \end{bmatrix}, \quad AB = \begin{bmatrix} 7 & 6 \\ 9 & 8 \end{bmatrix}$$

Applying the formula in Theorem 1.4.5, we obtain

$$A^{-1} = \begin{bmatrix} 3 & -2 \\ -1 & 1 \end{bmatrix}, \quad B^{-1} = \begin{bmatrix} 1 & -1 \\ -1 & \frac{3}{2} \end{bmatrix}, \quad (AB)^{-1} = \begin{bmatrix} 4 & -3 \\ -\frac{9}{2} & \frac{7}{2} \end{bmatrix}$$

Also,

$$B^{-1}A^{-1} = \begin{bmatrix} 1 & -1 \\ -1 & \frac{3}{2} \end{bmatrix}\begin{bmatrix} 3 & -2 \\ -1 & 1 \end{bmatrix} = \begin{bmatrix} 4 & -3 \\ -\frac{9}{2} & \frac{7}{2} \end{bmatrix}$$

Therefore, $(AB)^{-1} = B^{-1}A^{-1}$ as guaranteed by Theorem 1.4.6. ♦

Powers of a Matrix Next, we shall define powers of a square matrix and discuss their properties.

Definition

If A is a square matrix, then we define the nonnegative integer powers of A to be

$$A^0 = I \qquad A^n = \underbrace{AA \cdots A}_{n \text{ factors}} \qquad (n > 0)$$

Moreover, if A is invertible, then we define the negative integer powers to be

$$A^{-n} = (A^{-1})^n = \underbrace{A^{-1}A^{-1} \cdots A^{-1}}_{n \text{ factors}}$$

Because this definition parallels that for real numbers, the usual laws of exponents hold. (We omit the details.)

Theorem 1.4.7 Laws of Exponents

If A is a square matrix and r and s are integers, then

$$A^r A^s = A^{r+s}, \qquad (A^r)^s = A^{rs}$$

The next theorem provides some useful properties of negative exponents.

Theorem 1.4.8 Laws of Exponents

If A is an invertible matrix, then:

(a) A^{-1} *is invertible and* $(A^{-1})^{-1} = A$.
(b) A^n *is invertible and* $(A^n)^{-1} = (A^{-1})^n$ *for* $n = 0, 1, 2, \ldots$.
(c) For any nonzero scalar k, the matrix kA is invertible and $(kA)^{-1} = \frac{1}{k}A^{-1}$.

Proof.

(a) Since $AA^{-1} = A^{-1}A = I$, the matrix A^{-1} is invertible and $(A^{-1})^{-1} = A$.
(b) This part is left as an exercise.
(c) If k is any nonzero scalar, results (*l*) and (*m*) of Theorem 1.4.1 enable us to write

$$(kA)\left(\frac{1}{k}A^{-1}\right) = \frac{1}{k}(kA)A^{-1} = \left(\frac{1}{k}k\right)AA^{-1} = (1)I = I$$

Similarly, $\left(\frac{1}{k}A^{-1}\right)(kA) = I$ so that kA is invertible and $(kA)^{-1} = \frac{1}{k}A^{-1}$. ■

EXAMPLE 8 Powers of a Matrix

Let A and A^{-1} be as in Example 7, that is,

$$A = \begin{bmatrix} 1 & 2 \\ 1 & 3 \end{bmatrix} \quad \text{and} \quad A^{-1} = \begin{bmatrix} 3 & -2 \\ -1 & 1 \end{bmatrix}$$

Then

$$A^3 = \begin{bmatrix} 1 & 2 \\ 1 & 3 \end{bmatrix}\begin{bmatrix} 1 & 2 \\ 1 & 3 \end{bmatrix}\begin{bmatrix} 1 & 2 \\ 1 & 3 \end{bmatrix} = \begin{bmatrix} 11 & 30 \\ 15 & 41 \end{bmatrix}$$

$$A^{-3} = (A^{-1})^3 = \begin{bmatrix} 3 & -2 \\ -1 & 1 \end{bmatrix}\begin{bmatrix} 3 & -2 \\ -1 & 1 \end{bmatrix}\begin{bmatrix} 3 & -2 \\ -1 & 1 \end{bmatrix} = \begin{bmatrix} 41 & -30 \\ -15 & 11 \end{bmatrix}$$ ◆

Polynomial Expressions Involving Matrices If A is a square matrix, say $m \times m$, and if

$$p(x) = a_0 + a_1 x + \cdots + a_n x^n \tag{1}$$

is any polynomial, then we define

$$p(A) = a_0 I + a_1 A + \cdots + a_n A^n$$

where I is the $m \times m$ identity matrix. In words, $p(A)$ is the $m \times m$ matrix that results when A is substituted for x in (1) and a_0 is replaced by $a_0 I$.

EXAMPLE 9 Matrix Polynomial

If

$$p(x) = 2x^2 - 3x + 4 \quad \text{and} \quad A = \begin{bmatrix} -1 & 2 \\ 0 & 3 \end{bmatrix}$$

then

$$p(A) = 2A^2 - 3A + 4I = 2\begin{bmatrix} -1 & 2 \\ 0 & 3 \end{bmatrix}^2 - 3\begin{bmatrix} -1 & 2 \\ 0 & 3 \end{bmatrix} + 4\begin{bmatrix} 1 & 0 \\ 0 & 1 \end{bmatrix}$$

$$= \begin{bmatrix} 2 & 8 \\ 0 & 18 \end{bmatrix} - \begin{bmatrix} -3 & 6 \\ 0 & 9 \end{bmatrix} + \begin{bmatrix} 4 & 0 \\ 0 & 4 \end{bmatrix} = \begin{bmatrix} 9 & 2 \\ 0 & 13 \end{bmatrix}$$ ◆

Properties of the Transpose The next theorem lists the main properties of the transpose operation.

Theorem 1.4.9 Properties of the Transpose

If the sizes of the matrices are such that the stated operations can be performed, then

(*a*) $((A)^T)^T = A$
(*b*) $(A + B)^T = A^T + B^T$ *and* $(A - B)^T = A^T - B^T$
(*c*) $(kA)^T = kA^T$, *where k is any scalar*
(*d*) $(AB)^T = B^T A^T$

Keeping in mind that transposing a matrix interchanges its rows and columns, parts (*a*), (*b*), and (*c*) should be self-evident. For example, part (*a*) states that interchanging rows and columns twice leaves a matrix unchanged; part (*b*) asserts that adding and then interchanging rows and columns yields the same result as first interchanging rows and columns, then adding; and part (*c*) asserts that multiplying by a scalar and then interchanging rows and columns yields the same result as first interchanging rows and columns, then multiplying by the scalar. Part (*d*) is not so obvious, so we give its proof.

Proof (*d*). Let $A = [a_{ij}]_{m\times r}$ and $B = [b_{ij}]_{r\times n}$ so that the products AB and B^TA^T can both be formed. We leave it for the reader to check that $(AB)^T$ and B^TA^T have the same size, namely $n \times m$. Thus, it only remains to show that corresponding entries of $(AB)^T$ and B^TA^T are the same; that is,

$$\left((AB)^T\right)_{ij} = (B^TA^T)_{ij} \tag{2}$$

Applying Formula (11) of Section 1.3 to the left side of this equation and using the definition of matrix multiplication, we obtain

$$\left((AB)^T\right)_{ij} = (AB)_{ji} = a_{j1}b_{1i} + a_{j2}b_{2i} + \cdots + a_{jr}b_{ri} \tag{3}$$

To evaluate the right side of (2) it will be convenient to let a'_{ij} and b'_{ij} denote the ijth entries of A^T and B^T, respectively, so

$$a'_{ij} = a_{ji} \quad \text{and} \quad b'_{ij} = b_{ji}$$

From these relationships and the definition of matrix multiplication we obtain

$$\begin{aligned}(B^TA^T)_{ij} &= b'_{i1}a'_{1j} + b'_{i2}a'_{2j} + \cdots + b'_{ir}a'_{rj}\\ &= b_{1i}a_{j1} + b_{2i}a_{j2} + \cdots + b_{ri}a_{jr}\\ &= a_{j1}b_{1i} + a_{j2}b_{2i} + \cdots + a_{jr}b_{ri}\end{aligned}$$

This, together with (3), proves (2). ■

Although we shall not prove it, part (*d*) of this theorem can be extended to include three or more factors; that is,

The transpose of a product of any number of matrices is equal to the product of their transposes in the reverse order.

REMARK. Note the similarity between this result and the result following Theorem 1.4.6 about the inverse of a product of matrices.

Invertibility of a Transpose The following theorem establishes a relationship between the inverse of an invertible matrix and the inverse of its transpose.

Theorem 1.4.10

If A is an invertible matrix, then A^T is also invertible and

$$(A^T)^{-1} = (A^{-1})^T \tag{4}$$

Proof. We can prove the invertibility of A^T and obtain (4) by showing that

$$A^T(A^{-1})^T = (A^{-1})^TA^T = I$$

But from part (*d*) of Theorem 1.4.9 and the fact that $I^T = I$ we have

$$\begin{aligned}A^T(A^{-1})^T &= (A^{-1}A)^T = I^T = I\\ (A^{-1})^TA^T &= (AA^{-1})^T = I^T = I\end{aligned}$$

which completes the proof. ■

EXAMPLE 10 Verifying Theorem 1.4.10

Consider the matrices

$$A = \begin{bmatrix} -5 & -3 \\ 2 & 1 \end{bmatrix}, \quad A^T = \begin{bmatrix} -5 & 2 \\ -3 & 1 \end{bmatrix}$$

Applying Theorem 1.4.5 yields

$$A^{-1} = \begin{bmatrix} 1 & 3 \\ -2 & -5 \end{bmatrix}, \quad (A^{-1})^T = \begin{bmatrix} 1 & -2 \\ 3 & -5 \end{bmatrix}, \quad (A^T)^{-1} = \begin{bmatrix} 1 & -2 \\ 3 & -5 \end{bmatrix}$$

As guaranteed by Theorem 1.4.10, these matrices satisfy (4). ◆

Exercise Set 1.4

1. Let

$$A = \begin{bmatrix} 2 & -1 & 3 \\ 0 & 4 & 5 \\ -2 & 1 & 4 \end{bmatrix}, \quad B = \begin{bmatrix} 8 & -3 & -5 \\ 0 & 1 & 2 \\ 4 & -7 & 6 \end{bmatrix}, \quad C = \begin{bmatrix} 0 & -2 & 3 \\ 1 & 7 & 4 \\ 3 & 5 & 9 \end{bmatrix}, \quad a = 4, \quad b = -7$$

Show that

(a) $A + (B + C) = (A + B) + C$ (b) $(AB)C = A(BC)$ (c) $(a + b)C = aC + bC$
(d) $a(B - C) = aB - aC$

2. Using the matrices and scalars in Exercise 1, verify that

(a) $a(BC) = (aB)C = B(aC)$ (b) $A(B - C) = AB - AC$ (c) $(B + C)A = BA + CA$
(d) $a(bC) = (ab)C$

3. Using the matrices and scalars in Exercise 1, verify that

(a) $(A^T)^T = A$ (b) $(A + B)^T = A^T + B^T$ (c) $(aC)^T = aC^T$ (d) $(AB)^T = B^TA^T$

4. Use Theorem 1.4.5 to compute the inverses of the following matrices.

(a) $A = \begin{bmatrix} 3 & 1 \\ 5 & 2 \end{bmatrix}$ (b) $B = \begin{bmatrix} 2 & -3 \\ 4 & 4 \end{bmatrix}$ (c) $C = \begin{bmatrix} 6 & 4 \\ -2 & -1 \end{bmatrix}$ (d) $D = \begin{bmatrix} 2 & 0 \\ 0 & 3 \end{bmatrix}$

5. Use the matrices A and B in Exercise 4 to verify that

(a) $(A^{-1})^{-1} = A$ (b) $(B^T)^{-1} = (B^{-1})^T$

6. Use the matrices A, B, and C in Exercise 4 to verify that

(a) $(AB)^{-1} = B^{-1}A^{-1}$ (b) $(ABC)^{-1} = C^{-1}B^{-1}A^{-1}$

7. In each part use the given information to find A.

(a) $A^{-1} = \begin{bmatrix} 2 & -1 \\ 3 & 5 \end{bmatrix}$ (b) $(7A)^{-1} = \begin{bmatrix} -3 & 7 \\ 1 & -2 \end{bmatrix}$

(c) $(5A^T)^{-1} = \begin{bmatrix} -3 & -1 \\ 5 & 2 \end{bmatrix}$ (d) $(I + 2A)^{-1} = \begin{bmatrix} -1 & 2 \\ 4 & 5 \end{bmatrix}$

8. Let A be the matrix

$$\begin{bmatrix} 2 & 0 \\ 4 & 1 \end{bmatrix}$$

Compute A^3, A^{-3}, and $A^2 - 2A + I$.

9. Let A be the matrix

$$\begin{bmatrix} 3 & 1 \\ 2 & 1 \end{bmatrix}$$

In each part find $p(A)$.

(a) $p(x) = x - 2$ (b) $p(x) = 2x^2 - x + 1$ (c) $p(x) = x^3 - 2x + 4$

10. Let $p_1(x) = x^2 - 9$, $p_2(x) = x + 3$, and $p_3(x) = x - 3$.

(a) Show that $p_1(A) = p_2(A)p_3(A)$ for the matrix A in Exercise 9.

(b) Show that $p_1(A) = p_2(A)p_3(A)$ for any square matrix A.

11. Find the inverse of

$$\begin{bmatrix} \cos\theta & \sin\theta \\ -\sin\theta & \cos\theta \end{bmatrix}$$

12. Find the inverse of

$$\begin{bmatrix} \frac{1}{2}(e^x + e^{-x}) & \frac{1}{2}(e^x - e^{-x}) \\ \frac{1}{2}(e^x - e^{-x}) & \frac{1}{2}(e^x + e^{-x}) \end{bmatrix}$$

13. Consider the matrix

$$A = \begin{bmatrix} a_{11} & 0 & \cdots & 0 \\ 0 & a_{22} & \cdots & 0 \\ \vdots & \vdots & & \vdots \\ 0 & 0 & \cdots & a_{nn} \end{bmatrix}$$

where $a_{11}a_{22}\cdots a_{nn} \neq 0$. Show that A is invertible and find its inverse.

14. Show that if a square matrix A satisfies $A^2 - 3A + I = 0$, then $A^{-1} = 3I - A$.

15. (a) Show that a matrix with a row of zeros cannot have an inverse.

(b) Show that a matrix with a column of zeros cannot have an inverse.

16. Is the sum of two invertible matrices necessarily invertible?

17. Let A and B be square matrices such that $AB = 0$. Show that if A is invertible, then $B = 0$.

18. Let A, B, and 0 be 2×2 matrices. Assuming that A is invertible, find a matrix C so that

$$\left[\begin{array}{c|c} A^{-1} & 0 \\ \hline C & A^{-1} \end{array}\right]$$

is the inverse of the partitioned matrix

$$\left[\begin{array}{c|c} A & 0 \\ \hline B & A \end{array}\right]$$

(See Exercise 15 of the preceding section.)

19. Use the result in Exercise 18 to find the inverses of the following matrices.

(a) $\begin{bmatrix} 1 & 1 & 0 & 0 \\ -1 & 1 & 0 & 0 \\ 1 & 1 & 1 & 1 \\ 1 & 1 & -1 & 1 \end{bmatrix}$ (b) $\begin{bmatrix} 1 & 1 & 0 & 0 \\ 0 & 1 & 0 & 0 \\ 0 & 0 & 1 & 1 \\ 0 & 0 & 0 & 1 \end{bmatrix}$

20. (a) Find a nonzero 3×3 matrix A such that $A^T = A$.

(b) Find a nonzero 3×3 matrix A such that $A^T = -A$.

21. A square matrix A is called ***symmetric*** if $A^T = A$ and ***skew-symmetric*** if $A^T = -A$. Show that if B is a square matrix, then

(a) BB^T and $B + B^T$ are symmetric (b) $B - B^T$ is skew-symmetric

22. If A is a square matrix and n is a positive integer, is it true that $(A^n)^T = (A^T)^n$? Justify your answer.

23. Let A be the matrix

$$\begin{bmatrix} 1 & 0 & 1 \\ 1 & 1 & 0 \\ 0 & 1 & 1 \end{bmatrix}$$

Determine whether A is invertible, and if so, find its inverse. [***Hint.*** Solve $AX = I$ by equating corresponding entries on the two sides.]

24. Prove:

(a) part (b) of Theorem 1.4.1 (b) part (i) of Theorem 1.4.1 (c) part (m) of Theorem 1.4.1

25. Apply parts (d) and (m) of Theorem 1.4.1 to the matrices A, B, and $(-1)C$ to derive the result in part (f).

26. Prove Theorem 1.4.2.

27. Consider the laws of exponents $A^r A^s = A^{r+s}$ and $(A^r)^s = A^{rs}$.

(a) Show that if A is any square matrix, these laws are valid for all nonnegative integer values of r and s.

(b) Show that if A is invertible, these laws hold for all negative integer values of r and s.

28. Show that if A is invertible and k is any nonzero scalar, then $(kA)^n = k^n A^n$ for all integer values of n.

29. (a) Show that if A is invertible and $AB = AC$, then $B = C$.

(b) Explain why part (a) and Example 3 do not contradict one another.

30. Prove part (c) of Theorem 1.4.1. [***Hint.*** Assume that A is $m \times n$, B is $n \times p$, and C is $p \times q$. The ijth entry on the left side is $l_{ij} = a_{i1}[BC]_{1j} + a_{i2}[BC]_{2j} + \cdots + a_{in}[BC]_{nj}$ and the ijth entry on the right side is $r_{ij} = [AB]_{i1}c_{1j} + [AB]_{i2}c_{2j} + \cdots + [AB]_{ip}c_{pj}$. Verify that $l_{ij} = r_{ij}$.]

Discussion and Discovery

31. Let A and B be square matrices with the same size.

(a) Give an example in which $(A+B)^2 \neq A^2 + 2AB + B^2$.

(b) Fill in the blank to create a matrix identity that is valid for all choices of A and B.
$(A+B)^2 = A^2 + B^2 +$ ____________.

32. Let A and B be square matrices with the same size.

(a) Give an example in which $(A+B)(A-B) \neq A^2 - B^2$.

(b) Let A and B be square matrices with the same size. Fill in the blank to create a matrix identity that is valid for all choices of A and B. $(A+B)(A-B) =$ ____________.

33. In the real number system the equation $a^2 = 1$ has exactly two solutions. Find at least eight different 3×3 matrices that satisfy the equation $A^2 = I_3$. [***Hint.*** Look for solutions in which all entries off the main diagonal are zero.]

34. A statement of the form "If p, then q" is logically equivalent to the statement "If not q, then not p." (The second statement is called the ***logical contrapositive*** of the first.) For example, the logical contrapositive of the statement "If it is raining, then the ground is wet" is "If the ground is not wet, then it is not raining."

(a) Find the logical contrapositive of the following statement: If A^T is singular, then A is singular.

(b) Is the statement true or false? Explain.

35. Let A and B be $n \times n$ matrices. Indicate whether the statement is always true or sometimes false. Justify each answer.

(a) $(AB)^2 = A^2B^2$ (b) $(A-B)^2 = (B-A)^2$

(c) $(AB^{-1})(BA^{-1}) = I_n$ (d) $AB \neq BA$.

1.5 ELEMENTARY MATRICES AND A METHOD FOR FINDING A^{-1}

In this section we shall develop an algorithm for finding the inverse of an invertible matrix. We shall also discuss some of the basic properties of invertible matrices.

We begin with the definition of a special type of matrix that can be used to carry out an elementary row operation by matrix multiplication.

Definition

An $n \times n$ matrix is called an ***elementary matrix*** if it can be obtained from the $n \times n$ identity matrix I_n by performing a single elementary row operation.

EXAMPLE 1 Elementary Matrices and Row Operations

Listed below are four elementary matrices and the operations that produce them.

$$\begin{bmatrix} 1 & 0 \\ 0 & -3 \end{bmatrix} \quad \begin{bmatrix} 1 & 0 & 0 & 0 \\ 0 & 0 & 0 & 1 \\ 0 & 0 & 1 & 0 \\ 0 & 1 & 0 & 0 \end{bmatrix} \quad \begin{bmatrix} 1 & 0 & 3 \\ 0 & 1 & 0 \\ 0 & 0 & 1 \end{bmatrix} \quad \begin{bmatrix} 1 & 0 & 0 \\ 0 & 1 & 0 \\ 0 & 0 & 1 \end{bmatrix}$$

Multiply the second row of I_2 by -3.

Interchange the second and fourth rows of I_4.

Add 3 times the third row of I_3 to the first row.

Multiply the first row of I_3 by 1. ♦

When a matrix A is multiplied on the *left* by an elementary matrix E, the effect is to perform an elementary row operation on A. This is the content of the following theorem, the proof of which is left for the exercises.

Theorem 1.5.1 Row Operations by Matrix Multiplication

If the elementary matrix E results from performing a certain row operation on I_m and if A is an $m \times n$ matrix, then the product EA is the matrix that results when this same row operation is performed on A.

EXAMPLE 2 Using Elementary Matrices

Consider the matrix

$$A = \begin{bmatrix} 1 & 0 & 2 & 3 \\ 2 & -1 & 3 & 6 \\ 1 & 4 & 4 & 0 \end{bmatrix}$$

and consider the elementary matrix

$$E = \begin{bmatrix} 1 & 0 & 0 \\ 0 & 1 & 0 \\ 3 & 0 & 1 \end{bmatrix}$$

which results from adding 3 times the first row of I_3 to the third row. The product EA is

$$EA = \begin{bmatrix} 1 & 0 & 2 & 3 \\ 2 & -1 & 3 & 6 \\ 4 & 4 & 10 & 9 \end{bmatrix}$$

which is precisely the same matrix that results when we add 3 times the first row of A to the third row. ♦

REMARK. Theorem 1.5.1 is primarily of theoretical interest and will be used for developing some results about matrices and systems of linear equations. Computationally, it is preferable to perform row operations directly rather than multiplying on the left by an elementary matrix.

If an elementary row operation is applied to an identity matrix I to produce an elementary matrix E, then there is a second row operation that, when applied to E, produces I back again. For example, if E is obtained by multiplying the ith row of I by a nonzero constant c, then I can be recovered if the ith row of E is multiplied by $1/c$. The various possibilities are listed in Table 1. The operations on the right side of this table are called the ***inverse operations*** of the corresponding operations on the left.

TABLE 1

Row Operation on I That Produces E	Row Operation on E That Reproduces I
Multiply row i by $c \neq 0$	Multiply row i by $1/c$
Interchange rows i and j	Interchange rows i and j
Add c times row i to row j	Add $-c$ times row i to row j

EXAMPLE 3 Row Operations and Inverse Row Operations

In each of the following, an elementary row operation is applied to the 2×2 identity matrix to obtain an elementary matrix E, then E is restored to the identity matrix by applying the inverse row operation.

$$\begin{bmatrix} 1 & 0 \\ 0 & 1 \end{bmatrix} \rightarrow \begin{bmatrix} 1 & 0 \\ 0 & 7 \end{bmatrix} \rightarrow \begin{bmatrix} 1 & 0 \\ 0 & 1 \end{bmatrix}$$

Multiply the second row by 7.

Multiply the second row by $\frac{1}{7}$.

$$\begin{bmatrix} 1 & 0 \\ 0 & 1 \end{bmatrix} \rightarrow \begin{bmatrix} 0 & 1 \\ 1 & 0 \end{bmatrix} \rightarrow \begin{bmatrix} 1 & 0 \\ 0 & 1 \end{bmatrix}$$

Interchange the first and second rows. Interchange the first and second rows.

$$\begin{bmatrix} 1 & 0 \\ 0 & 1 \end{bmatrix} \rightarrow \begin{bmatrix} 1 & 5 \\ 0 & 1 \end{bmatrix} \rightarrow \begin{bmatrix} 1 & 0 \\ 0 & 1 \end{bmatrix}$$

Add 5 times the second row to the first. Add −5 times the second row to the first. ◆

The next theorem gives an important property of elementary matrices.

Theorem 1.5.2

Every elementary matrix is invertible, and the inverse is also an elementary matrix.

Proof. If E is an elementary matrix, then E results from performing some row operation on I. Let E_0 be the matrix that results when the inverse of this operation is performed on I. Applying Theorem 1.5.1 and using the fact that inverse row operations cancel the effect of each other, it follows that

$$E_0 E = I \quad \text{and} \quad E E_0 = I$$

Thus, the elementary matrix E_0 is the inverse of E. ■

The next theorem establishes some fundamental relationships between invertibility, homogeneous linear systems, reduced row-echelon forms, and elementary matrices. These results are extremely important and will be used many times in later sections.

Theorem 1.5.3 Equivalent Statements

If A is an $n \times n$ matrix, then the following statements are equivalent, that is, all true or all false.

(*a*) *A is invertible.*
(*b*) *$A\mathbf{x} = \mathbf{0}$ has only the trivial solution.*
(*c*) *The reduced row-echelon form of A is I_n.*
(*d*) *A is expressible as a product of elementary matrices.*

Proof. We shall prove the equivalence by establishing the chain of implications: $(a) \Rightarrow (b) \Rightarrow (c) \Rightarrow (d) \Rightarrow (a)$.

$(a) \Rightarrow (b)$. Assume A is invertible and let $\mathbf{x}_0$ be any solution of $A\mathbf{x} = \mathbf{0}$; thus, $A\mathbf{x}_0 = \mathbf{0}$. Multiplying both sides of this equation by the matrix A^{-1} gives $A^{-1}(A\mathbf{x}_0) = A^{-1}\mathbf{0}$, or $(A^{-1}A)\mathbf{x}_0 = \mathbf{0}$, or $I\mathbf{x}_0 = \mathbf{0}$, or $\mathbf{x}_0 = \mathbf{0}$. Thus, $A\mathbf{x} = \mathbf{0}$ has only the trivial solution.

(b) ⇒ *(c)*. Let $A\mathbf{x} = \mathbf{0}$ be the matrix form of the system

$$\begin{aligned} a_{11}x_1 + a_{12}x_2 + \cdots + a_{1n}x_n &= 0 \\ a_{21}x_1 + a_{22}x_2 + \cdots + a_{2n}x_n &= 0 \\ \vdots \qquad \vdots \qquad\qquad \vdots \quad &\;\; \vdots \\ a_{n1}x_1 + a_{n2}x_2 + \cdots + a_{nn}x_n &= 0 \end{aligned} \tag{1}$$

and assume that the system has only the trivial solution. If we solve by Gauss–Jordan elimination, then the system of equations corresponding to the reduced row-echelon form of the augmented matrix will be

$$\begin{aligned} x_1 \qquad\qquad\quad &= 0 \\ x_2 \qquad\quad &= 0 \\ \ddots \quad & \\ x_n &= 0 \end{aligned} \tag{2}$$

Thus, the augmented matrix

$$\begin{bmatrix} a_{11} & a_{12} & \cdots & a_{1n} & 0 \\ a_{21} & a_{22} & \cdots & a_{2n} & 0 \\ \vdots & \vdots & & \vdots & \vdots \\ a_{n1} & a_{n2} & \cdots & a_{nn} & 0 \end{bmatrix}$$

for (1) can be reduced to the augmented matrix

$$\begin{bmatrix} 1 & 0 & 0 & \cdots & 0 & 0 \\ 0 & 1 & 0 & \cdots & 0 & 0 \\ 0 & 0 & 1 & \cdots & 0 & 0 \\ \vdots & \vdots & \vdots & & \vdots & \vdots \\ 0 & 0 & 0 & \cdots & 1 & 0 \end{bmatrix}$$

for (2) by a sequence of elementary row operations. If we disregard the last column (of zeros) in each of these matrices, we can conclude that the reduced row-echelon form of A is I_n.

(c) ⇒ *(d)*. Assume that the reduced row-echelon form of A is I_n, so that A can be reduced to I_n by a finite sequence of elementary row operations. By Theorem 1.5.1 each of these operations can be accomplished by multiplying on the left by an appropriate elementary matrix. Thus, we can find elementary matrices $E_1, E_2, \ldots, E_k$ such that

$$E_k \cdots E_2 E_1 A = I_n \tag{3}$$

By Theorem 1.5.2, $E_1, E_2, \ldots, E_k$ are invertible. Multiplying both sides of Equation (3) on the left successively by $E_k^{-1}, \ldots, E_2^{-1}, E_1^{-1}$ we obtain

$$A = E_1^{-1} E_2^{-1} \cdots E_k^{-1} I_n = E_1^{-1} E_2^{-1} \cdots E_k^{-1} \tag{4}$$

By Theorem 1.5.2, this equation expresses A as a product of elementary matrices.

(d) ⇒ *(a)*. If A is a product of elementary matrices, then from Theorems 1.4.6 and 1.5.2 the matrix A is a product of invertible matrices, and hence is invertible. ■

Row Equivalence If a matrix B can be obtained from a matrix A by performing a finite sequence of elementary row operations, then obviously we can get from B back to A by performing the inverses of these elementary row operations in reverse order. Matrices that can be obtained from one another by a finite sequence of elementary row operations are said to be ***row equivalent***. With this terminology it follows from parts (*a*) and (*c*) of Theorem 1.5.3 that an $n \times n$ matrix A is invertible if and only if it is row equivalent to the $n \times n$ identity matrix.

A Method for Inverting Matrices As our first application of Theorem 1.5.3, we shall establish a method for determining the inverse of an invertible matrix. Multiplying (3) on the right by A^{-1} yields

$$A^{-1} = E_k \cdots E_2 E_1 I_n \tag{5}$$

which tells us that A^{-1} can be obtained by multiplying I_n successively on the left by the elementary matrices $E_1, E_2, \ldots, E_k$. Since each multiplication on the left by one of these elementary matrices performs a row operation, it follows, by comparing Equations (3) and (5), that *the sequence of row operations that reduces A to I_n will reduce I_n to A^{-1}*. Thus, we have the following result:

To find the inverse of an invertible matrix A, we must find a sequence of elementary row operations that reduces A to the identity and then perform this same sequence of operations on I_n to obtain A^{-1}.

A simple method for carrying out this procedure is given in the following example.

EXAMPLE 4 Using Row Operations to Find A^{-1}

Find the inverse of

$$A = \begin{bmatrix} 1 & 2 & 3 \\ 2 & 5 & 3 \\ 1 & 0 & 8 \end{bmatrix}$$

Solution.

We want to reduce A to the identity matrix by row operations and simultaneously apply these operations to I to produce A^{-1}. To accomplish this we shall adjoin the identity matrix to the right side of A, thereby producing a matrix of the form

$$[A \mid I]$$

Then we shall apply row operations to this matrix until the left side is reduced to I; these operations will convert the right side to A^{-1}, so that the final matrix will have the form

$$[I \mid A^{-1}]$$

The computations are as follows:

$$\left[\begin{array}{rrr|rrr} 1 & 2 & 3 & 1 & 0 & 0 \\ 2 & 5 & 3 & 0 & 1 & 0 \\ 1 & 0 & 8 & 0 & 0 & 1 \end{array}\right]$$

$$\left[\begin{array}{rrr|rrr} 1 & 2 & 3 & 1 & 0 & 0 \\ 0 & 1 & -3 & -2 & 1 & 0 \\ 0 & -2 & 5 & -1 & 0 & 1 \end{array}\right]$$

← We added −2 times the first row to the second and −1 times the first row to the third.

$$\left[\begin{array}{rrr|rrr} 1 & 2 & 3 & 1 & 0 & 0 \\ 0 & 1 & -3 & -2 & 1 & 0 \\ 0 & 0 & -1 & -5 & 2 & 1 \end{array}\right]$$

← We added 2 times the second row to the third.

$$\left[\begin{array}{ccc|ccc} 1 & 2 & 3 & 1 & 0 & 0 \\ 0 & 1 & -3 & -2 & 1 & 0 \\ 0 & 0 & 1 & 5 & -2 & -1 \end{array}\right]$$

← We multiplied the third row by −1.

$$\left[\begin{array}{ccc|ccc} 1 & 2 & 0 & -14 & 6 & 3 \\ 0 & 1 & 0 & 13 & -5 & -3 \\ 0 & 0 & 1 & 5 & -2 & -1 \end{array}\right]$$

← We added 3 times the third row to the second and −3 times the third row to the first.

$$\left[\begin{array}{ccc|ccc} 1 & 0 & 0 & -40 & 16 & 9 \\ 0 & 1 & 0 & 13 & -5 & -3 \\ 0 & 0 & 1 & 5 & -2 & -1 \end{array}\right]$$

← We added −2 times the second row to the first.

Thus,

$$A^{-1} = \begin{bmatrix} -40 & 16 & 9 \\ 13 & -5 & -3 \\ 5 & -2 & -1 \end{bmatrix}$$ ♦

Often it will not be known in advance whether a given matrix is invertible. If an $n \times n$ matrix A is not invertible, then it cannot be reduced to I_n by elementary row operations [part (c) of Theorem 1.5.3]. Stated another way, the reduced row-echelon form of A has at least one row of zeros. Thus, if the procedure in the last example is attempted on a matrix that is not invertible, then at some point in the computations a row of zeros will occur on the *left side*. It can then be concluded that the given matrix is not invertible, and the computations can be stopped.

EXAMPLE 5 Showing That a Matrix Is Not Invertible

Consider the matrix

$$A = \begin{bmatrix} 1 & 6 & 4 \\ 2 & 4 & -1 \\ -1 & 2 & 5 \end{bmatrix}$$

Applying the procedure of Example 4 yields

$$\left[\begin{array}{ccc|ccc} 1 & 6 & 4 & 1 & 0 & 0 \\ 2 & 4 & -1 & 0 & 1 & 0 \\ -1 & 2 & 5 & 0 & 0 & 1 \end{array}\right]$$

$$\left[\begin{array}{ccc|ccc} 1 & 6 & 4 & 1 & 0 & 0 \\ 0 & -8 & -9 & -2 & 1 & 0 \\ 0 & 8 & 9 & 1 & 0 & 1 \end{array}\right]$$

← We added −2 times the first row to the second and added the first row to the third.

$$\left[\begin{array}{ccc|ccc} 1 & 6 & 4 & 1 & 0 & 0 \\ 0 & -8 & -9 & -2 & 1 & 0 \\ 0 & 0 & 0 & -1 & 1 & 1 \end{array}\right]$$

← We added the second row to the third.

Since we have obtained a row of zeros on the left side, A is not invertible. ♦

EXAMPLE 6 A Consequence of Invertibility

In Example 4 we showed that

$$A = \begin{bmatrix} 1 & 2 & 3 \\ 2 & 5 & 3 \\ 1 & 0 & 8 \end{bmatrix}$$

is an invertible matrix. From Theorem 1.5.3 it follows that the homogeneous system

$$\begin{aligned} x_1 + 2x_2 + 3x_3 &= 0 \\ 2x_1 + 5x_2 + 3x_3 &= 0 \\ x_1 \qquad\quad + 8x_3 &= 0 \end{aligned}$$

has only the trivial solution. ♦

Exercise Set 1.5

1. Which of the following are elementary matrices?

(a) $\begin{bmatrix} 1 & 0 \\ -5 & 1 \end{bmatrix}$ (b) $\begin{bmatrix} -5 & 1 \\ 1 & 0 \end{bmatrix}$ (c) $\begin{bmatrix} 1 & 0 \\ 0 & \sqrt{3} \end{bmatrix}$ (d) $\begin{bmatrix} 0 & 0 & 1 \\ 0 & 1 & 0 \\ 1 & 0 & 0 \end{bmatrix}$

(e) $\begin{bmatrix} 1 & 1 & 0 \\ 0 & 0 & 1 \\ 0 & 0 & 0 \end{bmatrix}$ (f) $\begin{bmatrix} 1 & 0 & 0 \\ 0 & 1 & 9 \\ 0 & 0 & 1 \end{bmatrix}$ (g) $\begin{bmatrix} 2 & 0 & 0 & 2 \\ 0 & 1 & 0 & 0 \\ 0 & 0 & 1 & 0 \\ 0 & 0 & 0 & 1 \end{bmatrix}$

2. Find a row operation that will restore the given elementary matrix to an identity matrix.

(a) $\begin{bmatrix} 1 & 0 \\ -3 & 1 \end{bmatrix}$ (b) $\begin{bmatrix} 1 & 0 & 0 \\ 0 & 1 & 0 \\ 0 & 0 & 3 \end{bmatrix}$ (c) $\begin{bmatrix} 0 & 0 & 0 & 1 \\ 0 & 1 & 0 & 0 \\ 0 & 0 & 1 & 0 \\ 1 & 0 & 0 & 0 \end{bmatrix}$ (d) $\begin{bmatrix} 1 & 0 & -\frac{1}{7} & 0 \\ 0 & 1 & 0 & 0 \\ 0 & 0 & 1 & 0 \\ 0 & 0 & 0 & 1 \end{bmatrix}$

3. Consider the matrices

$$A = \begin{bmatrix} 3 & 4 & 1 \\ 2 & -7 & -1 \\ 8 & 1 & 5 \end{bmatrix}, \quad B = \begin{bmatrix} 8 & 1 & 5 \\ 2 & -7 & -1 \\ 3 & 4 & 1 \end{bmatrix}, \quad C = \begin{bmatrix} 3 & 4 & 1 \\ 2 & -7 & -1 \\ 2 & -7 & 3 \end{bmatrix}$$

Find elementary matrices, E_1, E_2, E_3, and E_4 such that

(a) $E_1A = B$ (b) $E_2B = A$ (c) $E_3A = C$ (d) $E_4C = A$

4. In Exercise 3 is it possible to find an elementary matrix E such that $EB = C$? Justify your answer.

In Exercises 5–7 use the method shown in Examples 4 and 5 to find the inverse of the given matrix if the matrix is invertible and check your answer by multiplication.

5. (a) $\begin{bmatrix} 1 & 4 \\ 2 & 7 \end{bmatrix}$ (b) $\begin{bmatrix} -3 & 6 \\ 4 & 5 \end{bmatrix}$ (c) $\begin{bmatrix} 6 & -4 \\ -3 & 2 \end{bmatrix}$

6. (a) $\begin{bmatrix} 3 & 4 & -1 \\ 1 & 0 & 3 \\ 2 & 5 & -4 \end{bmatrix}$ (b) $\begin{bmatrix} -1 & 3 & -4 \\ 2 & 4 & 1 \\ -4 & 2 & -9 \end{bmatrix}$ (c) $\begin{bmatrix} 1 & 0 & 1 \\ 0 & 1 & 1 \\ 1 & 1 & 0 \end{bmatrix}$ (d) $\begin{bmatrix} 2 & 6 & 6 \\ 2 & 7 & 6 \\ 2 & 7 & 7 \end{bmatrix}$ (e) $\begin{bmatrix} 1 & 0 & 1 \\ -1 & 1 & 1 \\ 0 & 1 & 0 \end{bmatrix}$

7. (a) $\begin{bmatrix} \frac{1}{5} & \frac{1}{5} & -\frac{2}{5} \\ \frac{1}{5} & \frac{1}{5} & \frac{1}{10} \\ \frac{1}{5} & -\frac{4}{5} & \frac{1}{10} \end{bmatrix}$ (b) $\begin{bmatrix} \sqrt{2} & 3\sqrt{2} & 0 \\ -4\sqrt{2} & \sqrt{2} & 0 \\ 0 & 0 & 1 \end{bmatrix}$ (c) $\begin{bmatrix} 1 & 0 & 0 & 0 \\ 1 & 3 & 0 & 0 \\ 1 & 3 & 5 & 0 \\ 1 & 3 & 5 & 7 \end{bmatrix}$

(d) $\begin{bmatrix} -8 & 17 & 2 & \frac{1}{3} \\ 4 & 0 & \frac{2}{5} & -9 \\ 0 & 0 & 0 & 0 \\ -1 & 13 & 4 & 2 \end{bmatrix}$ (e) $\begin{bmatrix} 0 & 0 & 2 & 0 \\ 1 & 0 & 0 & 1 \\ 0 & -1 & 3 & 0 \\ 2 & 1 & 5 & -3 \end{bmatrix}$

8. Find the inverse of each of the following 4×4 matrices, where k_1, k_2, k_3, k_4, and k are all nonzero.

(a) $\begin{bmatrix} k_1 & 0 & 0 & 0 \\ 0 & k_2 & 0 & 0 \\ 0 & 0 & k_3 & 0 \\ 0 & 0 & 0 & k_4 \end{bmatrix}$ (b) $\begin{bmatrix} 0 & 0 & 0 & k_1 \\ 0 & 0 & k_2 & 0 \\ 0 & k_3 & 0 & 0 \\ k_4 & 0 & 0 & 0 \end{bmatrix}$ (c) $\begin{bmatrix} k & 0 & 0 & 0 \\ 1 & k & 0 & 0 \\ 0 & 1 & k & 0 \\ 0 & 0 & 1 & k \end{bmatrix}$

9. Consider the matrix

$$A = \begin{bmatrix} 1 & 0 \\ -5 & 2 \end{bmatrix}$$

(a) Find elementary matrices E_1 and E_2 such that $E_2 E_1 A = I$.
(b) Write A^{-1} as a product of two elementary matrices.
(c) Write A as a product of two elementary matrices.

10. In each part perform the stated row operation on

$$\begin{bmatrix} 2 & -1 & 0 \\ 4 & 5 & -3 \\ 1 & -4 & 7 \end{bmatrix}$$

by multiplying A on the left by a suitable elementary matrix. Check your answer in each case by performing the row operation directly on A.

(a) Interchange the first and third rows.
(b) Multiply the second row by $\frac{1}{3}$.
(c) Add twice the second row to the first row.

11. Express the matrix

$$A = \begin{bmatrix} 0 & 1 & 7 & 8 \\ 1 & 3 & 3 & 8 \\ -2 & -5 & 1 & -8 \end{bmatrix}$$

in the form $A = EFGR$, where E, F, and G are elementary matrices, and R is in row-echelon form.

12. Show that if

$$A = \begin{bmatrix} 1 & 0 & 0 \\ 0 & 1 & 0 \\ a & b & c \end{bmatrix}$$

is an elementary matrix, then at least one entry in the third row must be a zero.

13. Show that

$$A = \begin{bmatrix} 0 & a & 0 & 0 & 0 \\ b & 0 & c & 0 & 0 \\ 0 & d & 0 & e & 0 \\ 0 & 0 & f & 0 & g \\ 0 & 0 & 0 & h & 0 \end{bmatrix}$$

is not invertible for any values of the entries.

14. Prove that if A is an $m \times n$ matrix, there is an invertible matrix C such that CA is in reduced row-echelon form.

15. Prove that if A is an invertible matrix and B is row equivalent to A, then B is also invertible.

16. (a) Prove: If A and B are $m \times n$ matrices, then A and B are row equivalent if and only if A and B have the same reduced row-echelon form.

(b) Show that A and B are row equivalent, and find a sequence of elementary row operations that produces B from A.

$$A = \begin{bmatrix} 1 & 2 & 3 \\ 1 & 4 & 1 \\ 2 & 1 & 9 \end{bmatrix}, \quad B = \begin{bmatrix} 1 & 0 & 5 \\ 0 & 2 & -2 \\ 1 & 1 & 4 \end{bmatrix}$$

17. Prove Theorem 1.5.1.

Discussion and Discovery

18. Suppose that A is some unknown invertible matrix, but you know of a sequence of elementary row operations that produces the identity matrix when applied in succession to A. Explain how you can use the known information to find A.

19. Indicate whether the statement is always true or sometimes false. Justify your answer with a logical argument or a counterexample.

(a) Every square matrix can be expressed as a product of elementary matrices.

(b) The product of two elementary matrices is an elementary matrix.

(c) If A is invertible and a multiple of the first row of A is added to the second row, then the resulting matrix is invertible.

(d) If A is invertible and $AB = 0$, then it must be true that $B = 0$.

20. Indicate whether the statement is always true or sometimes false. Justify your answer with a logical argument or a counterexample.

(a) If A is a singular $n \times n$ matrix, then $A\mathbf{x} = \mathbf{0}$ has infinitely many solutions.

(b) If A is a singular $n \times n$ matrix, then the reduced row-echelon form of A has at least one row of zeros.

(c) If A^{-1} is expressible as a product of elementary matrices, then the homogeneous linear system $A\mathbf{x} = \mathbf{0}$ has only the trivial solution.

(d) If A is a singular $n \times n$ matrix, and B results by interchanging two rows of A, then B may or may not be singular.

21. Do you think that there is a 2×2 matrix A such that

$$A \begin{bmatrix} a & b \\ c & d \end{bmatrix} = \begin{bmatrix} b & d \\ a & c \end{bmatrix}$$

for all values of a, b, c, and d? Explain your reasoning.

1.6 FURTHER RESULTS ON SYSTEMS OF EQUATIONS AND INVERTIBILITY

In this section we shall establish more results about systems of linear equations and invertibility of matrices. Our work will lead to a new method for solving n equations in n unknowns.

A Basic Theorem In Section 1.1 we made the statement (based on Figure 1.1.1) that every linear system has either no solutions, one solution, or infinitely many solutions. We are now in a position to prove this fundamental result.

Theorem 1.6.1

Every system of linear equations has either no solutions, exactly one solution, or infinitely many solutions.

Proof. If $A\mathbf{x} = \mathbf{b}$ is a system of linear equations, exactly one of the following is true: (a) the system has no solutions, (b) the system has exactly one solution, or (c) the system has more than one solution. The proof will be complete if we can show that the system has infinitely many solutions in case (c).

Assume that $A\mathbf{x} = \mathbf{b}$ has more than one solution, and let $\mathbf{x}_0 = \mathbf{x}_1 - \mathbf{x}_2$, where $\mathbf{x}_1$ and $\mathbf{x}_2$ are any two distinct solutions. Because $\mathbf{x}_1$ and $\mathbf{x}_2$ are distinct, the matrix $\mathbf{x}_0$ is nonzero; moreover,

$$A\mathbf{x}_0 = A(\mathbf{x}_1 - \mathbf{x}_2) = A\mathbf{x}_1 - A\mathbf{x}_2 = \mathbf{b} - \mathbf{b} = \mathbf{0}$$

If we now let k be any scalar, then

$$\begin{aligned} A(\mathbf{x}_1 + k\mathbf{x}_0) &= A\mathbf{x}_1 + A(k\mathbf{x}_0) = A\mathbf{x}_1 + k(A\mathbf{x}_0) \\ &= \mathbf{b} + k\mathbf{0} = \mathbf{b} + \mathbf{0} = \mathbf{b} \end{aligned}$$

But this says that $\mathbf{x}_1 + k\mathbf{x}_0$ is a solution of $A\mathbf{x} = \mathbf{b}$. Since $\mathbf{x}_0$ is nonzero and there are infinitely many choices for k, the system $A\mathbf{x} = \mathbf{b}$ has infinitely many solutions. ■

Solving Linear Systems by Matrix Inversion Thus far, we have studied two methods for solving linear systems: Gaussian elimination and Gauss–Jordan elimination. The following theorem provides a new method for solving certain linear systems.

Theorem 1.6.2

If A is an invertible $n \times n$ matrix, then for each $n \times 1$ matrix $\mathbf{b}$, the system of equations $A\mathbf{x} = \mathbf{b}$ has exactly one solution, namely, $\mathbf{x} = A^{-1}\mathbf{b}$.

Proof. Since $A(A^{-1}\mathbf{b}) = \mathbf{b}$, it follows that $\mathbf{x} = A^{-1}\mathbf{b}$ is a solution of $A\mathbf{x} = \mathbf{b}$. To show that this is the only solution, we will assume that $\mathbf{x}_0$ is an arbitrary solution and then show that $\mathbf{x}_0$ must be the solution $A^{-1}\mathbf{b}$.

If $\mathbf{x}_0$ is any solution, then $A\mathbf{x}_0 = \mathbf{b}$. Multiplying both sides by A^{-1}, we obtain $\mathbf{x}_0 = A^{-1}\mathbf{b}$. ■

EXAMPLE 1 Solution of a Linear System Using A^{-1}

Consider the system of linear equations

$$\begin{aligned} x_1 + 2x_2 + 3x_3 &= 5 \\ 2x_1 + 5x_2 + 3x_3 &= 3 \\ x_1 \qquad\quad + 8x_3 &= 17 \end{aligned}$$

In matrix form this system can be written as $A\mathbf{x} = \mathbf{b}$, where

$$A = \begin{bmatrix} 1 & 2 & 3 \\ 2 & 5 & 3 \\ 1 & 0 & 8 \end{bmatrix}, \quad \mathbf{x} = \begin{bmatrix} x_1 \\ x_2 \\ x_3 \end{bmatrix}, \quad \mathbf{b} = \begin{bmatrix} 5 \\ 3 \\ 17 \end{bmatrix}$$

In Example 4 of the preceding section we showed that A is invertible and

$$A^{-1} = \begin{bmatrix} -40 & 16 & 9 \\ 13 & -5 & -3 \\ 5 & -2 & -1 \end{bmatrix}$$

By Theorem 1.6.2 the solution of the system is

$$\mathbf{x} = A^{-1}\mathbf{b} = \begin{bmatrix} -40 & 16 & 9 \\ 13 & -5 & -3 \\ 5 & -2 & -1 \end{bmatrix} \begin{bmatrix} 5 \\ 3 \\ 17 \end{bmatrix} = \begin{bmatrix} 1 \\ -1 \\ 2 \end{bmatrix}$$

or $x_1 = 1, x_2 = -1, x_3 = 2$. ◆

REMARK. Note that the method of Example 1 applies only when the system has as many equations as unknowns and the coefficient matrix is invertible.

Linear Systems with a Common Coefficient Matrix

Frequently, one is concerned with solving a sequence of systems

$$A\mathbf{x} = \mathbf{b}_1, \quad A\mathbf{x} = \mathbf{b}_2, \quad A\mathbf{x} = \mathbf{b}_3, \ldots, \quad A\mathbf{x} = \mathbf{b}_k$$

each of which has the same square coefficient matrix A. If A is invertible, then the solutions

$$\mathbf{x}_1 = A^{-1}\mathbf{b}_1, \quad \mathbf{x}_2 = A^{-1}\mathbf{b}_2, \quad \mathbf{x}_3 = A^{-1}\mathbf{b}_3, \ldots, \quad \mathbf{x}_k = A^{-1}\mathbf{b}_k$$

can be obtained with one matrix inversion and k matrix multiplications. However, a more efficient method is to form the matrix

$$[A \mid \mathbf{b}_1 \mid \mathbf{b}_2 \mid \cdots \mid \mathbf{b}_k] \tag{1}$$

in which the coefficient matrix A is "augmented" by all k of the matrices $\mathbf{b}_1, \mathbf{b}_2, \ldots, \mathbf{b}_k$. By reducing (1) to reduced row-echelon form we can solve all k systems at once by Gauss–Jordan elimination. This method has the added advantage that it applies even when A is not invertible.

EXAMPLE 2 Solving Two Linear Systems at Once

Solve the systems

$$\text{(a)} \quad \begin{aligned} x_1 + 2x_2 + 3x_3 &= 4 \\ 2x_1 + 5x_2 + 3x_3 &= 5 \\ x_1 \qquad\quad + 8x_3 &= 9 \end{aligned} \qquad \text{(b)} \quad \begin{aligned} x_1 + 2x_2 + 3x_3 &= 1 \\ 2x_1 + 5x_2 + 3x_3 &= 6 \\ x_1 \qquad\quad + 8x_3 &= -6 \end{aligned}$$

Solution.

The two systems have the same coefficient matrix. If we augment this coefficient matrix with the columns of constants on the right sides of these systems, we obtain

$$\left[\begin{array}{ccc|c|c} 1 & 2 & 3 & 4 & 1 \\ 2 & 5 & 3 & 5 & 6 \\ 1 & 0 & 8 & 9 & -6 \end{array}\right]$$

Reducing this matrix to reduced row-echelon form yields (verify)

$$\left[\begin{array}{ccc|c|c} 1 & 0 & 0 & 1 & 2 \\ 0 & 1 & 0 & 0 & 1 \\ 0 & 0 & 1 & 1 & -1 \end{array}\right]$$

It follows from the last two columns that the solution of system (a) is $x_1 = 1, x_2 = 0, x_3 = 1$ and of system (b) is $x_1 = 2, x_2 = 1, x_3 = -1$. ♦

Properties of Invertible Matrices Up to now, to show that an $n \times n$ matrix A is invertible, it has been necessary to find an $n \times n$ matrix B such that

$$AB = I \quad \text{and} \quad BA = I$$

The next theorem shows that if we produce an $n \times n$ matrix B satisfying *either* condition, then the other condition holds automatically.

Theorem 1.6.3

Let A be a square matrix.

(a) If B is a square matrix satisfying $BA = I$, then $B = A^{-1}$.
(b) If B is a square matrix satisfying $AB = I$, then $B = A^{-1}$.

We shall prove part (a) and leave part (b) as an exercise.

Proof (a). Assume that $BA = I$. If we can show that A is invertible, the proof can be completed by multiplying $BA = I$ on both sides by A^{-1} to obtain

$$BAA^{-1} = IA^{-1} \quad \text{or} \quad BI = IA^{-1} \quad \text{or} \quad B = A^{-1}$$

To show that A is invertible, it suffices to show that the system $A\mathbf{x} = \mathbf{0}$ has only the trivial solution (see Theorem 1.5.3). Let $\mathbf{x}_0$ be any solution of this system. If we multiply both sides of $A\mathbf{x}_0 = \mathbf{0}$ on the left by B, we obtain $BA\mathbf{x}_0 = B\mathbf{0}$ or $I\mathbf{x}_0 = \mathbf{0}$ or $\mathbf{x}_0 = \mathbf{0}$. Thus, the system of equations $A\mathbf{x} = \mathbf{0}$ has only the trivial solution. ■

We are now in a position to add two more statements that are equivalent to the four given in Theorem 1.5.3.

Theorem 1.6.4 **Equivalent Statements**

If A is an $n \times n$ matrix, then the following are equivalent.

(a) A is invertible.
(b) $A\mathbf{x} = \mathbf{0}$ has only the trivial solution.
(c) The reduced row-echelon form of A is I_n.
(d) A is expressible as a product of elementary matrices.
(e) $A\mathbf{x} = \mathbf{b}$ is consistent for every $n \times 1$ matrix $\mathbf{b}$.
(f) $A\mathbf{x} = \mathbf{b}$ has exactly one solution for every $n \times 1$ matrix $\mathbf{b}$.

Proof. Since we proved in Theorem 1.5.3 that (a), (b), (c), and (d) are equivalent, it will be sufficient to prove that $(a) \Rightarrow (f) \Rightarrow (e) \Rightarrow (a)$.

$(a) \Rightarrow (f)$. This was already proved in Theorem 1.6.2.

$(f) \Rightarrow (e)$. This is self-evident: If $A\mathbf{x} = \mathbf{b}$ has exactly one solution for every $n \times 1$ matrix $\mathbf{b}$, then $A\mathbf{x} = \mathbf{b}$ is consistent for every $n \times 1$ matrix $\mathbf{b}$.

$(e) \Rightarrow (a)$. If the system $A\mathbf{x} = \mathbf{b}$ is consistent for every $n \times 1$ matrix $\mathbf{b}$, then in particular, the systems

$$A\mathbf{x} = \begin{bmatrix} 1 \\ 0 \\ 0 \\ \vdots \\ 0 \end{bmatrix}, \quad A\mathbf{x} = \begin{bmatrix} 0 \\ 1 \\ 0 \\ \vdots \\ 0 \end{bmatrix}, \ldots, \quad A\mathbf{x} = \begin{bmatrix} 0 \\ 0 \\ 0 \\ \vdots \\ 1 \end{bmatrix}$$

are consistent. Let $\mathbf{x}_1, \mathbf{x}_2, \ldots, \mathbf{x}_n$ be solutions of the respective systems, and let us form an $n \times n$ matrix C having these solutions as columns. Thus, C has the form

$$C = [\mathbf{x}_1 \mid \mathbf{x}_2 \mid \cdots \mid \mathbf{x}_n]$$

As discussed in Section 1.3, the successive columns of the product AC will be

$$A\mathbf{x}_1, A\mathbf{x}_2, \ldots, A\mathbf{x}_n$$

Thus,

$$AC = [A\mathbf{x}_1 \mid A\mathbf{x}_2 \mid \cdots \mid A\mathbf{x}_n] = \begin{bmatrix} 1 & 0 & \cdots & 0 \\ 0 & 1 & \cdots & 0 \\ 0 & 0 & \cdots & 0 \\ \vdots & \vdots & & \vdots \\ 0 & 0 & \cdots & 1 \end{bmatrix} = I$$

By part (b) of Theorem 1.6.3 it follows that $C = A^{-1}$. Thus, A is invertible. ■

We know from earlier work that invertible matrix factors produce an invertible product. The following theorem, which will be proved later, looks at the converse: It shows that if the product of square matrices is invertible, then the factors themselves must be invertible.

Theorem 1.6.5

Let A and B be square matrices of the same size. If AB is invertible, then A and B must also be invertible.

In our later work the following fundamental problem will occur frequently in various contexts.

A Fundamental Problem. Let A be a fixed $m \times n$ matrix. Find all $m \times 1$ matrices $\mathbf{b}$ such that the system of equations $A\mathbf{x} = \mathbf{b}$ is consistent.

If A is an invertible matrix, Theorem 1.6.2 completely solves this problem by asserting that for *every* $m \times 1$ matrix $\mathbf{b}$, the linear system $A\mathbf{x} = \mathbf{b}$ has the unique solution $\mathbf{x} = A^{-1}\mathbf{b}$. If A is not square, or if A is square but not invertible, then Theorem 1.6.2 does not apply. In these cases the matrix $\mathbf{b}$ must usually satisfy certain conditions in order for $A\mathbf{x} = \mathbf{b}$ to be consistent. The following example illustrates how the elimination methods of Section 1.2 can be used to determine such conditions.

EXAMPLE 3 Determining Consistency by Elimination

What conditions must b_1, b_2, and b_3 satisfy in order for the system of equations

$$\begin{aligned} x_1 + x_2 + 2x_3 &= b_1 \\ x_1 \qquad + x_3 &= b_2 \\ 2x_1 + x_2 + 3x_3 &= b_3 \end{aligned}$$

to be consistent?

Solution.

The augmented matrix is

$$\begin{bmatrix} 1 & 1 & 2 & b_1 \\ 1 & 0 & 1 & b_2 \\ 2 & 1 & 3 & b_3 \end{bmatrix}$$

which can be reduced to row-echelon form as follows.

$$\begin{bmatrix} 1 & 1 & 2 & b_1 \\ 0 & -1 & -1 & b_2 - b_1 \\ 0 & -1 & -1 & b_3 - 2b_1 \end{bmatrix}$$ ← -1 times the first row was added to the second and -2 times the first row was added to the third.

$$\begin{bmatrix} 1 & 1 & 2 & b_1 \\ 0 & 1 & 1 & b_1 - b_2 \\ 0 & -1 & -1 & b_3 - 2b_1 \end{bmatrix}$$ ← The second row was multiplied by -1.

$$\begin{bmatrix} 1 & 1 & 2 & b_1 \\ 0 & 1 & 1 & b_1 - b_2 \\ 0 & 0 & 0 & b_3 - b_2 - b_1 \end{bmatrix}$$ ← The second row was added to the third.

It is now evident from the third row in the matrix that the system has a solution if and only if b_1, b_2, and b_3 satisfy the condition

$$b_3 - b_2 - b_1 = 0 \quad \text{or} \quad b_3 = b_1 + b_2$$

To express this condition another way, $A\mathbf{x} = \mathbf{b}$ is consistent if and only if $\mathbf{b}$ is a matrix of the form

$$\mathbf{b} = \begin{bmatrix} b_1 \\ b_2 \\ b_1 + b_2 \end{bmatrix}$$

where b_1 and b_2 are arbitrary. ♦

EXAMPLE 4 Determining Consistency by Elimination

What conditions must b_1, b_2, and b_3 satisfy in order for the system of equations

$$\begin{aligned} x_1 + 2x_2 + 3x_3 &= b_1 \\ 2x_1 + 5x_2 + 3x_3 &= b_2 \\ x_1 \qquad + 8x_3 &= b_3 \end{aligned}$$

to be consistent?

Solution.

The augmented matrix is

$$\begin{bmatrix} 1 & 2 & 3 & b_1 \\ 2 & 5 & 3 & b_2 \\ 1 & 0 & 8 & b_3 \end{bmatrix}$$

Reducing this to reduced row-echelon form yields (verify)

$$\begin{bmatrix} 1 & 0 & 0 & -40b_1 + 16b_2 + 9b_3 \\ 0 & 1 & 0 & 13b_1 - 5b_2 - 3b_3 \\ 0 & 0 & 1 & 5b_1 - 2b_2 - b_3 \end{bmatrix} \tag{2}$$

In this case there are no restrictions on b_1, b_2, and b_3; that is, the given system $A\mathbf{x} = \mathbf{b}$ has the unique solution

$$x_1 = -40b_1 + 16b_2 + 9b_3, \quad x_2 = 13b_1 - 5b_2 - 3b_3, \quad x_3 = 5b_1 - 2b_2 - b_3 \tag{3}$$

for all **b**. ♦

REMARK. Because the system $A\mathbf{x} = \mathbf{b}$ in the preceding example is consistent for all **b**, it follows from Theorem 1.6.4 that A is invertible. We leave it for the reader to verify that the formulas in (3) can also be obtained by calculating $\mathbf{x} = A^{-1}\mathbf{b}$.

Exercise Set 1.6

In Exercises 1–8 solve the system by inverting the coefficient matrix and using Theorem 1.6.2.

1. $\begin{aligned} x_1 + x_2 &= 2 \\ 5x_1 + 6x_2 &= 9 \end{aligned}$

2. $\begin{aligned} 4x_1 - 3x_2 &= -3 \\ 2x_1 - 5x_2 &= 9 \end{aligned}$

3. $\begin{aligned} x_1 + 3x_2 + x_3 &= 4 \\ 2x_1 + 2x_2 + x_3 &= -1 \\ 2x_1 + 3x_2 + x_3 &= 3 \end{aligned}$

4. $\begin{aligned} 5x_1 + 3x_2 + 2x_3 &= 4 \\ 3x_1 + 3x_2 + 2x_3 &= 2 \\ x_2 + x_3 &= 5 \end{aligned}$

5. $\begin{aligned} x + y + z &= 5 \\ x + y - 4z &= 10 \\ -4x + y + z &= 0 \end{aligned}$

6. $\begin{aligned} -x - 2y - 3z &= 0 \\ w + x + 4y + 4z &= 7 \\ w + 3x + 7y + 9z &= 4 \\ -w - 2x - 4y - 6z &= 6 \end{aligned}$

7. $\begin{aligned} 3x_1 + 5x_2 &= b_1 \\ x_1 + 2x_2 &= b_2 \end{aligned}$

8. $\begin{aligned} x_1 + 2x_2 + 3x_3 &= b_1 \\ 2x_1 + 5x_2 + 5x_3 &= b_2 \\ 3x_1 + 5x_2 + 8x_3 &= b_3 \end{aligned}$

9. Solve the following general system by inverting the coefficient matrix and using Theorem 1.6.2.

$$\begin{aligned} x_1 + 2x_2 + x_3 &= b_1 \\ x_1 - x_2 + x_3 &= b_2 \\ x_1 + x_2 &= b_3 \end{aligned}$$

Use the resulting formulas to find the solution if

(a) $b_1 = -1,\ b_2 = 3,\ b_3 = 4$ (b) $b_1 = 5,\ b_2 = 0,\ b_3 = 0$ (c) $b_1 = -1,\ b_2 = -1,\ b_3 = 3$

10. Solve the three systems in Exercise 9 using the method of Example 2.

In Exercises 11–14 use the method of Example 2 to solve the systems in all parts simultaneously.

11. $$\begin{aligned} x_1 - 5x_2 &= b_1 \\ 3x_1 + 2x_2 &= b_2 \end{aligned}$$

(a) $b_1 = 1,\ \ b_2 = 4$
(b) $b_1 = -2,\ \ b_2 = 5$

12. $$\begin{aligned} -x_1 + 4x_2 + x_3 &= b_1 \\ x_1 + 9x_2 - 2x_3 &= b_2 \\ 6x_1 + 4x_2 - 8x_3 &= b_3 \end{aligned}$$

(a) $b_1 = 0,\ \ b_2 = 1,\ \ b_3 = 0$
(b) $b_1 = -3,\ \ b_2 = 4,\ \ b_3 = -5$

13. $$\begin{aligned} 4x_1 - 7x_2 &= b_1 \\ x_1 + 2x_2 &= b_2 \end{aligned}$$

(a) $b_1 = 0,\ \ b_2 = 1$
(b) $b_1 = -4,\ \ b_2 = 6$
(c) $b_1 = -1,\ \ b_2 = 3$
(d) $b_1 = -5,\ \ b_2 = 1$

14. $$\begin{aligned} x_1 + 3x_2 + 5x_3 &= b_1 \\ -x_1 - 2x_2 \qquad &= b_2 \\ 2x_1 + 5x_2 + 4x_3 &= b_3 \end{aligned}$$

(a) $b_1 = 1,\ \ b_2 = 0,\ \ b_3 = -1$
(b) $b_1 = 0,\ \ b_2 = 1,\ \ b_3 = 1$
(c) $b_1 = -1,\ \ b_2 = -1,\ \ b_3 = 0$

15. The method of Example 2 can be used for linear systems with infinitely many solutions. Use that method to solve the systems in both parts at the same time.

(a) $$\begin{aligned} x_1 - 2x_2 + x_3 &= -2 \\ 2x_1 - 5x_2 + x_3 &= 1 \\ 3x_1 - 7x_2 + 2x_3 &= -1 \end{aligned}$$

(b) $$\begin{aligned} x_1 - 2x_2 + x_3 &= 1 \\ 2x_1 - 5x_2 + x_3 &= -1 \\ 3x_1 - 7x_2 + 2x_3 &= 0 \end{aligned}$$

In Exercises 16–19 find conditions that b's must satisfy for the system to be consistent.

16. $$\begin{aligned} 6x_1 - 4x_2 &= b_1 \\ 3x_1 - 2x_2 &= b_2 \end{aligned}$$

17. $$\begin{aligned} x_1 - 2x_2 + 5x_3 &= b_1 \\ 4x_1 - 5x_2 + 8x_3 &= b_2 \\ -3x_1 + 3x_2 - 3x_3 &= b_3 \end{aligned}$$

18. $$\begin{aligned} x_1 - 2x_2 - x_3 &= b_1 \\ -4x_1 + 5x_2 + 2x_3 &= b_2 \\ -4x_1 + 7x_2 + 4x_3 &= b_3 \end{aligned}$$

19. $$\begin{aligned} x_1 - x_2 + 3x_3 + 2x_1 &= b_1 \\ -2x_1 + x_2 + 5x_3 + x_1 &= b_2 \\ -3x_1 + 2x_2 + 2x_3 - x_1 &= b_3 \\ 4x_1 - 3x_2 + x_3 + 3x_1 &= b_4 \end{aligned}$$

20. Consider the matrices

$$A = \begin{bmatrix} 2 & 1 & 2 \\ 2 & 2 & -2 \\ 3 & 1 & 1 \end{bmatrix} \quad \text{and} \quad \mathbf{x} = \begin{bmatrix} x_1 \\ x_2 \\ x_3 \end{bmatrix}$$

(a) Show that the equation $A\mathbf{x} = \mathbf{x}$ can be rewritten as $(A - I)\mathbf{x} = \mathbf{0}$ and use this result to solve $A\mathbf{x} = \mathbf{x}$ for $\mathbf{x}$.

(b) Solve $A\mathbf{x} = 4\mathbf{x}$.

21. Solve the following matrix equation for X.

$$\begin{bmatrix} 1 & -1 & 1 \\ 2 & 3 & 0 \\ 0 & 2 & -1 \end{bmatrix} X = \begin{bmatrix} 2 & -1 & 5 & 7 & 8 \\ 4 & 0 & -3 & 0 & 1 \\ 3 & 5 & -7 & 2 & 1 \end{bmatrix}$$

22. In each part determine whether the homogeneous system has a nontrivial solution (without using pencil and paper); then state whether the given matrix is invertible.

(a) $$\begin{aligned} 2x_1 + x_2 - 3x_3 + x_4 &= 0 \\ 5x_2 + 4x_3 + 3x_4 &= 0 \\ x_3 + 2x_4 &= 0 \\ 3x_4 &= 0 \end{aligned} \qquad \begin{bmatrix} 2 & 1 & -3 & 1 \\ 0 & 5 & 4 & 3 \\ 0 & 0 & 1 & 2 \\ 0 & 0 & 0 & 3 \end{bmatrix}$$

(b) $$\begin{aligned} 5x_1 + x_2 + 4x_3 + x_4 &= 0 \\ 2x_3 - x_4 &= 0 \\ x_3 + x_4 &= 0 \\ 7x_4 &= 0 \end{aligned} \qquad \begin{bmatrix} 5 & 1 & 4 & 1 \\ 0 & 0 & 2 & -1 \\ 0 & 0 & 1 & 1 \\ 0 & 0 & 0 & 7 \end{bmatrix}$$

23. Let $A\mathbf{x} = \mathbf{0}$ be a homogeneous system of n linear equations in n unknowns that has only the trivial solution. Show that if k is any positive integer, then the system $A^k\mathbf{x} = \mathbf{0}$ also has only the trivial solution.

24. Let $A\mathbf{x} = \mathbf{0}$ be a homogeneous system of n linear equations in n unknowns, and let Q be an invertible $n \times n$ matrix. Show that $A\mathbf{x} = \mathbf{0}$ has just the trivial solution if and only if $(QA)\mathbf{x} = \mathbf{0}$ has just the trivial solution.

25. Let $A\mathbf{x} = \mathbf{b}$ be any consistent system of linear equations, and let $\mathbf{x}_1$ be a fixed solution. Show that every solution to the system can be written in the form $\mathbf{x} = \mathbf{x}_1 + \mathbf{x}_0$, where $\mathbf{x}_0$ is a solution to $A\mathbf{x} = \mathbf{0}$. Show also that every matrix of this form is a solution.

26. Use part (a) of Theorem 1.6.3 to prove part (b).

Discussion and Discovery

27. (a) If A is an $n \times n$ matrix and if $\mathbf{b}$ is an $n \times 1$ matrix, what conditions would you impose to ensure that the equation $\mathbf{x} = A\mathbf{x} + \mathbf{b}$ has a unique solution for $\mathbf{x}$?

(b) Assuming that your conditions are satisfied, find a formula for the solution in terms of an appropriate inverse.

28. Suppose that A is an invertible $n \times n$ matrix. Must the system of equations $A\mathbf{x} = \mathbf{x}$ have a unique solution? Explain your reasoning.

29. Is it possible to have $AB = I$ without B being the inverse of A? Explain your reasoning.

30. Create a theorem by rewriting Theorem 1.6.5 in contrapositive form (see Exercise 34 of Section 1.4).

1.7 DIAGONAL, TRIANGULAR, AND SYMMETRIC MATRICES

In this section we shall consider certain classes of matrices that have special forms. The matrices that we study in this section are among the most important kinds of matrices encountered in linear algebra and will arise in many different settings throughout the text.

Diagonal Matrices A square matrix in which all the entries off the main diagonal are zero is called a ***diagonal matrix***. Here are some examples.

$$\begin{bmatrix} 2 & 0 \\ 0 & -5 \end{bmatrix}, \quad \begin{bmatrix} 1 & 0 & 0 \\ 0 & 1 & 0 \\ 0 & 0 & 1 \end{bmatrix}, \quad \begin{bmatrix} 6 & 0 & 0 & 0 \\ 0 & -4 & 0 & 0 \\ 0 & 0 & 0 & 0 \\ 0 & 0 & 0 & 8 \end{bmatrix}$$

A general $n \times n$ diagonal matrix D can be written as

$$D = \begin{bmatrix} d_1 & 0 & \cdots & 0 \\ 0 & d_2 & \cdots & 0 \\ \vdots & \vdots & & \vdots \\ 0 & 0 & \cdots & d_n \end{bmatrix} \tag{1}$$

A diagonal matrix is invertible if and only if all of its diagonal entries are nonzero; in this case the inverse of (1) is

$$D^{-1} = \begin{bmatrix} 1/d_1 & 0 & \cdots & 0 \\ 0 & 1/d_2 & \cdots & 0 \\ \vdots & \vdots & & \vdots \\ 0 & 0 & \cdots & 1/d_n \end{bmatrix}$$

The reader should verify that $DD^{-1} = D^{-1}D = I$.

Powers of diagonal matrices are easy to compute; we leave it for the reader to verify that if D is the diagonal matrix (1) and k is a positive integer, then

$$D^k = \begin{bmatrix} d_1{}^k & 0 & \cdots & 0 \\ 0 & d_2{}^k & \cdots & 0 \\ \vdots & \vdots & & \vdots \\ 0 & 0 & \cdots & d_n{}^k \end{bmatrix}$$

EXAMPLE 1 Inverses and Powers of Diagonal Matrices

If

$$A = \begin{bmatrix} 1 & 0 & 0 \\ 0 & -3 & 0 \\ 0 & 0 & 2 \end{bmatrix}$$

then

$$A^{-1} = \begin{bmatrix} 1 & 0 & 0 \\ 0 & -\frac{1}{3} & 0 \\ 0 & 0 & \frac{1}{2} \end{bmatrix}, \quad A^5 = \begin{bmatrix} 1 & 0 & 0 \\ 0 & -243 & 0 \\ 0 & 0 & 32 \end{bmatrix}, \quad A^{-5} = \begin{bmatrix} 1 & 0 & 0 \\ 0 & -\frac{1}{243} & 0 \\ 0 & 0 & \frac{1}{32} \end{bmatrix}$$

◆

Matrix products that involve diagonal factors are especially easy to compute. For example,

$$\begin{bmatrix} d_1 & 0 & 0 \\ 0 & d_2 & 0 \\ 0 & 0 & d_3 \end{bmatrix} \begin{bmatrix} a_{11} & a_{12} & a_{13} & a_{14} \\ a_{21} & a_{22} & a_{23} & a_{24} \\ a_{31} & a_{32} & a_{33} & a_{34} \end{bmatrix} = \begin{bmatrix} d_1a_{11} & d_1a_{12} & d_1a_{13} & d_1a_{14} \\ d_2a_{21} & d_2a_{22} & d_2a_{23} & d_2a_{24} \\ d_3a_{31} & d_3a_{32} & d_3a_{33} & d_3a_{34} \end{bmatrix}$$

$$\begin{bmatrix} a_{11} & a_{12} & a_{13} \\ a_{21} & a_{22} & a_{23} \\ a_{31} & a_{32} & a_{33} \\ a_{41} & a_{42} & a_{43} \end{bmatrix} \begin{bmatrix} d_1 & 0 & 0 \\ 0 & d_2 & 0 \\ 0 & 0 & d_3 \end{bmatrix} = \begin{bmatrix} d_1a_{11} & d_2a_{12} & d_3a_{13} \\ d_1a_{21} & d_2a_{22} & d_3a_{23} \\ d_1a_{31} & d_2a_{32} & d_3a_{33} \\ d_1a_{41} & d_2a_{42} & d_3a_{43} \end{bmatrix}$$

In words, *to multiply a matrix A on the left by a diagonal matrix D, one can multiply successive rows of A by the successive diagonal entries of D, and to multiply A on the right by D one can multiply successive columns of A by the successive diagonal entries of D.*

Triangular Matrices A square matrix in which all the entries above the main diagonal are zero is called ***lower triangular***, and a square matrix in which all the entries below the main diagonal are zero is called ***upper triangular***. A matrix that is either upper triangular or lower triangular is called ***triangular***.

EXAMPLE 2 Upper and Lower Triangular Matrices

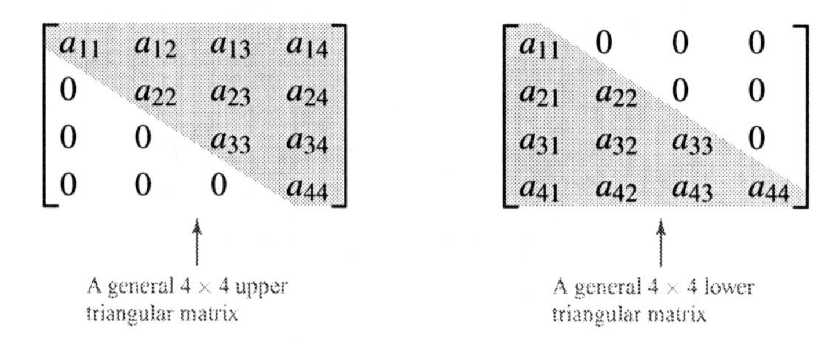

◆

REMARK. Observe that diagonal matrices are both upper triangular and lower triangular since they have zeros below and above the main diagonal. Observe also that a *square* matrix in row-echelon form is upper triangular since it has zeros below the main diagonal.

The following are four useful characterizations of triangular matrices. The reader will find it instructive to verify that the matrices in Example 2 have the stated properties.

- A square matrix $A = [a_{ij}]$ is upper triangular if and only if the ith row starts with at least $i - 1$ zeros.
- A square matrix $A = [a_{ij}]$ is lower triangular if and only if the jth column starts with at least $j - 1$ zeros.
- A square matrix $A = [a_{ij}]$ is upper triangular if and only if $a_{ij} = 0$ for $i > j$.
- A square matrix $A = [a_{ij}]$ is lower triangular if and only if $a_{ij} = 0$ for $i < j$.

The following theorem lists some of the basic properties of triangular matrices.

Theorem 1.7.1

(*a*) *The transpose of a lower triangular matrix is upper triangular, and the transpose of an upper triangular matrix is lower triangular.*

(*b*) *The product of lower triangular matrices is lower triangular, and the product of upper triangular matrices is upper triangular.*

(*c*) *A triangular matrix is invertible if and only if its diagonal entries are all nonzero.*

(*d*) *The inverse of an invertible lower triangular matrix is lower triangular, and the inverse of an invertible upper triangular matrix is upper triangular.*

Part (*a*) is evident from the fact that transposing a square matrix can be accomplished by reflecting the entries about the main diagonal; we omit the formal proof. We will prove (*b*), but we will defer the proofs of (*c*) and (*d*) to the next chapter, where we will have the tools to prove those results more efficiently.

Proof (*b*). We will prove the result for lower triangular matrices; the proof for upper triangular matrices is similar. Let $A = [a_{ij}]$ and $B = [b_{ij}]$ be lower triangular $n \times n$ matrices, and let $C = [c_{ij}]$ be the product $C = AB$. From the remark preceding this theorem, we can prove that C is lower triangular by showing that $c_{ij} = 0$ for $i < j$. But from the definition of matrix multiplication,

$$c_{ij} = a_{i1}b_{1j} + a_{i2}b_{2j} + \cdots + a_{in}b_{nj}$$

If we assume that $i < j$, then the terms in this expression can be grouped as follows:

$$c_{ij} = \underbrace{a_{i1}b_{1j} + a_{i2}b_{2j} + \cdots + a_{i(j-1)}b_{(j-1)j}}_{\substack{\text{Terms in which the row}\\ \text{number of } b \text{ is less than the}\\ \text{column number of } b}} + \underbrace{a_{ij}b_{jj} + \cdots + a_{in}b_{nj}}_{\substack{\text{Terms in which the row}\\ \text{number of } a \text{ is less than}\\ \text{the column number of } a}}$$

In the first grouping all of the b factors are zero since B is lower triangular, and in the second grouping all of the a factors are zero since A is lower triangular. Thus, $c_{ij} = 0$, which is what we wanted to prove. ■

EXAMPLE 3 Upper Triangular Matrices

Consider the upper triangular matrices

$$A = \begin{bmatrix} 1 & 3 & -1 \\ 0 & 2 & 4 \\ 0 & 0 & 5 \end{bmatrix}, \quad B = \begin{bmatrix} 3 & -2 & 2 \\ 0 & 0 & -1 \\ 0 & 0 & 1 \end{bmatrix}$$

The matrix A is invertible, since its diagonal entries are nonzero, but the matrix B is not. We leave it for the reader to calculate the inverse of A by the method of Section 1.5 and show that

$$A^{-1} = \begin{bmatrix} 1 & -\frac{3}{2} & \frac{7}{5} \\ 0 & \frac{1}{2} & -\frac{2}{5} \\ 0 & 0 & \frac{1}{5} \end{bmatrix}$$

This inverse is upper triangular, as guaranteed by part (d) of Theorem 1.7.1. We also leave it for the reader to check that the product AB is

$$AB = \begin{bmatrix} 3 & -2 & -2 \\ 0 & 0 & 2 \\ 0 & 0 & 5 \end{bmatrix}$$

This product is upper triangular, as guaranteed by part (b) of Theorem 1.7.1. ♦

Symmetric Matrices A square matrix A is called ***symmetric*** if $A = A^T$.

EXAMPLE 4 Symmetric Matrices

The following matrices are symmetric, since each is equal to its own transpose (verify).

$$\begin{bmatrix} 7 & -3 \\ -3 & 5 \end{bmatrix}, \quad \begin{bmatrix} 1 & 4 & 5 \\ 4 & -3 & 0 \\ 5 & 0 & 7 \end{bmatrix}, \quad \begin{bmatrix} d_1 & 0 & 0 & 0 \\ 0 & d_2 & 0 & 0 \\ 0 & 0 & d_3 & 0 \\ 0 & 0 & 0 & d_4 \end{bmatrix}$$

♦

It is easy to recognize symmetric matrices by inspection: The entries on the main diagonal may be arbitrary, but as shown in (2), "mirror images" of entries across the main diagonal must be equal.

$$\begin{bmatrix} 1 & 4 & 5 \\ 4 & -3 & 0 \\ 5 & 0 & 7 \end{bmatrix} \tag{2}$$

This follows from the fact that transposing a square matrix can be accomplished by interchanging entries that are symmetrically positioned about the main diagonal. Expressed in terms of the individual entries, a matrix $A = [a_{ij}]$ is symmetric if and only if $a_{ij} = a_{ji}$ for all values of i and j. As illustrated in Example 4, all diagonal matrices are symmetric.

The following theorem lists the main algebraic properties of symmetric matrices. The proofs are direct consequences of Theorem 1.4.9 and are left for the reader.

Theorem 1.7.2

If A and B are symmetric matrices with the same size, and if k is any scalar, then:

(a) A^T is symmetric.
(b) $A + B$ and $A - B$ are symmetric.
(c) kA is symmetric.

REMARK. It is not true, in general, that the product of symmetric matrices is symmetric. To see why this is so, let A and B be symmetric matrices with the same size. Then from part (d) of Theorem 1.4.9 and the symmetry we have

$$(AB)^T = B^T A^T = BA$$

Since AB and BA are not usually equal, it follows that AB will not usually be symmetric. However, in the special case where $AB = BA$, the product AB will be symmetric. If A and B are matrices such that $AB = BA$, then we say that A and B ***commute***. In summary: *The product of two symmetric matrices is symmetric if and only if the matrices commute.*

EXAMPLE 5 Products of Symmetric Matrices

The first of the following equations shows a product of symmetric matrices that *is not* symmetric, and the second shows a product of symmetric matrices that *is* symmetric. We conclude that the factors in the first equation do not commute, but those in the second equation do. We leave it for the reader to verify that this is so.

$$\begin{bmatrix} 1 & 2 \\ 2 & 3 \end{bmatrix} \begin{bmatrix} -4 & 1 \\ 1 & 0 \end{bmatrix} = \begin{bmatrix} -2 & 1 \\ -5 & 2 \end{bmatrix}$$

$$\begin{bmatrix} 1 & 2 \\ 2 & 3 \end{bmatrix} \begin{bmatrix} -4 & 3 \\ 3 & -1 \end{bmatrix} = \begin{bmatrix} 2 & 1 \\ 1 & 3 \end{bmatrix}$$ ♦

In general, a symmetric matrix need not be invertible; for example, a square zero matrix is symmetric, but not invertible. However, if a symmetric matrix is invertible, then that inverse is also symmetric.

Theorem 1.7.3

If A is an invertible symmetric matrix, then A^{-1} is symmetric.

Proof. Assume that A is symmetric and invertible. From Theorem 1.4.10 and the fact that $A = A^T$ we have

$$(A^{-1})^T = (A^T)^{-1} = A^{-1}$$

which proves that A^{-1} is symmetric. ■

Products AA^T and A^TA Matrix products of the form AA^T and A^TA arise in a variety of applications. If A is an $m \times n$ matrix, then A^T is an $n \times m$ matrix, so the products AA^T and A^TA are both square matrices—the matrix AA^T has size $m \times m$ and the matrix A^TA has size $n \times n$. Such products are always symmetric since

$$(AA^T)^T = (A^T)^T A^T = AA^T \quad \text{and} \quad (A^TA)^T = A^T(A^T)^T = A^TA$$

EXAMPLE 6 The Product of a Matrix and Its Transpose Is Symmetric

Let A be the 2×3 matrix

$$A = \begin{bmatrix} 1 & -2 & 4 \\ 3 & 0 & -5 \end{bmatrix}$$

Then

$$A^TA = \begin{bmatrix} 1 & 3 \\ -2 & 0 \\ 4 & -5 \end{bmatrix} \begin{bmatrix} 1 & -2 & 4 \\ 3 & 0 & -5 \end{bmatrix} = \begin{bmatrix} 10 & -2 & -11 \\ -2 & 4 & -8 \\ -11 & -8 & 41 \end{bmatrix}$$

$$AA^T = \begin{bmatrix} 1 & -2 & 4 \\ 3 & 0 & -5 \end{bmatrix} \begin{bmatrix} 1 & 3 \\ -2 & 0 \\ 4 & -5 \end{bmatrix} = \begin{bmatrix} 21 & -17 \\ -17 & 34 \end{bmatrix}$$

Observe that A^TA and AA^T are symmetric as expected. ◆

Later in this text, we will obtain general conditions on A under which AA^T and A^TA are invertible. However, in the special case where A is *square* we have the following result.

Theorem 1.7.4

If A is an invertible matrix, then AA^T and A^TA are also invertible.

Proof. Since A is invertible, so is A^T by Theorem 1.4.10. Thus, AA^T and A^TA are invertible, since they are the products of invertible matrices. ■

Exercise Set 1.7

1. Determine whether the matrix is invertible; if so, find the inverse by inspection.

(a) $\begin{bmatrix} 2 & 0 \\ 0 & -5 \end{bmatrix}$ (b) $\begin{bmatrix} 4 & 0 & 0 \\ 0 & 0 & 0 \\ 0 & 0 & 5 \end{bmatrix}$ (c) $\begin{bmatrix} -1 & 0 & 0 \\ 0 & 2 & 0 \\ 0 & 0 & \frac{1}{3} \end{bmatrix}$

2. Compute the product by inspection.

(a) $\begin{bmatrix} 3 & 0 & 0 \\ 0 & -1 & 0 \\ 0 & 0 & 2 \end{bmatrix} \begin{bmatrix} 2 & 1 \\ -4 & 1 \\ 2 & 5 \end{bmatrix}$ (b) $\begin{bmatrix} 2 & 0 & 0 \\ 0 & -1 & 0 \\ 0 & 0 & 4 \end{bmatrix} \begin{bmatrix} 4 & -1 & 3 \\ 1 & 2 & 0 \\ -5 & 1 & -2 \end{bmatrix} \begin{bmatrix} -3 & 0 & 0 \\ 0 & 5 & 0 \\ 0 & 0 & 2 \end{bmatrix}$

3. Find A^2, A^{-2}, and A^{-k} by inspection.

(a) $A = \begin{bmatrix} 1 & 0 \\ 0 & -2 \end{bmatrix}$ (b) $A = \begin{bmatrix} \frac{1}{2} & 0 & 0 \\ 0 & \frac{1}{3} & 0 \\ 0 & 0 & \frac{1}{4} \end{bmatrix}$

4. Which of the following matrices are symmetric?

(a) $\begin{bmatrix} 2 & -1 \\ 1 & 2 \end{bmatrix}$ (b) $\begin{bmatrix} 3 & 4 \\ 4 & 0 \end{bmatrix}$ (c) $\begin{bmatrix} 2 & -1 & 3 \\ -1 & 5 & 1 \\ 3 & 1 & 7 \end{bmatrix}$ (d) $\begin{bmatrix} 0 & 0 & 1 \\ 0 & 2 & 0 \\ 3 & 0 & 0 \end{bmatrix}$

5. By inspection, determine whether the given triangular matrix is invertible.

(a) $\begin{bmatrix} -1 & 2 & 4 \\ 0 & 3 & 0 \\ 0 & 0 & 5 \end{bmatrix}$ (b) $\begin{bmatrix} 0 & 1 & -2 & 5 \\ 0 & 1 & 5 & 6 \\ 0 & 0 & -3 & 1 \\ 0 & 0 & 0 & 5 \end{bmatrix}$

6. Find all values of a, b, and c for which A is symmetric.

$$A = \begin{bmatrix} 2 & a-2b+2c & 2a+b+c \\ 3 & 5 & a+c \\ 0 & -2 & 7 \end{bmatrix}$$

7. Find all values of a and b for which A and B are both not invertible.

$$A = \begin{bmatrix} a+b-1 & 0 \\ 0 & 3 \end{bmatrix}, \quad B = \begin{bmatrix} 5 & 0 \\ 0 & 2a-3b-7 \end{bmatrix}$$

8. Use the given equation to determine by inspection whether the matrices on the left commute.

(a) $\begin{bmatrix} 1 & -3 \\ -3 & 2 \end{bmatrix}\begin{bmatrix} 4 & 1 \\ 1 & 2 \end{bmatrix} = \begin{bmatrix} 1 & -5 \\ -10 & 1 \end{bmatrix}$ (b) $\begin{bmatrix} 2 & -1 \\ -1 & 3 \end{bmatrix}\begin{bmatrix} 3 & 2 \\ 2 & 1 \end{bmatrix} = \begin{bmatrix} 4 & 3 \\ 3 & 1 \end{bmatrix}$

9. Show that A and B commute if $a - d = 7b$.

$$A = \begin{bmatrix} 2 & 1 \\ 1 & -5 \end{bmatrix}, \quad B = \begin{bmatrix} a & b \\ b & d \end{bmatrix}$$

10. Find a diagonal matrix A that satisfies

(a) $A^5 = \begin{bmatrix} 1 & 0 & 0 \\ 0 & -1 & 0 \\ 0 & 0 & -1 \end{bmatrix}$ (b) $A^{-2} = \begin{bmatrix} 9 & 0 & 0 \\ 0 & 4 & 0 \\ 0 & 0 & 1 \end{bmatrix}$

11. (a) Factor A into the form $A = BD$, where D is a diagonal matrix.

$$A = \begin{bmatrix} 3a_{11} & 5a_{12} & 7a_{13} \\ 3a_{21} & 5a_{22} & 7a_{23} \\ 3a_{31} & 5a_{32} & 7a_{33} \end{bmatrix}$$

(b) Is your factorization the only one possible? Explain.

12. Verify Theorem 1.7.1b for the product AB, where

$$A = \begin{bmatrix} -1 & 2 & 5 \\ 0 & 1 & 3 \\ 0 & 0 & -4 \end{bmatrix}, \quad B = \begin{bmatrix} 2 & -8 & 0 \\ 0 & 2 & 1 \\ 0 & 0 & 3 \end{bmatrix}$$

13. Verify Theorem 1.7.1d for the matrices A and B in Exercise 12.

14. Verify Theorem 1.7.3 for the given matrix A.

(a) $A = \begin{bmatrix} 2 & -1 \\ -1 & 3 \end{bmatrix}$ (b) $A = \begin{bmatrix} 1 & -2 & 3 \\ -2 & 1 & -7 \\ 3 & -7 & 4 \end{bmatrix}$

15. Let A be a symmetric matrix.

(a) Show that A^2 is symmetric.

(b) Show that $2A^2 - 3A + I$ is symmetric.

16. Let A be a symmetric matrix.

(a) Show that A^k is symmetric if k is any nonnegative integer.

(b) If $p(x)$ is a polynomial, is $p(A)$ necessarily symmetric? Explain.

17. Let A be an upper triangular matrix and let $p(x)$ be a polynomial. Is $p(A)$ necessarily upper triangular? Explain.

18. Prove: If $A^TA = A$, then A is symmetric and $A = A^2$.

19. Find all 3×3 diagonal matrices A that satisfy $A^2 - 3A - 4I = 0$.

20. Let $A = [a_{ij}]$ be an $n \times n$ matrix. Determine whether A is symmetric.

(a) $a_{ij} = i^2 + j^2$ (b) $a_{ij} = i^2 - j^2$

(c) $a_{ij} = 2i + 2j$ (d) $a_{ij} = 2i^2 + 2j^3$

21. Based on your experience with Exercise 20, devise a general test that can be applied to a formula for a_{ij} to determine whether $A = [a_{ij}]$ is symmetric.

22. A square matrix A is called ***skew-symmetric*** if $A^T = -A$. Prove:

(a) If A is an invertible skew-symmetric matrix, then A^{-1} is skew-symmetric.

(b) If A and B are skew-symmetric, then so are A^T, $A + B$, $A - B$, and kA for any scalar k.

(c) Every square matrix A can be expressed as the sum of a symmetric matrix and a skew-symmetric matrix. [***Hint.*** Note the identity $A = \frac{1}{2}(A + A^T) + \frac{1}{2}(A - A^T)$.]

23. We showed in the text that the product of symmetric matrices is symmetric if and only if the matrices commute. Is the product of commuting skew-symmetric matrices skew-symmetric? Explain. [***Note.*** See Exercise 22 for terminology.]

24. If the $n \times n$ matrix A can be expressed as $A = LU$, where L is a lower triangular matrix and U is an upper triangular matrix, then the linear system $A\mathbf{x} = \mathbf{b}$ can be expressed as $LU\mathbf{x} = \mathbf{b}$ and can be solved in two steps:

Step 1. Let $U\mathbf{x} = \mathbf{y}$, so that $LU\mathbf{x} = \mathbf{b}$ can be expressed as $L\mathbf{y} = \mathbf{b}$. Solve this system.

Step 2. Solve the system $U\mathbf{x} = \mathbf{y}$ for $\mathbf{x}$.

In each part use this two-step method to solve the given system.

(a) $$\begin{bmatrix} 1 & 0 & 0 \\ -2 & 3 & 0 \\ 2 & 4 & 1 \end{bmatrix} \begin{bmatrix} 2 & -1 & 3 \\ 0 & 1 & 2 \\ 0 & 0 & 4 \end{bmatrix} \begin{bmatrix} x_1 \\ x_2 \\ x_3 \end{bmatrix} = \begin{bmatrix} 1 \\ -2 \\ 0 \end{bmatrix}$$

(b) $$\begin{bmatrix} 2 & 0 & 0 \\ 4 & 1 & 0 \\ -3 & -2 & 3 \end{bmatrix} \begin{bmatrix} 3 & -5 & 2 \\ 0 & 4 & 1 \\ 0 & 0 & 2 \end{bmatrix} \begin{bmatrix} x_1 \\ x_2 \\ x_3 \end{bmatrix} = \begin{bmatrix} 4 \\ -5 \\ 2 \end{bmatrix}$$

25. Find an upper triangular matrix that satisfies

$$A^3 = \begin{bmatrix} 1 & 30 \\ 0 & -8 \end{bmatrix}$$

Discussion and Discovery

26. What is the maximum number of distinct entries that an $n \times n$ symmetric matrix can have? Explain your reasoning.

27. Invent and prove a theorem that describes how to multiply two diagonal matrices.

28. Suppose that A is a square matrix and D is a diagonal matrix such that $AD = I$. What can you say about the matrix A? Explain your reasoning.

29. (a) Make up a consistent linear system of five equations in five unknowns that has a lower triangular coefficient matrix with no zeros on or below the main diagonal.
(b) Devise an efficient procedure for solving your system by hand.
(c) Invent an appropriate name for your procedure.

30. Indicate whether the statement is always true or sometimes false. Justify each answer.

(a) If AA^T is singular, then so is A.
(b) If $A + B$ is symmetric, then so are A and B.
(c) If A is an $n \times n$ matrix and $A\mathbf{x} = \mathbf{0}$ has only the trivial solution, then so does $A^T\mathbf{x} = \mathbf{0}$.
(d) If A^2 is symmetric, then so is A.

Chapter 1 Supplementary Exercises

1. Use Gauss–Jordan elimination to solve for x' and y' in terms of x and y.

$$\begin{aligned} x &= \tfrac{3}{5}x' - \tfrac{4}{5}y' \\ y &= \tfrac{4}{5}x' + \tfrac{3}{5}y' \end{aligned}$$

2. Use Gauss–Jordan elimination to solve for x' and y' in terms of x and y.

$$\begin{aligned} x &= x'\cos\theta - y'\sin\theta \\ y &= x'\sin\theta + y'\cos\theta \end{aligned}$$

3. Find a homogeneous linear system with two equations that are not multiples of one another and such that

$$x_1 = 1, \quad x_2 = -1, \quad x_3 = 1, \quad x_4 = 2$$

and

$$x_1 = 2, \quad x_2 = 0, \quad x_3 = 3, \quad x_4 = -1$$

are solutions of the system.

4. A box containing pennies, nickels, and dimes has 13 coins with a total value of 83 cents. How many coins of each type are in the box?

5. Find positive integers that satisfy

$$\begin{aligned} x + y + z &= 9 \\ x + 5y + 10z &= 44 \end{aligned}$$

6. For which value(s) of a does the following system have zero, one, infinitely many solutions?

$$\begin{aligned} x_1 + x_2 + x_3 &= 4 \\ x_3 &= 2 \\ (a^2 - 4)x_3 &= a - 2 \end{aligned}$$

7. Let

$$\begin{bmatrix} a & 0 & b & 2 \\ a & a & 4 & 4 \\ 0 & a & 2 & b \end{bmatrix}$$

be the augmented matrix for a linear system. For what values of a and b does the system have

(a) a unique solution, (b) a one-parameter solution,
(c) a two-parameter solution, (d) no solution?

8. Solve for x, y, and z.

$$\begin{aligned} xy - 2\sqrt{y} + 3zy &= 8 \\ 2xy - 3\sqrt{y} + 2zy &= 7 \\ -xy + \sqrt{y} + 2zy &= 4 \end{aligned}$$

9. Find a matrix K such that $AKB = C$ given that

$$A = \begin{bmatrix} 1 & 4 \\ -2 & 3 \\ 1 & -2 \end{bmatrix}, \quad B = \begin{bmatrix} 2 & 0 & 0 \\ 0 & 1 & -1 \end{bmatrix}, \quad C = \begin{bmatrix} 8 & 6 & -6 \\ 6 & -1 & 1 \\ -4 & 0 & 0 \end{bmatrix}$$

10. How should the coefficients a, b, and c be chosen so that the system

$$\begin{aligned} ax + by - 3z &= -3 \\ -2x - by + cz &= -1 \\ ax + 3y - cz &= -3 \end{aligned}$$

has the solution $x = 1$, $y = -1$, and $z = 2$?

11. In each part solve the matrix equation for X.

(a) $X\begin{bmatrix} -1 & 0 & 1 \\ 1 & 1 & 0 \\ 3 & 1 & -1 \end{bmatrix} = \begin{bmatrix} 1 & 2 & 0 \\ -3 & 1 & 5 \end{bmatrix}$ (b) $X\begin{bmatrix} 1 & -1 & 2 \\ 3 & 0 & 1 \end{bmatrix} = \begin{bmatrix} -5 & -1 & 0 \\ 6 & -3 & 7 \end{bmatrix}$

(c) $\begin{bmatrix} 3 & 1 \\ -1 & 2 \end{bmatrix}X - X\begin{bmatrix} 1 & 4 \\ 2 & 0 \end{bmatrix} = \begin{bmatrix} 2 & -2 \\ 5 & 4 \end{bmatrix}$

12. (a) Express the equations

$$\begin{aligned} y_1 &= x_1 - x_2 + x_3 \\ y_2 &= 3x_1 + x_2 - 4x_3 \\ y_3 &= -2x_1 - 2x_2 + 3x_3 \end{aligned} \quad \text{and} \quad \begin{aligned} z_1 &= 4y_1 - y_2 + y_3 \\ z_2 &= -3y_1 + 5y_2 - y_3 \end{aligned}$$

in the matrix forms $Y = AX$ and $Z = BY$. Then use these to obtain a direct relationship $Z = CX$ between Z and X.

(b) Use the equation $Z = CX$ obtained in (a) to express z_1 and z_2 in terms of x_1, x_2, and x_3.

(c) Check the result in (b) by directly substituting the equations for y_1, y_2, and y_3 into the equations for z_1 and z_2 and then simplifying.

13. If A is $m \times n$ and B is $n \times p$, how many multiplication operations and how many addition operations are needed to calculate the matrix product AB?

14. Let A be a square matrix.

(a) Show that $(I - A)^{-1} = I + A + A^2 + A^3$ if $A^4 = 0$.

(b) Show that $(I - A)^{-1} = I + A + A^2 + \cdots + A^n$ if $A^{n+1} = 0$.

15. Find values of a, b, and c so that the graph of the polynomial $p(x) = ax^2 + bx + c$ passes through the points $(1, 2)$, $(-1, 6)$, and $(2, 3)$.

16. ***(For readers who have studied calculus.)*** Find values of a, b, and c so that the graph of the polynomial $p(x) = ax^2 + bx + c$ passes through the point $(-1, 0)$ and has a horizontal tangent at $(2, -9)$.

17. Let J_n be the $n \times n$ matrix each of whose entries is 1. Show that if $n > 1$, then

$$(I - J_n)^{-1} = I - \frac{1}{n-1}J_n$$

18. Show that if a square matrix A satisfies $A^3 + 4A^2 - 2A + 7I = 0$, then so does A^T.

19. Prove: If B is invertible, then $AB^{-1} = B^{-1}A$ if and only if $AB = BA$.

20. Prove: If A is invertible, then $A + B$ and $I + BA^{-1}$ are both invertible or both not invertible.

21. Prove that if A and B are $n \times n$ matrices, then

(a) $\operatorname{tr}(A+B) = \operatorname{tr}(A) + \operatorname{tr}(B)$ (b) $\operatorname{tr}(kA) = k\operatorname{tr}(A)$ (c) $\operatorname{tr}(A^T) = \operatorname{tr}(A)$ (d) $\operatorname{tr}(AB) = \operatorname{tr}(BA)$

22. Use Exercise 21 to show that there are no square matrices A and B such that

$$AB - BA = I$$

23. Prove: If A is an $m \times n$ matrix and B is the $n \times 1$ matrix each of whose entries is $1/n$, then

$$AB = \begin{bmatrix} \overline{r}_1 \\ \overline{r}_2 \\ \vdots \\ \overline{r}_m \end{bmatrix}$$

where $\overline{r}_i$ is the average of the entries in the ith row of A.

24. (***For readers who have studied calculus.***) If the entries of the matrix

$$C = \begin{bmatrix} c_{11}(x) & c_{12}(x) & \cdots & c_{1n}(x) \\ c_{21}(x) & c_{22}(x) & \cdots & c_{2n}(x) \\ \vdots & \vdots & & \vdots \\ c_{m1}(x) & c_{m2}(x) & \cdots & c_{mn}(x) \end{bmatrix}$$

are differentiable functions of x, then we define

$$\frac{dC}{dx} = \begin{bmatrix} c'_{11}(x) & c'_{12}(x) & \cdots & c'_{1n}(x) \\ c'_{21}(x) & c'_{22}(x) & \cdots & c'_{2n}(x) \\ \vdots & \vdots & & \vdots \\ c'_{m1}(x) & c'_{m2}(x) & \cdots & c'_{mn}(x) \end{bmatrix}$$

Show that if the entries in A and B are differentiable functions of x and the sizes of the matrices are such that the stated operations can be performed, then

(a) $\dfrac{d}{dx}(kA) = k\dfrac{dA}{dx}$ (b) $\dfrac{d}{dx}(A+B) = \dfrac{dA}{dx} + \dfrac{dB}{dx}$ (c) $\dfrac{d}{dx}(AB) = \dfrac{dA}{dx}B + A\dfrac{dB}{dx}$

25. (***For readers who have studied calculus.***) Use part (c) of Exercise 24 to show that

$$\frac{dA^{-1}}{dx} = -A^{-1}\frac{dA}{dx}A^{-1}$$

State all the assumptions you make in obtaining this formula.

26. Find the values of A, B, and C that will make the equation

$$\frac{x^2 + x - 2}{(3x-1)(x^2+1)} = \frac{A}{3x-1} + \frac{Bx + C}{x^2+1}$$

an identity. [***Hint.*** Multiply through by $(3x-1)(x^2+1)$ and equate the corresponding coefficients of the polynomials on each side of the resulting equation.]

27. If P is an $n \times 1$ matrix such that $P^T P = 1$, then $H = I - 2PP^T$ is called the corresponding ***Householder matrix*** (named after the American mathematician A. S. Householder).

(a) Verify that $P^T P = 1$ if $P^T = \left[\frac{3}{4}\ \frac{1}{6}\ \frac{1}{4}\ \frac{5}{12}\ \frac{5}{12}\right]$ and compute the corresponding Householder matrix.

(b) Prove that if H is any Householder matrix, then $H = H^T$ and $H^T H = I$.

(c) Verify that the Householder matrix found in part (a) satisfies the conditions proved in part (b).

28. Assuming that the stated inverses exist, prove the following equalities.

(a) $(C^{-1} + D^{-1})^{-1} = C(C+D)^{-1}D$ (b) $(I + CD)^{-1}C = C(I + DC)^{-1}$

(c) $(C + DD^T)^{-1}D = C^{-1}D(I + D^TC^{-1}D)^{-1}$

2

Determinants

Chapter Contents

INTRODUCTION: We are all familiar with functions such as $f(x) = \sin x$ and $f(x) = x^2$, which associate a real number $f(x)$ with a real value of the variable x. Since both x and $f(x)$ assume only real values, such functions are described as real-valued functions of a real variable. In this section we shall study the "determinant function," which is a real-valued function of a matrix variable in the sense that it associates a real number $f(X)$ with a square matrix X. Our work on determinant functions will have important applications to the theory of systems of linear equations and will also lead us to an explicit formula for the inverse of an invertible matrix.

2.2 EVALUATING DETERMINANTS BY ROW REDUCTION

In this section we shall show that the determinant of a square matrix can be evaluated by reducing the matrix to row-echelon form. This method is important since it avoids the lengthy computations involved in directly applying the determinant definition.

A Basic Theorem As discussed at the end of the last section, the definition of a determinant is helpful for proving theorems *about* determinants, but it does not provide a practical means for evaluating them, especially determinants of matrices larger than 3×3. Accordingly, we begin with a fundamental theorem that will lead us to an efficient procedure for evaluating the determinant of a matrix of any order n.

Theorem 2.2.1

Let A be a square matrix.

(*a*) *If A has a row of zeros or a column of zeros, then* $\det(A) = 0$.

(*b*) $\det(A) = \det(A^T)$.

Proof (*a*). Since every signed elementary product from A has one factor from each row and one factor from each column, every signed elementary product would of necessity have a factor from a zero row or a factor from a zero column. In such cases, every signed elementary product is zero, and $\det(A)$, which is the sum of all the signed elementary products, is zero. ■

We omit the proof of part (*b*), but recall that an elementary product has one factor from each row and each column, so it is evident that A and A^T have precisely the same set of elementary products. With the help of some theorems on permutations, which would take us too far afield to discuss, it can be shown that A and A^T actually have the same set of *signed* elementary products. This implies that $\det(A) = \det(A^T)$.

REMARK. Because of Theorem 2.2.1*b*, nearly every theorem about determinants that contains the word "row" in its statement is also true when the word "column" is substituted for "row." To prove a column statement one need only transpose the matrix in question to convert the column statement to a row statement, and then apply the corresponding known result for rows.

Triangular Matrices The following theorem makes it easy to evaluate the determinant of a triangular matrix, regardless of its size.

Theorem 2.2.2

If A is an $n \times n$ *triangular matrix (upper triangular, lower triangular, or diagonal), then* $\det(A)$ *is the product of the entries on the main diagonal of the matrix; that is,* $\det(A) = a_{11}a_{22}\cdots a_{nn}$.

For simplicity of notation, we will prove the result for a 4×4 lower triangular matrix

$$A = \begin{bmatrix} a_{11} & 0 & 0 & 0 \\ a_{21} & a_{22} & 0 & 0 \\ a_{31} & a_{32} & a_{33} & 0 \\ a_{41} & a_{42} & a_{43} & a_{44} \end{bmatrix}$$

The argument in the $n \times n$ case is similar. A proof for upper triangular matrices can be obtained by applying Theorem 2.2.1*b* and observing that the transpose of an upper triangular matrix is a lower triangular matrix with the same diagonal entries.

Proof of Theorem 2.2.2 (4 × 4 lower triangular case). The only elementary product from A that can be nonzero is $a_{11}a_{22}a_{33}a_{44}$. To see that this is so, consider a typical elementary product $a_{1j_1}a_{2j_2}a_{3j_3}a_{4j_4}$. Since $a_{12} = a_{13} = a_{14} = 0$, we must have $j_1 = 1$ in order to have a nonzero elementary product. If $j_1 = 1$, we must have $j_2 \neq 1$, since no two factors come from the same column. Further, since $a_{23} = a_{24} = 0$, we must have $j_2 = 2$ in order to have a nonzero product. Continuing in this way, we obtain $j_3 = 3$ and $j_4 = 4$. Since $a_{11}a_{22}a_{33}a_{44}$ is multiplied by $+1$ in forming the signed elementary product, we obtain

$$\det(A) = a_{11}a_{22}a_{33}a_{44}$$ ■

EXAMPLE 1 Determinant of an Upper Triangular Matrix

$$\begin{vmatrix} 2 & 7 & -3 & 8 & 3 \\ 0 & -3 & 7 & 5 & 1 \\ 0 & 0 & 6 & 7 & 6 \\ 0 & 0 & 0 & 9 & 8 \\ 0 & 0 & 0 & 0 & 4 \end{vmatrix} = (2)(-3)(6)(9)(4) = -1296$$ ♦

Elementary Row Operations The next theorem shows how an elementary row operation on a matrix affects the value of its determinant.

Theorem 2.2.3

Let A be an $n \times n$ matrix.

(*a*) *If B is the matrix that results when a single row or single column of A is multiplied by a scalar k, then* $\det(B) = k\det(A)$.

(*b*) *If B is the matrix that results when two rows or two columns of A are interchanged, then* $\det(B) = -\det(A)$.

(*c*) *If B is the matrix that results when a multiple of one row of A is added to another row or when a multiple of one column is added to another column, then* $\det(B) = \det(A)$.

A proof of this theorem can be obtained by using Formula (1) of Section 2.1 to compute the determinants involved, and then verifying the equalities. We omit the proof but give the following example that illustrates the theorem for 3×3 determinants.

EXAMPLE 2 Theorem 2.2.3 Applied to 3 × 3 Determinants

Relationship	Operation
$\begin{vmatrix} ka_{11} & ka_{12} & ka_{13} \\ a_{21} & a_{22} & a_{23} \\ a_{31} & a_{32} & a_{33} \end{vmatrix} = k \begin{vmatrix} a_{11} & a_{12} & a_{13} \\ a_{21} & a_{22} & a_{23} \\ a_{31} & a_{32} & a_{33} \end{vmatrix}$ $\det(B) = k\det(A)$	The first row of A is multiplied by k.
$\begin{vmatrix} a_{21} & a_{22} & a_{23} \\ a_{11} & a_{12} & a_{13} \\ a_{31} & a_{32} & a_{33} \end{vmatrix} = - \begin{vmatrix} a_{11} & a_{12} & a_{13} \\ a_{21} & a_{22} & a_{23} \\ a_{31} & a_{32} & a_{33} \end{vmatrix}$ $\det(B) = -\det(A)$	The first and second rows of A are interchanged.
$\begin{vmatrix} a_{11}+ka_{21} & a_{12}+ka_{22} & a_{13}+ka_{23} \\ a_{21} & a_{22} & a_{23} \\ a_{31} & a_{32} & a_{33} \end{vmatrix} = \begin{vmatrix} a_{11} & a_{12} & a_{13} \\ a_{21} & a_{22} & a_{23} \\ a_{31} & a_{32} & a_{33} \end{vmatrix}$ $\det(B) = \det(A)$	A multiple of the second row of A is added to the first row.

We will verify the equation in the last row of the table and leave the first two for the reader. With the help of Example 7 in Section 2.1 we obtain

$$\begin{aligned} \det(B) &= (a_{11} + ka_{21})a_{22}a_{33} + (a_{12} + ka_{22})a_{23}a_{31} + (a_{13} + ka_{23})a_{21}a_{32} \\ &\quad - a_{31}a_{22}(a_{13} + ka_{23}) - a_{33}a_{21}(a_{12} + ka_{22}) - a_{32}a_{23}(a_{11} + ka_{21}) \\ &= \det(A) + k(a_{21}a_{22}a_{33} + a_{22}a_{23}a_{31} + a_{23}a_{21}a_{32} \\ &\qquad - a_{31}a_{22}a_{23} - a_{33}a_{21}a_{22} - a_{32}a_{23}a_{21}) \\ &= \det(A) + 0 = \det(A) \end{aligned}$$

◆

REMARK. As illustrated by the first equation in Example 2, part (*a*) of Theorem 2.2.3 allows us to bring a "common factor" from any row (or column) through the determinant sign.

Elementary Matrices Recall that an elementary matrix results from performing a single elementary row operation on an identity matrix; thus, if we let $A = I_n$ in Theorem 2.2.3 [so that we have $\det(A) = \det(I_n) = 1$], then the matrix B is an elementary matrix, and the theorem yields the following result about determinants of elementary matrices.

Theorem 2.2.4

Let E be an $n \times n$ elementary matrix.

(*a*) *If E results from multiplying a row of I_n by k, then* $\det(E) = k$.
(*b*) *If E results from interchanging two rows of I_n, then* $\det(E) = -1$.
(*c*) *If E results from adding a multiple of one row of I_n to another, then* $\det(E) = 1$.

EXAMPLE 3 Determinants of Elementary Matrices

The following determinants of elementary matrices, which are evaluated by inspection, illustrate Theorem 2.2.4.

$$\begin{vmatrix} 1 & 0 & 0 & 0 \\ 0 & 3 & 0 & 0 \\ 0 & 0 & 1 & 0 \\ 0 & 0 & 0 & 1 \end{vmatrix} = 3, \quad \begin{vmatrix} 0 & 0 & 0 & 1 \\ 0 & 1 & 0 & 0 \\ 0 & 0 & 1 & 0 \\ 1 & 0 & 0 & 0 \end{vmatrix} = -1, \quad \begin{vmatrix} 1 & 0 & 0 & 7 \\ 0 & 1 & 0 & 0 \\ 0 & 0 & 1 & 0 \\ 0 & 0 & 0 & 1 \end{vmatrix} = 1$$

The second row of I_4 was multiplied by 3.

The first and last rows of I_4 were interchanged.

7 times the last row of I_4 was added to the first row. ♦

Matrices with Proportional Rows or Columns If a square matrix A has two proportional rows, then a row of zeros can be introduced by adding a suitable multiple of one of the rows to the other. Similarly for columns. But adding a multiple of one row or column to another does not change the determinant, so from Theorem 2.2.1a, we must have $\det(A) = 0$. This proves the following theorem.

Theorem 2.2.5

If A is a square matrix with two proportional rows or two proportional columns, then $\det(A) = 0$.

EXAMPLE 4 Introducing Zero Rows

The following computation illustrates the introduction of a row of zeros when there are two proportional rows:

$$\begin{vmatrix} 1 & 3 & -2 & 4 \\ 2 & 6 & -4 & 8 \\ 3 & 9 & 1 & 5 \\ 1 & 1 & 4 & 8 \end{vmatrix} = \begin{vmatrix} 1 & 3 & -2 & 4 \\ 0 & 0 & 0 & 0 \\ 3 & 9 & 1 & 5 \\ 1 & 1 & 4 & 8 \end{vmatrix} = 0$$

← The second row is 2 times the first, so we added −2 times the first row to the second to introduce a row of zeros.

Each of the following matrices has two proportional rows or columns; thus, each has a determinant of zero.

$$\begin{bmatrix} -1 & 4 \\ -2 & 8 \end{bmatrix}, \quad \begin{bmatrix} 1 & -2 & 7 \\ -4 & 8 & 5 \\ 2 & -4 & 3 \end{bmatrix}, \quad \begin{bmatrix} 3 & -1 & 4 & -5 \\ 6 & -2 & 5 & 2 \\ 5 & 8 & 1 & 4 \\ -9 & 3 & -12 & 15 \end{bmatrix}$$ ♦

Evaluating Determinants by Row Reduction We shall now give a method for evaluating determinants that involves substantially less computation than applying the determinant definition directly. The idea of the method is to reduce the given matrix to upper triangular form by elementary row operations, then compute the determinant of the upper triangular matrix (an easy computation), then relate that determinant to that of the original matrix. Here is an example.

EXAMPLE 5 Using Row Reduction to Evaluate a Determinant

Evaluate $\det(A)$ where

$$A = \begin{bmatrix} 0 & 1 & 5 \\ 3 & -6 & 9 \\ 2 & 6 & 1 \end{bmatrix}$$

Solution.

We will reduce A to row-echelon form (which is upper triangular) and apply Theorem 2.2.3:

$$\det(A) = \begin{vmatrix} 0 & 1 & 5 \\ 3 & -6 & 9 \\ 2 & 6 & 1 \end{vmatrix} = -\begin{vmatrix} 3 & -6 & 9 \\ 0 & 1 & 5 \\ 2 & 6 & 1 \end{vmatrix}$$

← The first and second rows of A were interchanged.

$$= -3\begin{vmatrix} 1 & -2 & 3 \\ 0 & 1 & 5 \\ 2 & 6 & 1 \end{vmatrix}$$

← A common factor of 3 from the first row was taken through the determinant sign.

$$= -3\begin{vmatrix} 1 & -2 & 3 \\ 0 & 1 & 5 \\ 0 & 10 & -5 \end{vmatrix}$$

← -2 times the first row was added to the third row.

$$= -3\begin{vmatrix} 1 & -2 & 3 \\ 0 & 1 & 5 \\ 0 & 0 & -55 \end{vmatrix}$$

← -10 times the second row was added to the third row.

$$= (-3)(-55)\begin{vmatrix} 1 & -2 & 3 \\ 0 & 1 & 5 \\ 0 & 0 & 1 \end{vmatrix}$$

← A common factor of -55 from the last row was taken through the determinant sign.

$$= (-3)(-55)(1) = 165$$

◆

REMARK. The method of row reduction is well suited for computer evaluation of determinants because it is systematic and easily programmed. However, in subsequent sections we will develop methods that are often easier for hand computation.

EXAMPLE 6 Using Column Operations to Evaluate a Determinant

Compute the determinant of

$$A = \begin{bmatrix} 1 & 0 & 0 & 3 \\ 2 & 7 & 0 & 6 \\ 0 & 6 & 3 & 0 \\ 7 & 3 & 1 & -5 \end{bmatrix}$$

Solution.

This determinant could be computed as above by using elementary row operations to reduce A to row-echelon form, but we can put A in lower triangular form in one step by adding -3 times the first column to the fourth to obtain

$$\det(A) = \det\begin{bmatrix} 1 & 0 & 0 & 0 \\ 2 & 7 & 0 & 0 \\ 0 & 6 & 3 & 0 \\ 7 & 3 & 1 & -26 \end{bmatrix} = (1)(7)(3)(-26) = -546$$

This example points out the utility of keeping an eye open for column operations that can shorten computations. ♦

Exercise Set 2.2

1. Verify that $\det(A) = \det(A^T)$ for

(a) $A = \begin{bmatrix} -2 & 3 \\ 1 & 4 \end{bmatrix}$ (b) $A = \begin{bmatrix} 2 & -1 & 3 \\ 1 & 2 & 4 \\ 5 & -3 & 6 \end{bmatrix}$

2. Evaluate the following determinants by inspection.

(a) $\begin{vmatrix} 3 & -17 & 4 \\ 0 & 5 & 1 \\ 0 & 0 & -2 \end{vmatrix}$ (b) $\begin{vmatrix} \sqrt{2} & 0 & 0 & 0 \\ -8 & \sqrt{2} & 0 & 0 \\ 7 & 0 & -1 & 0 \\ 9 & 5 & 6 & 1 \end{vmatrix}$ (c) $\begin{vmatrix} -2 & 1 & 3 \\ 1 & -7 & 4 \\ -2 & 1 & 3 \end{vmatrix}$ (d) $\begin{vmatrix} 1 & -2 & 3 \\ 2 & -4 & 6 \\ 5 & -8 & 1 \end{vmatrix}$

3. Find the determinants of the following elementary matrices by inspection.

(a) $\begin{bmatrix} 1 & 0 & 0 & 0 \\ 0 & 1 & 0 & 0 \\ 0 & 0 & -5 & 0 \\ 0 & 0 & 0 & 1 \end{bmatrix}$ (b) $\begin{bmatrix} 1 & 0 & 0 & 0 \\ 0 & 0 & 1 & 0 \\ 0 & 1 & 0 & 0 \\ 0 & 0 & 0 & 1 \end{bmatrix}$ (c) $\begin{bmatrix} 1 & 0 & 0 & 0 \\ 0 & 1 & 0 & -9 \\ 0 & 0 & 1 & 0 \\ 0 & 0 & 0 & 1 \end{bmatrix}$

In Exercises 4–11 evaluate the determinant of the given matrix by reducing the matrix to row-echelon form.

4. $\begin{bmatrix} 3 & 6 & -9 \\ 0 & 0 & -2 \\ -2 & 1 & 5 \end{bmatrix}$ **5.** $\begin{bmatrix} 0 & 3 & 1 \\ 1 & 1 & 2 \\ 3 & 2 & 4 \end{bmatrix}$ **6.** $\begin{bmatrix} 1 & -3 & 0 \\ -2 & 4 & 1 \\ 5 & -2 & 2 \end{bmatrix}$ **7.** $\begin{bmatrix} 3 & -6 & 9 \\ -2 & 7 & -2 \\ 0 & 1 & 5 \end{bmatrix}$

8. $\begin{bmatrix} 1 & -2 & 3 & 1 \\ 5 & -9 & 6 & 3 \\ -1 & 2 & -6 & -2 \\ 2 & 8 & 6 & 1 \end{bmatrix}$ **9.** $\begin{bmatrix} 2 & 1 & 3 & 1 \\ 1 & 0 & 1 & 1 \\ 0 & 2 & 1 & 0 \\ 0 & 1 & 2 & 3 \end{bmatrix}$ **10.** $\begin{bmatrix} 0 & 1 & 1 & 1 \\ \frac{1}{2} & \frac{1}{2} & 1 & \frac{1}{2} \\ \frac{2}{3} & \frac{1}{3} & \frac{1}{3} & 0 \\ -\frac{1}{3} & \frac{2}{3} & 0 & 0 \end{bmatrix}$ **11.** $\begin{bmatrix} 1 & 3 & 1 & 5 & 3 \\ -2 & -7 & 0 & -4 & 2 \\ 0 & 0 & 1 & 0 & 1 \\ 0 & 0 & 2 & 1 & 1 \\ 0 & 0 & 0 & 1 & 1 \end{bmatrix}$

12. Given that $\begin{vmatrix} a & b & c \\ d & e & f \\ g & h & i \end{vmatrix} = -6$, find

(a) $\begin{vmatrix} d & e & f \\ g & h & i \\ a & b & c \end{vmatrix}$ (b) $\begin{vmatrix} 3a & 3b & 3c \\ -d & -e & -f \\ 4g & 4h & 4i \end{vmatrix}$ (c) $\begin{vmatrix} a+g & b+h & c+i \\ d & e & f \\ g & h & i \end{vmatrix}$ (d) $\begin{vmatrix} -3a & -3b & -3c \\ d & e & f \\ g-4d & h-4e & i-4f \end{vmatrix}$

13. Use row reduction to show that

$$\begin{vmatrix} 1 & 1 & 1 \\ a & b & c \\ a^2 & b^2 & c^2 \end{vmatrix} = (b-a)(c-a)(c-b)$$

14. Use an argument like that in the proof of Theorem 2.2.2 to show that

(a) $\det\begin{bmatrix} 0 & 0 & a_{13} \\ 0 & a_{22} & a_{23} \\ a_{31} & a_{32} & a_{33} \end{bmatrix} = -a_{13}a_{22}a_{31}$ (b) $\det\begin{bmatrix} 0 & 0 & 0 & a_{14} \\ 0 & 0 & a_{23} & a_{24} \\ 0 & a_{32} & a_{33} & a_{34} \\ a_{41} & a_{42} & a_{43} & a_{44} \end{bmatrix} = a_{14}a_{23}a_{32}a_{41}$

15. Prove the following special cases of Theorem 2.2.3.

(a) $\begin{vmatrix} ka_{11} & ka_{12} & ka_{13} \\ a_{21} & a_{22} & a_{23} \\ a_{31} & a_{32} & a_{33} \end{vmatrix} = k\begin{vmatrix} a_{11} & a_{12} & a_{13} \\ a_{21} & a_{22} & a_{23} \\ a_{31} & a_{32} & a_{33} \end{vmatrix}$ (b) $\begin{vmatrix} a_{21} & a_{22} & a_{23} \\ a_{11} & a_{12} & a_{13} \\ a_{31} & a_{32} & a_{33} \end{vmatrix} = -\begin{vmatrix} a_{11} & a_{12} & a_{13} \\ a_{21} & a_{22} & a_{23} \\ a_{31} & a_{32} & a_{33} \end{vmatrix}$

Discussion and Discovery

16. In each part, find $\det(A)$ by inspection, and explain your reasoning.

(a) $A = \begin{bmatrix} 0 & 0 & 1 \\ 0 & 1 & 0 \\ 1 & 0 & 0 \end{bmatrix}$ (b) $A = \begin{bmatrix} 0 & 0 & 0 & 1 \\ 0 & 0 & 1 & 0 \\ 0 & 1 & 0 & 0 \\ 1 & 0 & 0 & 0 \end{bmatrix}$

17. By inspection, solve the equation

$$\begin{vmatrix} x & 5 & 7 \\ 0 & x+1 & 6 \\ 0 & 0 & 2x-1 \end{vmatrix} = 0$$

Explain your reasoning.

18. (a) By inspection, find two solutions of the equation

$$\begin{vmatrix} 1 & x & x^2 \\ 1 & 1 & 1 \\ 1 & -3 & 9 \end{vmatrix} = 0$$

(b) Is it possible that there are other solutions? Justify your answer.

2.3 PROPERTIES OF THE DETERMINANT FUNCTION

In this section we shall develop some of the fundamental properties of the determinant function. Our work here will give us some further insight into the relationship between a square matrix and its determinant. One of the immediate consequences of this material will be an important determinant test for the invertibility of a matrix.

Basic Properties of Determinants Suppose that A and B are $n \times n$ matrices and k is any scalar. We begin by considering possible relationships between $\det(A)$, $\det(B)$, and

$$\det(kA), \quad \det(A+B), \quad \text{and} \quad \det(AB)$$

Since a common factor of any row of a matrix can be moved through the det sign, and since each of the n rows in kA has a common factor of k, we obtain

$$\det(kA) = k^n \det(A) \tag{1}$$

For example,

$$\begin{vmatrix} ka_{11} & ka_{12} & ka_{13} \\ ka_{21} & ka_{22} & ka_{23} \\ ka_{31} & ka_{32} & ka_{33} \end{vmatrix} = k^3 \begin{vmatrix} a_{11} & a_{12} & a_{13} \\ a_{21} & a_{22} & a_{23} \\ a_{31} & a_{32} & a_{33} \end{vmatrix}$$

Unfortunately, no simple relationship exists between $\det(A)$, $\det(B)$, and $\det(A+B)$ in general. In particular, we emphasize that $\det(A+B)$ is usually *not* equal to $\det(A)+\det(B)$. The following example illustrates this fact.

EXAMPLE 1 $\det[A+B] \neq \det[A] + \det[B]$

Consider

$$A = \begin{bmatrix} 1 & 2 \\ 2 & 5 \end{bmatrix}, \quad B = \begin{bmatrix} 3 & 1 \\ 1 & 3 \end{bmatrix}, \quad A+B = \begin{bmatrix} 4 & 3 \\ 3 & 8 \end{bmatrix}$$

We have $\det(A) = 1$, $\det(B) = 8$, and $\det(A+B) = 23$; thus

$$\det(A+B) \neq \det(A) + \det(B)$$

◆

In spite of the negative tone of the preceding example, there is one important relationship concerning sums of determinants that is often useful. To obtain it, consider two 2×2 matrices that differ only in the second row:

$$A = \begin{bmatrix} a_{11} & a_{12} \\ a_{21} & a_{22} \end{bmatrix} \quad \text{and} \quad B = \begin{bmatrix} a_{11} & a_{12} \\ b_{21} & b_{22} \end{bmatrix}$$

We have

$$\begin{aligned} \det(A) + \det(B) &= (a_{11}a_{22} - a_{12}a_{21}) + (a_{11}b_{22} - a_{12}b_{21}) \\ &= a_{11}(a_{22}+b_{22}) - a_{12}(a_{21}+b_{21}) \\ &= \det \begin{bmatrix} a_{11} & a_{12} \\ a_{21}+b_{21} & a_{22}+b_{22} \end{bmatrix} \end{aligned}$$

Thus,

$$\det \begin{bmatrix} a_{11} & a_{12} \\ a_{21} & a_{22} \end{bmatrix} + \det \begin{bmatrix} a_{11} & a_{12} \\ b_{21} & b_{22} \end{bmatrix} = \det \begin{bmatrix} a_{11} & a_{12} \\ a_{21}+b_{21} & a_{22}+b_{22} \end{bmatrix}$$

This is a special case of the following general result.

Theorem 2.3.1

Let A, B, and C be $n \times n$ matrices that differ only in a single row, say the rth, and assume that the rth row of C can be obtained by adding corresponding entries in the rth rows of A and B. Then

$$\det(C) = \det(A) + \det(B)$$

The same result holds for columns.

EXAMPLE 2 Using Theorem 2.3.1

By evaluating the determinants, the reader can check that

$$\det\begin{bmatrix} 1 & 7 & 5 \\ 2 & 0 & 3 \\ 1+0 & 4+1 & 7+(-1) \end{bmatrix} = \det\begin{bmatrix} 1 & 7 & 5 \\ 2 & 0 & 3 \\ 1 & 4 & 7 \end{bmatrix} + \det\begin{bmatrix} 1 & 7 & 5 \\ 2 & 0 & 3 \\ 0 & 1 & -1 \end{bmatrix}$$ ♦

Determinant of a Matrix Product When one considers the complexity of the definitions of matrix multiplication and determinants, it would seem unlikely that any simple relationship should exist between them. This is what makes the elegant simplicity of the following result so surprising: We will show that if A and B are square matrices of the same size, then

$$\det(AB) = \det(A)\det(B) \tag{2}$$

The proof of this theorem is fairly intricate, so we will have to develop some preliminary results first. We begin with the special case of (2) in which A is an elementary matrix. Because this special case is only a prelude to (2), we call it a lemma.

Lemma 2.3.2

If B is an $n \times n$ matrix and E is an $n \times n$ elementary matrix, then

$$\det(EB) = \det(E)\det(B)$$

Proof. We shall consider three cases, each depending on the row operation that produces matrix E.

Case 1. If E results from multiplying a row of I_n by k, then by Theorem 1.5.1 EB results from B by multiplying a row by k; so from Theorem 2.2.3a we have

$$\det(EB) = k\det(B)$$

But from Theorem 2.2.4a we have $\det(E) = k$, so

$$\det(EB) = \det(E)\det(B)$$

Cases 2 and 3. The proofs of the cases where E results from interchanging two rows of I_n or from adding a multiple of one row to another follow the same pattern as Case 1 and are left as exercises. ■

REMARK. It follows by repeated applications of Lemma 2.3.2 that if B is an $n \times n$ matrix and $E_1, E_2, \ldots, E_r$ are $n \times n$ elementary matrices, then

$$\det(E_1E_2\cdots E_rB) = \det(E_1)\det(E_2)\cdots\det(E_r)\det(B) \tag{3}$$

For example,

$$\det(E_1E_2B) = \det(E_1)\det(E_2B) = \det(E_1)\det(E_2)\det(B)$$

Determinant Test for Invertibility The next theorem is one of the most fundamental in linear algebra; it provides an important criterion for invertibility in terms of determinants, and it will be used in proving (2).

Theorem 2.3.3

A square matrix A is invertible if and only if $\det(A) \neq 0$.

Proof. Let R be the reduced row-echelon form of A. As a preliminary step, we will show that $\det(A)$ and $\det(R)$ are both zero or both nonzero: Let $E_1, E_2, \ldots, E_r$ be the elementary matrices that correspond to the elementary row operations that produce R from A. Thus,

$$R = E_r \cdots E_2 E_1 A$$

and from (3)

$$\det(R) = \det(E_r) \cdots \det(E_2)\det(E_1)\det(A) \tag{4}$$

But from Theorem 2.2.4 the determinants of the elementary matrices are all nonzero. (Keep in mind that multiplying a row by zero is *not* an allowable elementary row operation, so $k \neq 0$ in this application of Theorem 2.2.4.) Thus, it follows from (4) that $\det(A)$ and $\det(R)$ are both zero or both nonzero. Now to the main body of the proof.

If A is invertible, then by Theorem 1.6.4 we have $R = I$, so $\det(R) = 1 \neq 0$ and consequently $\det(A) \neq 0$. Conversely, if $\det(A) \neq 0$, then $\det(R) \neq 0$, so R cannot have a row of zeros. It follows from Theorem 1.4.3 that $R = I$, so A is invertible by Theorem 1.6.4. ■

It follows from Theorems 2.3.3 and 2.2.5 that a square matrix with two proportional rows or columns is not invertible.

EXAMPLE 3 Determinant Test for Invertibility

Since the first and third rows of

$$A = \begin{bmatrix} 1 & 2 & 3 \\ 1 & 0 & 1 \\ 2 & 4 & 6 \end{bmatrix}$$

are proportional, $\det(A) = 0$. Thus, A is not invertible. ♦

We are now ready for the main result in this section.

Theorem 2.3.4

If A and B are square matrices of the same size, then

$$\det(AB) = \det(A)\det(B)$$

Proof. We divide the proof into two cases that depend on whether or not A is invertible. If the matrix A is not invertible, then by Theorem 1.6.5 neither is the product AB. Thus, from Theorem 2.3.3, we have $\det(AB) = 0$ and $\det(A) = 0$, so it follows that $\det(AB) = \det(A)\det(B)$.

Now assume that A is invertible. By Theorem 1.6.4, the matrix A is expressible as a product of elementary matrices, say

$$A = E_1 E_2 \cdots E_r \tag{5}$$

so

$$AB = E_1 E_2 \cdots E_r B$$

Applying (3) to this equation yields

$$\det(AB) = \det(E_1)\det(E_2)\cdots\det(E_r)\det(B)$$

and applying (3) again yields

$$\det(AB) = \det(E_1 E_2 \cdots E_r)\det(B)$$

which, from (5), can be written as $\det(AB) = \det(A)\det(B)$. ■

EXAMPLE 4 Verifying That $\det[AB] = \det[A]\det[B]$

Consider the matrices

$$A = \begin{bmatrix} 3 & 1 \\ 2 & 1 \end{bmatrix}, \quad B = \begin{bmatrix} -1 & 3 \\ 5 & 8 \end{bmatrix}, \quad AB = \begin{bmatrix} 2 & 17 \\ 3 & 14 \end{bmatrix}$$

We leave it for the reader to verify that

$$\det(A) = 1, \quad \det(B) = -23, \quad \text{and} \quad \det(AB) = -23$$

Thus, $\det(AB) = \det(A)\det(B)$ as guaranteed by Theorem 2.3.4. ♦

The following theorem gives a useful relationship between the determinant of an invertible matrix and the determinant of its inverse.

Theorem 2.3.5

If A is invertible, then

$$\det(A^{-1}) = \frac{1}{\det(A)}$$

Proof. Since $A^{-1}A = I$, it follows that $\det(A^{-1}A) = \det(I)$. Therefore, we must have $\det(A^{-1})\det(A) = 1$. Since $\det(A) \neq 0$, the proof can be completed by dividing through by $\det(A)$. ■

Linear Systems of the Form $A\mathbf{x} = \lambda\mathbf{x}$ Many applications of linear algebra are concerned with systems of n linear equations in n unknowns that are expressed in the form

$$A\mathbf{x} = \lambda\mathbf{x} \tag{6}$$

where λ is a scalar. Such systems are really homogeneous linear systems in disguise, since (6) can be rewritten as $\lambda\mathbf{x} - A\mathbf{x} = \mathbf{0}$ or, by inserting an identity matrix and factoring, as

$$(\lambda I - A)\mathbf{x} = \mathbf{0} \tag{7}$$

Here is an example.

EXAMPLE 5 Finding $\lambda I - A$

The linear system

$$\begin{aligned} x_1 + 3x_2 &= \lambda x_1 \\ 4x_1 + 2x_2 &= \lambda x_2 \end{aligned}$$

can be written in matrix form as

$$\begin{bmatrix} 1 & 3 \\ 4 & 2 \end{bmatrix}\begin{bmatrix} x_1 \\ x_2 \end{bmatrix} = \lambda \begin{bmatrix} x_1 \\ x_2 \end{bmatrix}$$

which is of form (6) with

$$A = \begin{bmatrix} 1 & 3 \\ 4 & 2 \end{bmatrix} \quad \text{and} \quad \mathbf{x} = \begin{bmatrix} x_1 \\ x_2 \end{bmatrix}$$

This system can be rewritten as

$$\lambda \begin{bmatrix} x_1 \\ x_2 \end{bmatrix} - \begin{bmatrix} 1 & 3 \\ 4 & 2 \end{bmatrix}\begin{bmatrix} x_1 \\ x_2 \end{bmatrix} = \begin{bmatrix} 0 \\ 0 \end{bmatrix}$$

or

$$\lambda \begin{bmatrix} 1 & 0 \\ 0 & 1 \end{bmatrix}\begin{bmatrix} x_1 \\ x_2 \end{bmatrix} - \begin{bmatrix} 1 & 3 \\ 4 & 2 \end{bmatrix}\begin{bmatrix} x_1 \\ x_2 \end{bmatrix} = \begin{bmatrix} 0 \\ 0 \end{bmatrix}$$

or

$$\begin{bmatrix} \lambda - 1 & -3 \\ -4 & \lambda - 2 \end{bmatrix}\begin{bmatrix} x_1 \\ x_2 \end{bmatrix} = \begin{bmatrix} 0 \\ 0 \end{bmatrix}$$

which is of form (7) with

$$\lambda I - A = \begin{bmatrix} \lambda - 1 & -3 \\ -4 & \lambda - 2 \end{bmatrix}$$ ♦

The primary problem of interest for linear systems of the form (7) is to determine those values of λ for which the system has a nontrivial solution; such a value of λ is called a ***characteristic value*** or an ***eigenvalue***† of A. If λ is an eigenvalue of A, then the nontrivial solutions of (7) are called the ***eigenvectors*** of A corresponding to λ.

It follows from Theorem 2.3.3 that the system $(\lambda I - A)\mathbf{x} = \mathbf{0}$ has a nontrivial solution if and only if

$$\det(\lambda I - A) = 0 \tag{8}$$

This is called the ***characteristic equation*** of A; the eigenvalues of A can be found by solving this equation for λ.

Eigenvalues and eigenvectors will be studied again in subsequent chapters, where we will discuss their geometric interpretation and develop their properties in more depth.

EXAMPLE 6 Eigenvalues and Eigenvectors

Find the eigenvalues and corresponding eigenvectors of the matrix A in Example 5.

Solution.

The characteristic equation of A is

$$\det(\lambda I - A) = \begin{vmatrix} \lambda - 1 & -3 \\ -4 & \lambda - 2 \end{vmatrix} = 0 \quad \text{or} \quad \lambda^2 - 3\lambda - 10 = 0$$

† The word *eigenvalue* is a mixture of German and English. The German prefix *eigen* can be translated as "proper," which stems from the older literature where eigenvalues were known as *proper values*; they were also called *latent roots*.

The factored form of this equation is $(\lambda + 2)(\lambda - 5) = 0$, so the eigenvalues of A are $\lambda = -2$ and $\lambda = 5$.

By definition

$$\mathbf{x} = \begin{bmatrix} x_1 \\ x_2 \end{bmatrix}$$

is an eigenvector of A if and only if $\mathbf{x}$ is a nontrivial solution of $(\lambda I - A)\mathbf{x} = \mathbf{0}$; that is,

$$\begin{bmatrix} \lambda - 1 & -3 \\ -4 & \lambda - 2 \end{bmatrix} \begin{bmatrix} x_1 \\ x_2 \end{bmatrix} = \begin{bmatrix} 0 \\ 0 \end{bmatrix} \tag{9}$$

If $\lambda = -2$, then (9) becomes

$$\begin{bmatrix} -3 & -3 \\ -4 & -4 \end{bmatrix} \begin{bmatrix} x_1 \\ x_2 \end{bmatrix} = \begin{bmatrix} 0 \\ 0 \end{bmatrix}$$

Solving this system yields (verify) $x_1 = -t,\ x_2 = t$, so the eigenvectors corresponding to $\lambda = -2$ are the nonzero solutions of the form

$$\mathbf{x} = \begin{bmatrix} x_1 \\ x_2 \end{bmatrix} = \begin{bmatrix} -t \\ t \end{bmatrix}$$

Again from (9), the eigenvectors of A corresponding to $\lambda = 5$ are the nontrivial solutions of

$$\begin{bmatrix} 4 & -3 \\ -4 & 3 \end{bmatrix} \begin{bmatrix} x_1 \\ x_2 \end{bmatrix} = \begin{bmatrix} 0 \\ 0 \end{bmatrix}$$

We leave it for the reader to solve this system and show that the eigenvectors of A corresponding to $\lambda = 5$ are the nonzero solutions of the form

$$\mathbf{x} = \begin{bmatrix} \frac{3}{4}t \\ t \end{bmatrix}$$ ◆

Summary In Theorem 1.6.4 we listed five results that are equivalent to the invertibility of a matrix A. We conclude this section by merging Theorem 2.3.3 with that list to produce the following theorem that relates all of the major topics we have studied thus far.

Theorem 2.3.6 Equivalent Statements

If A is an $n \times n$ matrix, then the following are equivalent.

(a) *A is invertible.*
(b) *$A\mathbf{x} = \mathbf{0}$ has only the trivial solution.*
(c) *The reduced row-echelon form of A is I_n.*
(d) *A is expressible as a product of elementary matrices.*
(e) *$A\mathbf{x} = \mathbf{b}$ is consistent for every $n \times 1$ matrix $\mathbf{b}$.*
(f) *$A\mathbf{x} = \mathbf{b}$ has exactly one solution for every $n \times 1$ matrix $\mathbf{b}$.*
(g) $\det(A) \neq 0$.

Exercise Set 2.3

1. Verify that $\det(kA) = k^n \det(A)$ for

(a) $A = \begin{bmatrix} -1 & 2 \\ 3 & 4 \end{bmatrix}$; $k = 2$ (b) $A = \begin{bmatrix} 2 & -1 & 3 \\ 3 & 2 & 1 \\ 1 & 4 & 5 \end{bmatrix}$; $k = -2$

2. Verify that $\det(AB) = \det(A)\det(B)$ for

$$A = \begin{bmatrix} 2 & 1 & 0 \\ 3 & 4 & 0 \\ 0 & 0 & 2 \end{bmatrix} \quad \text{and} \quad B = \begin{bmatrix} 1 & -1 & 3 \\ 7 & 1 & 2 \\ 5 & 0 & 1 \end{bmatrix}$$

3. By inspection, explain why $\det(A) = 0$.

$$A = \begin{bmatrix} -2 & 8 & 1 & 4 \\ 3 & 2 & 5 & 1 \\ 1 & 10 & 6 & 5 \\ 4 & -6 & 4 & -3 \end{bmatrix}$$

4. Use Theorem 2.3.3 to determine which of the following matrices are invertible.

(a) $\begin{bmatrix} 1 & 0 & -1 \\ 9 & -1 & 4 \\ 8 & 9 & -1 \end{bmatrix}$ (b) $\begin{bmatrix} 4 & 2 & 8 \\ -2 & 1 & -4 \\ 3 & 1 & 6 \end{bmatrix}$ (c) $\begin{bmatrix} \sqrt{2} & -\sqrt{7} & 0 \\ 3\sqrt{2} & -3\sqrt{7} & 0 \\ 5 & -9 & 0 \end{bmatrix}$ (d) $\begin{bmatrix} -3 & 0 & 1 \\ 5 & 0 & 6 \\ 8 & 0 & 3 \end{bmatrix}$

5. Let

$$A = \begin{bmatrix} a & b & c \\ d & e & f \\ g & h & i \end{bmatrix}$$

Assuming that $\det(A) = -7$, find

(a) $\det(3A)$ (b) $\det(A^{-1})$ (c) $\det(2A^{-1})$ (d) $\det((2A)^{-1})$ (e) $\det \begin{bmatrix} a & g & d \\ b & h & e \\ c & i & f \end{bmatrix}$

6. Without directly evaluating, show that $x = 0$ and $x = 2$ satisfy

$$\begin{vmatrix} x^2 & x & 2 \\ 2 & 1 & 1 \\ 0 & 0 & -5 \end{vmatrix} = 0$$

7. Without directly evaluating, show that

$$\det \begin{bmatrix} b+c & c+a & b+a \\ a & b & c \\ 1 & 1 & 1 \end{bmatrix} = 0$$

In Exercises 8–11 prove the identity without evaluating the determinants.

8. $\begin{vmatrix} a_1 & b_1 & a_1+b_1+c_1 \\ a_2 & b_2 & a_2+b_2+c_2 \\ a_3 & b_3 & a_3+b_3+c_3 \end{vmatrix} = \begin{vmatrix} a_1 & b_1 & c_1 \\ a_2 & b_2 & c_2 \\ a_3 & b_3 & c_3 \end{vmatrix}$

9. $\begin{vmatrix} a_1+b_1 & a_1-b_1 & c_1 \\ a_2+b_2 & a_2-b_2 & c_2 \\ a_3+b_3 & a_3-b_3 & c_3 \end{vmatrix} = -2 \begin{vmatrix} a_1 & b_1 & c_1 \\ a_2 & b_2 & c_2 \\ a_3 & b_3 & c_3 \end{vmatrix}$

10. $$\begin{vmatrix} a_1+b_1t & a_2+b_2t & a_3+b_3t \\ a_1t+b_1 & a_2t+b_2 & a_3t+b_3 \\ c_1 & c_2 & c_3 \end{vmatrix} = (1-t^2)\begin{vmatrix} a_1 & a_2 & a_3 \\ b_1 & b_2 & b_3 \\ c_1 & c_2 & c_3 \end{vmatrix}$$

11. $$\begin{vmatrix} a_1 & b_1+ta_1 & c_1+rb_1+sa_1 \\ a_2 & b_2+ta_2 & c_2+rb_2+sa_2 \\ a_3 & b_3+ta_3 & c_3+rb_3+sa_3 \end{vmatrix} = \begin{vmatrix} a_1 & a_2 & a_3 \\ b_1 & b_2 & b_3 \\ c_1 & c_2 & c_3 \end{vmatrix}$$

12. For which value(s) of k does A fail to be invertible?

(a) $A = \begin{bmatrix} k-3 & -2 \\ -2 & k-2 \end{bmatrix}$ (b) $A = \begin{bmatrix} 1 & 2 & 4 \\ 3 & 1 & 6 \\ k & 3 & 2 \end{bmatrix}$

13. Use Theorem 2.3.3 to show that

$$\begin{bmatrix} \sin^2\alpha & \sin^2\beta & \sin^2\gamma \\ \cos^2\alpha & \cos^2\beta & \cos^2\gamma \\ 1 & 1 & 1 \end{bmatrix}$$

is not invertible for any values of α, β, and γ.

14. Express the following linear systems in the form $(\lambda I - A)\mathbf{x} = \mathbf{0}$.

(a) $\begin{aligned} x_1 + 2x_2 &= \lambda x_1 \\ 2x_1 + x_2 &= \lambda x_2 \end{aligned}$ (b) $\begin{aligned} 2x_1 + 3x_2 &= \lambda x_1 \\ 4x_1 + 3x_2 &= \lambda x_2 \end{aligned}$ (c) $\begin{aligned} 3x_1 + x_2 &= \lambda x_1 \\ -5x_1 - 3x_2 &= \lambda x_2 \end{aligned}$

15. For each of the systems in Exercise 14, find

(i) the characteristic equation;
(ii) the eigenvalues;
(iii) the eigenvectors corresponding to each of the eigenvalues.

16. Let A and B be $n \times n$ matrices. Show that if A is invertible, then $\det(B) = \det(A^{-1}BA)$.

17. (a) Express

$$\begin{vmatrix} a_1+b_1 & c_1+d_1 \\ a_2+b_2 & c_2+d_2 \end{vmatrix}$$

as a sum of four determinants whose entries contain no sums.

(b) Express

$$\begin{vmatrix} a_1+b_1 & c_1+d_1 & e_1+f_1 \\ a_2+b_2 & c_2+d_2 & e_2+f_2 \\ a_3+b_3 & c_3+d_3 & e_3+f_3 \end{vmatrix}$$

as a sum of eight determinants whose entries contain no sums.

18. Prove that a square matrix A is invertible if and only if A^TA is invertible.

19. Prove Cases 2 and 3 of Lemma 2.3.2.

Discussion and Discovery

20. Let A and B be $n \times n$ matrices. You know from earlier work that AB and BA need not be equal. Is the same true for $\det(AB)$ and $\det(BA)$? Explain your reasoning.

21. Let A and B be $n \times n$ matrices. You know from earlier work that AB is invertible if A and B are invertible. What can you say about the invertibility of AB if one or both of the factors are singular? Explain your reasoning.

22. Indicate whether the statement is always true or sometimes false. Justify each answer by giving a logical argument or a counterexample.

(a) $\det(2A) = 2\det(A)$

(b) $|A^2| = |A|^2$

(c) $\det(I + A) = 1 + \det(A)$

(d) If $\det(A) = 0$, then the homogeneous system $A\mathbf{x} = \mathbf{0}$ has infinitely many solutions.

23. Indicate whether the statement is always true or sometimes false. Justify your answer by giving a logical argument or a counterexample.

(a) If $\det(A) = 0$, then A is not expressible as a product of elementary matrices.

(b) If the reduced row-echelon form of A has a row of zeros, then $\det(A) = 0$.

(c) The determinant of a matrix is unchanged if the columns are written in reverse order.

(d) There is no square matrix A such that $\det(AA^T) = -1$.

2.4 COFACTOR EXPANSION; CRAMER'S RULE

In this section we shall consider a method for evaluating determinants that is useful for hand computations and is also important theoretically. As a consequence of our work here, we will obtain a formula for the inverse of an invertible matrix as well as a formula for the solution to certain systems of linear equations in terms of determinants.

Minors and Cofactors In Example 7 of Section 2.1 we saw that the determinant of a 3×3 matrix

$$A = \begin{bmatrix} a_{11} & a_{12} & a_{13} \\ a_{21} & a_{22} & a_{23} \\ a_{33} & a_{32} & a_{33} \end{bmatrix}$$

is the number

$$\begin{aligned} \det(A) = {} & a_{11}a_{22}a_{33} + a_{12}a_{23}a_{31} + a_{13}a_{21}a_{32} \\ & - a_{13}a_{22}a_{31} - a_{12}a_{21}a_{33} - a_{11}a_{23}a_{32} \end{aligned} \tag{1}$$

By rearranging terms and factoring, (1) can be rewritten as

$$\det(A) = a_{11}(a_{22}a_{33} - a_{23}a_{32}) - a_{12}(a_{21}a_{33} - a_{23}a_{31}) + a_{13}(a_{21}a_{32} - a_{22}a_{31}) \tag{2}$$

The expressions highlighted in color in (2) are themselves determinants:

$$M_{11} = \begin{vmatrix} a_{22} & a_{23} \\ a_{32} & a_{33} \end{vmatrix}, \qquad M_{12} = \begin{vmatrix} a_{21} & a_{23} \\ a_{31} & a_{33} \end{vmatrix}, \qquad M_{13} = \begin{vmatrix} a_{21} & a_{22} \\ a_{31} & a_{32} \end{vmatrix}$$

The submatrices of A that appear in these determinants are given a special name:

Definition

If A is a square matrix, then the ***minor of entry a_{ij}*** is denoted by M_{ij} and is defined to be the determinant of the submatrix that remains after the ith row and jth column are deleted from A. The number $(-1)^{i+j}M_{ij}$ is denoted by C_{ij} and is called the ***cofactor of entry a_{ij}***.

EXAMPLE 1 Finding Minors and Cofactors

Let

$$A = \begin{bmatrix} 3 & 1 & -4 \\ 2 & 5 & 6 \\ 1 & 4 & 8 \end{bmatrix}$$

The minor of entry a_{11} is

$$M_{11} = \begin{vmatrix} 3 & 1 & -4 \\ 2 & 5 & 6 \\ 1 & 4 & 8 \end{vmatrix} = \begin{vmatrix} 5 & 6 \\ 4 & 8 \end{vmatrix} = 16$$

The cofactor of a_{11} is

$$C_{11} = (-1)^{1+1} M_{11} = M_{11} = 16$$

Similarly, the minor of entry a_{32} is

$$M_{32} = \begin{vmatrix} 3 & 1 & -4 \\ 2 & 5 & 6 \\ 1 & 4 & 8 \end{vmatrix} = \begin{vmatrix} 3 & -4 \\ 2 & 6 \end{vmatrix} = 26$$

The cofactor of a_{32} is

$$C_{32} = (-1)^{3+2} M_{32} = -M_{32} = -26$$

◆

Notice that the cofactor and the minor of an element a_{ij} differ only in sign, that is, $C_{ij} = \pm M_{ij}$. A quick way for determining whether to use the + or − is to use the fact that the sign relating C_{ij} and M_{ij} is in the ith row and jth column of the "checkerboard" array

$$\begin{bmatrix} + & - & + & - & + & \cdots \\ - & + & - & + & - & \cdots \\ + & - & + & - & + & \cdots \\ - & + & - & + & - & \cdots \\ \vdots & \vdots & \vdots & \vdots & \vdots & \end{bmatrix}$$

For example, $C_{11} = M_{11}$, $C_{21} = -M_{21}$, $C_{12} = -M_{12}$, $C_{22} = M_{22}$, and so on.

Cofactor Expansions In view of the definition on the preceding page, the expression in (2) can be written in terms of minors and cofactors as

$$\begin{aligned} \det(A) &= a_{11}M_{11} + a_{12}(-M_{12}) + a_{13}M_{13} \\ &= a_{11}C_{11} + a_{12}C_{12} + a_{13}C_{13} \end{aligned} \tag{3}$$

Equation (3) shows that the determinant of A can be computed by multiplying the entries in the first row of A by their corresponding cofactors and adding the resulting products. This method of evaluating $\det(A)$ is called ***cofactor expansion*** along the first row of A.

EXAMPLE 2 Cofactor Expansion Along the First Row

Let $A = \begin{bmatrix} 3 & 1 & 0 \\ -2 & -4 & 3 \\ 5 & 4 & -2 \end{bmatrix}$. Evaluate $\det(A)$ by cofactor expansion along the first row of A.

Solution.

From (3)

$$\det(A)=\begin{vmatrix}3 & 1 & 0\\ -2 & -4 & 3\\ 5 & 4 & -2\end{vmatrix}=3\begin{vmatrix}-4 & 3\\ 4 & -2\end{vmatrix}-1\begin{vmatrix}-2 & 3\\ 5 & -2\end{vmatrix}+0\begin{vmatrix}-2 & -4\\ 5 & 4\end{vmatrix}$$

$$=3(-4)-(1)(-11)+0=-1$$ ♦

By rearranging the terms in (1) in various ways, it is possible to obtain other formulas like (3). There should be no trouble checking that all of the following are correct (see Exercise 28):

$$\begin{aligned}\det(A)&=a_{11}C_{11}+a_{12}C_{12}+a_{13}C_{13}\\&=a_{11}C_{11}+a_{21}C_{21}+a_{31}C_{31}\\&=a_{21}C_{21}+a_{22}C_{22}+a_{23}C_{23}\\&=a_{12}C_{12}+a_{22}C_{22}+a_{32}C_{32}\\&=a_{31}C_{31}+a_{32}C_{32}+a_{33}C_{33}\\&=a_{13}C_{13}+a_{23}C_{23}+a_{33}C_{33}\end{aligned}\tag{4}$$

Notice that in each equation the entries and cofactors all come from the same row or column. These equations are called the ***cofactor expansions*** of $\det(A)$.

The results we have just given for 3×3 matrices form a special case of the following general theorem, which we state without proof.

Theorem 2.4.1 Expansions by Cofactors

The determinant of an $n\times n$ matrix A can be computed by multiplying the entries in any row (or column) by their cofactors and adding the resulting products; that is, for each $1\le i\le n$ and $1\le j\le n$,

$$\det(A)=a_{1j}C_{1j}+a_{2j}C_{2j}+\cdots+a_{nj}C_{nj}$$

(cofactor expansion along the jth column)

and

$$\det(A)=a_{i1}C_{i1}+a_{i2}C_{i2}+\cdots+a_{in}C_{in}$$

(cofactor expansion along the ith row)

EXAMPLE 3 Cofactor Expansion Along the First Column

Let A be the matrix in Example 2. Evaluate $\det(A)$ by cofactor expansion along the first column of A.

Solution.

From (4)

$$\det(A)=\begin{vmatrix}3 & 1 & 0\\ -2 & -4 & 3\\ 5 & 4 & -2\end{vmatrix}=3\begin{vmatrix}-4 & 3\\ 4 & -2\end{vmatrix}-(-2)\begin{vmatrix}1 & 0\\ 4 & -2\end{vmatrix}+5\begin{vmatrix}1 & 0\\ -4 & 3\end{vmatrix}$$

$$=3(-4)-(-2)(-2)+5(3)=-1$$

This agrees with the result obtained in Example 2. ♦

REMARK. In this example we had to compute three cofactors, but in Example 2 we only had to compute two of them, since the third was multiplied by zero. In general, the best strategy for evaluating a determinant by cofactor expansion is to expand along a row or column having the largest number of zeros.

Cofactor expansion and row or column operations can sometimes be used in combination to provide an effective method for evaluating determinants. The following example illustrates this idea.

EXAMPLE 4 Row Operations and Cofactor Expansion

Evaluate $\det(A)$ where

$$A = \begin{bmatrix} 3 & 5 & -2 & 6 \\ 1 & 2 & -1 & 1 \\ 2 & 4 & 1 & 5 \\ 3 & 7 & 5 & 3 \end{bmatrix}$$

Solution.

By adding suitable multiples of the second row to the remaining rows, we obtain

$$\det(A) = \begin{vmatrix} 0 & -1 & 1 & 3 \\ 1 & 2 & -1 & 1 \\ 0 & 0 & 3 & 3 \\ 0 & 1 & 8 & 0 \end{vmatrix}$$

$$= -\begin{vmatrix} -1 & 1 & 3 \\ 0 & 3 & 3 \\ 1 & 8 & 0 \end{vmatrix}$$ ← Cofactor expansion along the first column

$$= -\begin{vmatrix} -1 & 1 & 3 \\ 0 & 3 & 3 \\ 0 & 9 & 3 \end{vmatrix}$$ ← We added the first row to the third row.

$$= -(-1)\begin{vmatrix} 3 & 3 \\ 9 & 3 \end{vmatrix}$$ ← Cofactor expansion along the first column

$$= -18$$ ◆

Adjoint of a Matrix In a cofactor expansion we compute $\det(A)$ by multiplying the entries in a row or column by their cofactors and adding the resulting products. It turns out that if one multiplies the entries in any row by the corresponding cofactors from a *different* row, the sum of these products is always zero. (This result also holds for columns.) Although we omit the general proof, the next example illustrates the idea of the proof in a special case.

EXAMPLE 5 Entries and Cofactors from Different Rows

Let

$$A = \begin{bmatrix} a_{11} & a_{12} & a_{13} \\ a_{21} & a_{22} & a_{23} \\ a_{31} & a_{32} & a_{33} \end{bmatrix}$$

Consider the quantity

$$a_{11}C_{31} + a_{12}C_{32} + a_{13}C_{33}$$

that is formed by multiplying the entries in the first row by the cofactors of the corresponding entries in the third row and adding the resulting products. We now show that this quantity is equal to zero by the following trick. Construct a new matrix A' by replacing the third row of A with another copy of the first row. Thus,

$$A' = \begin{bmatrix} a_{11} & a_{12} & a_{13} \\ a_{21} & a_{22} & a_{23} \\ a_{11} & a_{12} & a_{13} \end{bmatrix}$$

Let C'_{31}, C'_{32}, C'_{33} be the cofactors of the entries in the third row of A'. Since the first two rows of A and A' are the same, and since the computations of C_{31}, C_{32}, C_{33}, C'_{31}, C'_{32}, and C'_{33} involve only entries from the first two rows of A and A', it follows that

$$C_{31} = C'_{31}, \qquad C_{32} = C'_{32}, \qquad C_{33} = C'_{33}$$

Since A' has two identical rows,

$$\det(A') = 0 \tag{5}$$

On the other hand, evaluating $\det(A')$ by cofactor expansion along the third row gives

$$\det(A') = a_{11}C'_{31} + a_{12}C'_{32} + a_{13}C'_{33} = a_{11}C_{31} + a_{12}C_{32} + a_{13}C_{33} \tag{6}$$

From (5) and (6) we obtain

$$a_{11}C_{31} + a_{12}C_{32} + a_{13}C_{33} = 0$$

♦

Definition

If A is any $n \times n$ matrix and C_{ij} is the cofactor of a_{ij}, then the matrix

$$\begin{bmatrix} C_{11} & C_{12} & \cdots & C_{1n} \\ C_{21} & C_{22} & \cdots & C_{2n} \\ \vdots & \vdots & & \vdots \\ C_{n1} & C_{n2} & \cdots & C_{nn} \end{bmatrix}$$

is called the ***matrix of cofactors from A***. The transpose of this matrix is called the ***adjoint of A*** and is denoted by $\text{adj}(A)$.

EXAMPLE 6 Adjoint of a 3 × 3 Matrix

Let

$$A = \begin{bmatrix} 3 & 2 & -1 \\ 1 & 6 & 3 \\ 2 & -4 & 0 \end{bmatrix}$$

The cofactors of A are

$$\begin{array}{lll} C_{11} = 12 & C_{12} = 6 & C_{13} = -16 \\ C_{21} = 4 & C_{22} = 2 & C_{23} = 16 \\ C_{31} = 12 & C_{32} = -10 & C_{33} = 16 \end{array}$$

so that the matrix of cofactors is

$$\begin{bmatrix} 12 & 6 & -16 \\ 4 & 2 & 16 \\ 12 & -10 & 16 \end{bmatrix}$$

and the adjoint of A is

$$\text{adj}(A) = \begin{bmatrix} 12 & 4 & 12 \\ 6 & 2 & -10 \\ -16 & 16 & 16 \end{bmatrix}$$

♦

We are now in a position to derive a formula for the inverse of an invertible matrix.

Theorem 2.4.2 Inverse of a Matrix Using Its Adjoint

If A is an invertible matrix, then

$$A^{-1} = \frac{1}{\det(A)}\text{adj}(A) \tag{7}$$

Proof. We show first that

$$A\,\text{adj}(A) = \det(A)I$$

Consider the product

$$A\,\text{adj}(A) = \begin{bmatrix} a_{11} & a_{12} & \cdots & a_{1n} \\ a_{21} & a_{22} & \cdots & a_{2n} \\ \vdots & \vdots & & \vdots \\ a_{i1} & a_{i2} & \cdots & a_{in} \\ \vdots & \vdots & & \vdots \\ a_{n1} & a_{n2} & \cdots & a_{nn} \end{bmatrix} \begin{bmatrix} C_{11} & C_{21} & \cdots & C_{j1} & \cdots & C_{n1} \\ C_{12} & C_{22} & \cdots & C_{j2} & \cdots & C_{n2} \\ \vdots & \vdots & & \vdots & & \vdots \\ C_{1n} & C_{2n} & \cdots & C_{jn} & \cdots & C_{nn} \end{bmatrix}$$

The entry in the ith row and jth column of the product $A\,\text{adj}(A)$ is

$$a_{i1}C_{j1} + a_{i2}C_{j2} + \cdots + a_{in}C_{jn} \tag{8}$$

(see the shaded lines above).

If $i = j$, then (8) is the cofactor expansion of $\det(A)$ along the ith row of A (Theorem 2.4.1), and if $i \neq j$, then the a's and the cofactors come from different rows of A, so the value of (8) is zero. Therefore,

$$A\,\text{adj}(A) = \begin{bmatrix} \det(A) & 0 & \cdots & 0 \\ 0 & \det(A) & \cdots & 0 \\ \vdots & \vdots & & \vdots \\ 0 & 0 & \cdots & \det(A) \end{bmatrix} = \det(A)I \tag{9}$$

Since A is invertible, $\det(A) \neq 0$. Therefore, Equation (9) can be rewritten as

$$\frac{1}{\det(A)}[A\,\text{adj}(A)] = I \quad \text{or} \quad A\left[\frac{1}{\det(A)}\text{adj}(A)\right] = I$$

Multiplying both sides on the left by A^{-1} yields

$$A^{-1} = \frac{1}{\det(A)}\text{adj}(A)$$

■

EXAMPLE 7 Using the Adjoint to Find an Inverse Matrix

Use (7) to find the inverse of the matrix A in Example 6.

Solution.

The reader can check that $\det(A) = 64$. Thus,

$$A^{-1} = \frac{1}{\det(A)}\operatorname{adj}(A) = \frac{1}{64}\begin{bmatrix} 12 & 4 & 12 \\ 6 & 2 & -10 \\ -16 & 16 & 16 \end{bmatrix} = \begin{bmatrix} \frac{12}{64} & \frac{4}{64} & \frac{12}{64} \\ \frac{6}{64} & \frac{2}{64} & -\frac{10}{64} \\ -\frac{16}{64} & \frac{16}{64} & \frac{16}{64} \end{bmatrix}$$ ◆

Applications of Formula (7) Although the method in the preceding example is reasonable for inverting 3×3 matrices by hand, the inversion algorithm discussed in Section 1.5 is more efficient for larger matrices. It should be kept in mind, however, that the method of Section 1.5 is just a computational procedure, whereas Formula (7) is an actual formula for the inverse. As we shall now see, this formula is useful for deriving properties of the inverse.

In Section 1.7 we stated two results about inverses without proof.

- **Theorem 1.7.1*c*:** A triangular matrix is invertible if and only if its diagonal entries are all nonzero.
- **Theorem 1.7.1*d*:** The inverse of an invertible lower triangular matrix is lower triangular, and the inverse of an invertible upper triangular matrix is upper triangular.

We will now prove these results using the adjoint formula for the inverse.

Proof of Theorem 1.7.1c. Let $A = [a_{ij}]$ be a triangular matrix, so that its diagonal entries are

$$a_{11}, a_{22}, \ldots, a_{nn}$$

From Theorems 2.2.2 and 2.3.3, the matrix A is invertible if and only if

$$\det(A) = a_{11}a_{22}\cdots a_{nn} \neq 0$$

which is true if and only if the diagonal entries are all nonzero. ◆

We leave it as an exercise for the reader to use the adjoint formula for A^{-1} to show that if $A = [a_{ij}]$ is an invertible triangular matrix, then the successive diagonal entries of A^{-1} are

$$\frac{1}{a_{11}}, \frac{1}{a_{22}}, \ldots, \frac{1}{a_{nn}}$$

(See Example 3 of Section 1.7.)

Proof of Theorem 1.7.1d. We will prove the result for upper triangular matrices and leave the lower triangular case as an exercise. Assume that A is upper triangular and invertible. Since

$$A^{-1} = \frac{1}{\det(A)}\operatorname{adj}(A)$$

we can prove that A^{-1} is upper triangular by showing that $\operatorname{adj}(A)$ is upper triangular, or equivalently, that the matrix of cofactors is lower triangular. We can do this by showing that every cofactor C_{ij} with $i < j$ (i.e., above the main diagonal) is zero. Since

$$C_{ij} = (-1)^{i+j}M_{ij}$$

Gabriel Cramer (1704–1752) was a Swiss mathematician. Although Cramer does not rank with the great mathematicians of his time, his contributions as a disseminator of mathematical ideas have earned him a well-deserved place in the history of mathematics. Cramer traveled extensively and met many of the leading mathematicians of his day.

Cramer's most widely known work, *Introduction à l'analyse des lignes courbes algébriques* (1750), was a study and classification of algebraic curves; Cramer's rule appeared in the appendix. Although the rule bears his name, variations of the idea were formulated earlier by various mathematicians. However, Cramer's superior notation helped clarify and popularize the technique.

Overwork combined with a fall from a carriage led to his death at the age of 48. Cramer was apparently a good-natured and pleasant person with broad interests. He wrote on philosophy of law and government and the history of mathematics. He served in public office, participated in artillery and fortifications activities for the government, instructed workers on techniques of cathedral repair, and undertook excavations of cathedral archives. Cramer received numerous honors for his activities.

it suffices to show that each minor M_{ij} with $i < j$ is zero. For this purpose let B_{ij} be the matrix that results when the ith row and jth column of A are deleted, so that

$$M_{ij} = \det(B_{ij}) \tag{10}$$

From the assumption that $i < j$ it follows that B_{ij} is upper triangular (Exercise 32). Since A is upper triangular, its $(i+1)$-st row begins with at least i zeros. But the ith row of B_{ij} is the $(i+1)$-st row of A with the entry in the jth column removed. Since $i < j$, none of the first i zeros is removed by deleting the jth column; thus, the ith row of B_{ij} starts with at least i zeros, which implies that this row has a zero on the main diagonal. It now follows from Theorem 2.2.2 that $\det(B_{ij}) = 0$ and from (10) that $M_{ij} = 0$. ■

Cramer's Rule

The next theorem provides a formula for the solution of certain linear systems of n equations in n unknowns. This formula, known as ***Cramer's rule*** is of marginal interest for computational purposes, but it is useful for studying the mathematical properties of a solution without the need for solving the system.

Theorem 2.4.3 Cramer's Rule

If $A\mathbf{x} = \mathbf{b}$ *is a system of* n *linear equations in* n *unknowns such that* $\det(A) \neq 0$, *then the system has a unique solution. This solution is*

$$x_1 = \frac{\det(A_1)}{\det(A)}, \quad x_2 = \frac{\det(A_2)}{\det(A)}, \ldots, \quad x_n = \frac{\det(A_n)}{\det(A)}$$

where A_j *is the matrix obtained by replacing the entries in the* jth *column of* A *by the entries in the matrix*

$$\mathbf{b} = \begin{bmatrix} b_1 \\ b_2 \\ \vdots \\ b_n \end{bmatrix}$$

Proof. If $\det(A) \neq 0$, then A is invertible and, by Theorem 1.6.2, $\mathbf{x} = A^{-1}\mathbf{b}$ is the unique solution of $A\mathbf{x} = \mathbf{b}$. Therefore, by Theorem 2.4.2 we have

$$\mathbf{x} = A^{-1}\mathbf{b} = \frac{1}{\det(A)}\operatorname{adj}(A)\mathbf{b} = \frac{1}{\det(A)}\begin{bmatrix} C_{11} & C_{21} & \cdots & C_{n1} \\ C_{12} & C_{22} & \cdots & C_{n2} \\ \vdots & \vdots & & \vdots \\ C_{1n} & C_{2n} & \cdots & C_{nn} \end{bmatrix}\begin{bmatrix} b_1 \\ b_2 \\ \vdots \\ b_n \end{bmatrix}$$

Multiplying the matrices out gives

$$\mathbf{x} = \frac{1}{\det(A)}\begin{bmatrix} b_1C_{11} + b_2C_{21} + \cdots + b_nC_{n1} \\ b_1C_{12} + b_2C_{22} + \cdots + b_nC_{n2} \\ \vdots \qquad \vdots \qquad \vdots \\ b_1C_{1n} + b_2C_{2n} + \cdots + b_nC_{nn} \end{bmatrix}$$

The entry in the jth row of $\mathbf{x}$ is therefore

$$x_j = \frac{b_1C_{1j} + b_2C_{2j} + \cdots + b_nC_{nj}}{\det(A)} \tag{11}$$

Now let

$$A_j = \begin{bmatrix} a_{11} & a_{12} & \cdots & a_{1j-1} & b_1 & a_{1j+1} & \cdots & a_{1n} \\ a_{21} & a_{22} & \cdots & a_{2j-1} & b_2 & a_{2j+1} & \cdots & a_{2n} \\ \vdots & \vdots & & \vdots & \vdots & \vdots & & \vdots \\ a_{n1} & a_{n2} & \cdots & a_{nj-1} & b_n & a_{nj+1} & \cdots & a_{nn} \end{bmatrix}$$

Since A_j differs from A only in the jth column, it follows that the cofactors of entries $b_1, b_2, \ldots, b_n$ in A_j are the same as the cofactors of the corresponding entries in the jth column of A. The cofactor expansion of $\det(A_j)$ along the jth column is therefore

$$\det(A_j) = b_1C_{1j} + b_2C_{2j} + \cdots + b_nC_{nj}$$

Substituting this result in (11) gives

$$x_j = \frac{\det(A_j)}{\det(A)}$$

■

EXAMPLE 8 Using Cramer's Rule to Solve a Linear System

Use Cramer's rule to solve

$$\begin{aligned} x_1 + \quad\;\; + 2x_3 &= 6 \\ -3x_1 + 4x_2 + 6x_3 &= 30 \\ -x_1 - 2x_2 + 3x_3 &= 8 \end{aligned}$$

Solution.

$$A = \begin{bmatrix} 1 & 0 & 2 \\ -3 & 4 & 6 \\ -1 & -2 & 3 \end{bmatrix}, \quad A_1 = \begin{bmatrix} 6 & 0 & 2 \\ 30 & 4 & 6 \\ 8 & -2 & 3 \end{bmatrix},$$

$$A_2 = \begin{bmatrix} 1 & 6 & 2 \\ -3 & 30 & 6 \\ -1 & 8 & 3 \end{bmatrix}, \quad A_3 = \begin{bmatrix} 1 & 0 & 6 \\ -3 & 4 & 30 \\ -1 & -2 & 8 \end{bmatrix}$$

Therefore,

$$x_1 = \frac{\det(A_1)}{\det(A)} = \frac{-40}{44} = \frac{-10}{11}, \quad x_2 = \frac{\det(A_2)}{\det(A)} = \frac{72}{44} = \frac{18}{11},$$

$$x_3 = \frac{\det(A_3)}{\det(A)} = \frac{152}{44} = \frac{38}{11}$$

◆

REMARK. To solve a system of n equations in n unknowns by Cramer's rule, it is necessary to evaluate $n+1$ determinants of $n \times n$ matrices. For systems with more than three equations, Gaussian elimination is far more efficient, since it is only necessary to reduce one $n \times (n+1)$ augmented matrix. However, Cramer's rule does give a formula for the solution if the determinant of the coefficient matrix is nonzero.

Exercise Set 2.4

1. Let

$$A = \begin{bmatrix} 1 & -2 & 3 \\ 6 & 7 & -1 \\ -3 & 1 & 4 \end{bmatrix}$$

(a) Find all the minors of A. (b) Find all the cofactors.

2. Let

$$A = \begin{bmatrix} 4 & -1 & 1 & 6 \\ 0 & 0 & -3 & 3 \\ 4 & 1 & 0 & 14 \\ 4 & 1 & 3 & 2 \end{bmatrix}$$

Find

(a) M_{13} and C_{13} (b) M_{23} and C_{23} (c) M_{22} and C_{22} (d) M_{21} and C_{21}

3. Evaluate the determinant of the matrix in Exercise 1 by a cofactor expansion along

(a) the first row (b) the first column (c) the second row
(d) the second column (e) the third row (f) the third column

4. For the matrix in Exercise 1, find

(a) $\text{adj}(A)$ (b) A^{-1} using Theorem 2.4.2

In Exercises 5–10 evaluate $\det(A)$ by a cofactor expansion along a row or column of your choice.

5. $A = \begin{bmatrix} -3 & 0 & 7 \\ 2 & 5 & 1 \\ -1 & 0 & 5 \end{bmatrix}$ **6.** $A = \begin{bmatrix} 3 & 3 & 1 \\ 1 & 0 & -4 \\ 1 & -3 & 5 \end{bmatrix}$ **7.** $A = \begin{bmatrix} 1 & k & k^2 \\ 1 & k & k^2 \\ 1 & k & k^2 \end{bmatrix}$

8. $A = \begin{bmatrix} k+1 & k-1 & 7 \\ 2 & k-3 & 4 \\ 5 & k+1 & k \end{bmatrix}$ **9.** $A = \begin{bmatrix} 3 & 3 & 0 & 5 \\ 2 & 2 & 0 & -2 \\ 4 & 1 & -3 & 0 \\ 2 & 10 & 3 & 2 \end{bmatrix}$ **10.** $A = \begin{bmatrix} 4 & 0 & 0 & 1 & 0 \\ 3 & 3 & 3 & -1 & 0 \\ 1 & 2 & 4 & 2 & 3 \\ 9 & 4 & 6 & 2 & 3 \\ 2 & 2 & 4 & 2 & 3 \end{bmatrix}$

In Exercises 11–14 find A^{-1} using Theorem 2.4.2.

11. $A = \begin{bmatrix} 2 & 5 & 5 \\ -1 & -1 & 0 \\ 2 & 4 & 3 \end{bmatrix}$ **12.** $A = \begin{bmatrix} 2 & 0 & 3 \\ 0 & 3 & 2 \\ -2 & 0 & -4 \end{bmatrix}$

13. $A = \begin{bmatrix} 2 & -3 & 5 \\ 0 & 1 & -3 \\ 0 & 0 & 2 \end{bmatrix}$ **14.** $A = \begin{bmatrix} 2 & 0 & 0 \\ 8 & 1 & 0 \\ -5 & 3 & 6 \end{bmatrix}$

15. Let

$$A = \begin{bmatrix} 1 & 3 & 1 & 1 \\ 2 & 5 & 2 & 2 \\ 1 & 3 & 8 & 9 \\ 1 & 3 & 2 & 2 \end{bmatrix}$$

(a) Evaluate A^{-1} using Theorem 2.4.2.
(b) Evaluate A^{-1} using the method of Example 4 in Section 1.5.
(c) Which method involves less computation?

In Exercises 16–21 solve by Cramer's rule, where it applies.

16. $\begin{aligned} 7x_1 - 2x_2 &= 3 \\ 3x_1 + x_2 &= 5 \end{aligned}$

17. $\begin{aligned} 4x + 5y &= 2 \\ 11x + y + 2z &= 3 \\ x + 5y + 2z &= 1 \end{aligned}$

18. $\begin{aligned} x - 4y + z &= 6 \\ 4x - y + 2z &= -1 \\ 2x + 2y - 3z &= -20 \end{aligned}$

19. $\begin{aligned} x_1 - 3x_2 + x_3 &= 4 \\ 2x_1 - x_2 &= -2 \\ 4x_1 - 3x_3 &= 0 \end{aligned}$

20. $\begin{aligned} -x_1 - 4x_2 + 2x_3 + x_4 &= -32 \\ 2x_1 - x_2 + 7x_3 + 9x_4 &= 14 \\ -x_1 + x_2 + 3x_3 + x_4 &= 11 \\ x_1 - 2x_2 + x_3 - 4x_4 &= -4 \end{aligned}$

21. $\begin{aligned} 3x_1 - x_2 + x_3 &= 4 \\ -x_1 + 7x_2 - 2x_3 &= 1 \\ 2x_1 + 6x_2 - x_3 &= 5 \end{aligned}$

22. Show that the matrix

$$A = \begin{bmatrix} \cos\theta & \sin\theta & 0 \\ -\sin\theta & \cos\theta & 0 \\ 0 & 0 & 1 \end{bmatrix}$$

is invertible for all values of θ; then find A^{-1} using Theorem 2.4.2.

23. Use Cramer's rule to solve for y without solving for x, z, and w.

$$\begin{aligned} 4x + y + z + w &= 6 \\ 3x + 7y - z + w &= 1 \\ 7x + 3y - 5z + 8w &= -3 \\ x + y + z + 2w &= 3 \end{aligned}$$

24. Let $A\mathbf{x} = \mathbf{b}$ be the system in Exercise 23.

(a) Solve by Cramer's rule. (b) Solve by Gauss–Jordan elimination.
(c) Which method involves fewer computations?

25. Prove that if $\det(A) = 1$ and all the entries in A are integers, then all the entries in A^{-1} are integers.

26. Let $A\mathbf{x} = \mathbf{b}$ be a system of n linear equations in n unknowns with integer coefficients and integer constants. Prove that if $\det(A) = 1$, the solution $\mathbf{x}$ has integer entries.

27. Prove that if A is an invertible lower triangular matrix, then A^{-1} is lower triangular.

28. Derive the last cofactor expansion listed in Formula (4).

29. Prove: The equation of the line through the distinct points (a_1, b_1) and (a_2, b_2) can be written as

$$\begin{vmatrix} x & y & 1 \\ a_1 & b_1 & 1 \\ a_2 & b_2 & 1 \end{vmatrix} = 0$$

30. Prove: (x_1, y_1), (x_2, y_2), and (x_3, y_3) are collinear points if and only if

$$\begin{vmatrix} x_1 & y_1 & 1 \\ x_2 & y_2 & 1 \\ x_3 & y_3 & 1 \end{vmatrix} = 0$$

31. (a) If $A = \left[\begin{array}{c|c} A_{11} & A_{12} \\ \hline 0 & A_{22} \end{array}\right]$ is an "upper triangular" block matrix, where A_{11} and A_{22} are square matrices, then $\det(A) = \det(A_{11})\det(A_{22})$. Use this result to evaluate $\det(A)$ for

$$\left[\begin{array}{cc|ccc} 2 & -1 & 2 & 5 & 6 \\ 4 & 3 & -1 & 3 & 4 \\ \hline 0 & 0 & 1 & 3 & 5 \\ 0 & 0 & -2 & 6 & 2 \\ 0 & 0 & 3 & 5 & 2 \end{array}\right]$$

(b) Verify your answer in part (a) by using a cofactor expansion to evaluate $\det(A)$.

32. Prove that if A is upper triangular and B_{ij} is the matrix that results when the ith row and jth column of A are deleted, then B_{ij} is upper triangular if $i < j$.

Discussion and Discovery

33. What is the maximum number of zeros that a 4×4 matrix can have without having a zero determinant? Explain your reasoning.

34. Let A be a matrix of the form

$$A = \begin{bmatrix} * & * & 0 & 0 & 0 \\ * & * & 0 & 0 & 0 \\ * & * & 0 & 0 & 0 \\ * & * & * & * & * \\ * & * & * & * & * \end{bmatrix}$$

How many different values can you obtain for $\det(A)$ by substituting numerical values (not necessarily all the same) for the $*$'s. Explain your reasoning.

35. Indicate whether the statement is always true or sometimes false. Justify your answer by giving a logical argument or a counterexample.

(a) $A\,\mathrm{adj}(A)$ is a diagonal matrix for every square matrix A.

(b) In theory, Cramer's rule can be used to solve any system of linear equations, though the amount of computation may be enormous.

(c) If A is invertible, then $\mathrm{adj}(A)$ must also be invertible.

(d) If A has a row of zeros, then so does $\mathrm{adj}(A)$.

Chapter 2 Supplementary Exercises

1. Use Cramer's rule to solve for x' and y' in terms of x and y.

$$\begin{aligned} x &= \tfrac{3}{5}x' - \tfrac{4}{5}y' \\ y &= \tfrac{4}{5}x' + \tfrac{3}{5}y' \end{aligned}$$

2. Use Cramer's rule to solve for x' and y' in terms of x and y.

$$\begin{aligned} x &= x'\cos\theta - y'\sin\theta \\ y &= x'\sin\theta + y'\cos\theta \end{aligned}$$

3. By examining the determinant of the coefficient matrix, show that the following system has a nontrivial solution if and only if $\alpha = \beta$.

$$\begin{aligned} x + y + \alpha z &= 0 \\ x + y + \beta z &= 0 \\ \alpha x + \beta y + z &= 0 \end{aligned}$$

4. Let A be a 3×3 matrix, each of whose entries is 1 or 0. What is the largest possible value for $\det(A)$?

5. (a) For the triangle in the accompanying figure, use trigonometry to show that

$$\begin{aligned} b\cos\gamma + c\cos\beta &= a \\ c\cos\alpha + a\cos\gamma &= b \\ a\cos\beta + b\cos\alpha &= c \end{aligned}$$

and then apply Cramer's rule to show that

$$\cos\alpha = \frac{b^2 + c^2 - a^2}{2bc}$$

(b) Use Cramer's rule to obtain similar formulas for $\cos\beta$ and $\cos\gamma$.

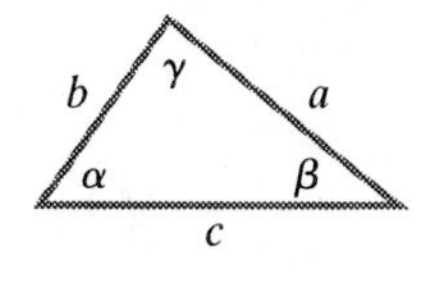

Figure Ex-5

3

Vectors in 2-Space and 3-Space

Chapter Contents

INTRODUCTION: Many physical quantities, such as area, length, mass, and temperature, are completely described once the magnitude of the quantity is given. Such quantities are called ***scalars***. Other physical quantities are not completely determined until both a magnitude and a direction are specified. These quantities are called ***vectors***. For example, wind movement is usually described by giving the speed and direction, say 20 mph northeast. The wind speed and wind direction form a vector called the wind ***velocity***. Other examples of vectors are ***force*** and ***displacement***. In this chapter our goal is to review some of the basic theory of vectors in two and three dimensions.

Note. *Readers already familiar with the contents of this chapter can go to Chapter 4 with no loss of continuity.*

3.1 INTRODUCTION TO VECTORS (GEOMETRIC)

In this section vectors in 2-space and 3-space will be introduced geometrically, arithmetic operations on vectors will be defined, and some basic properties of these arithmetic operations will be established.

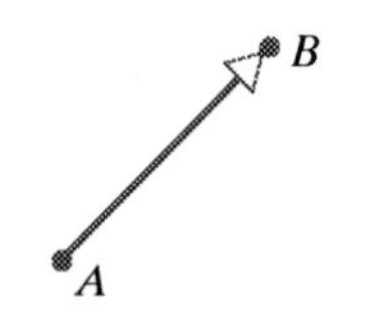

(a) The vector $\overrightarrow{AB}$

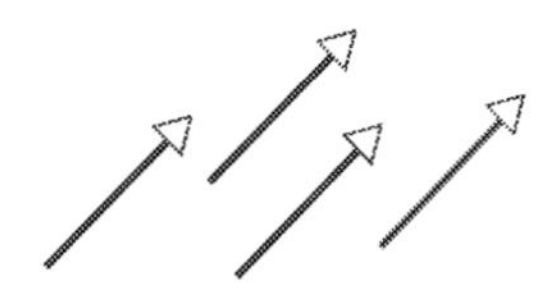

(b) Equivalent vectors

Figure 3.1.1

Geometric Vectors Vectors can be represented geometrically as directed line segments or arrows in 2-space or 3-space. The direction of the arrow specifies the direction of the vector, and the length of the arrow describes its magnitude. The tail of the arrow is called the ***initial point*** of the vector, and the tip of the arrow the ***terminal point***. Symbolically, we shall denote vectors in lowercase boldface type (for instance, **a**, **k**, **v**, **w**, and **x**). When discussing vectors, we shall refer to numbers as ***scalars***. For now, all our scalars will be real numbers and will be denoted in lowercase italic type (for instance, a, k, v, w, and x).

If, as in Figure 3.1.1a, the initial point of a vector **v** is A and the terminal point is B, we write

$$\mathbf{v} = \overrightarrow{AB}$$

Vectors with the same length and same direction, such as those in Figure 3.1.1b, are called ***equivalent***. Since we want a vector to be determined solely by its length and direction, equivalent vectors are regarded as ***equal*** even though they may be located in different positions. If **v** and **w** are equivalent, we write

$$\mathbf{v} = \mathbf{w}$$

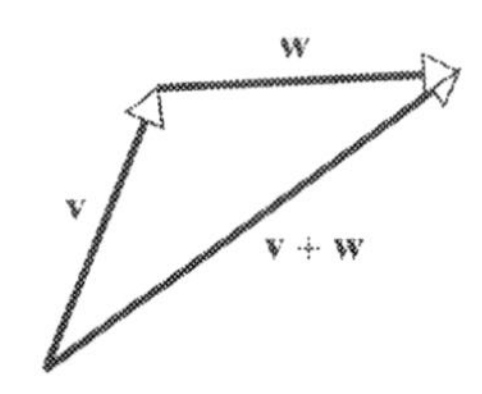

(a) The sum **v** + **w**

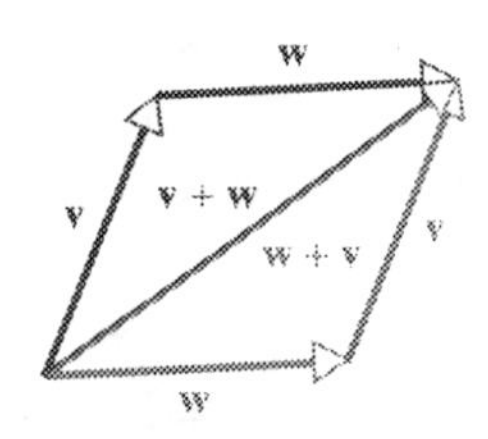

(b) **v** + **w** = **w** + **v**

Figure 3.1.2

Definition

If **v** and **w** are any two vectors, then the ***sum*** $\mathbf{v} + \mathbf{w}$ is the vector determined as follows: Position the vector **w** so that its initial point coincides with the terminal point of **v**. The vector $\mathbf{v} + \mathbf{w}$ is represented by the arrow from the initial point of **v** to the terminal point of **w** (Figure 3.1.2a).

In Figure 3.1.2b we have constructed two sums, $\mathbf{v} + \mathbf{w}$ (color arrows) and $\mathbf{w} + \mathbf{v}$ (gray arrows). It is evident that

$$\mathbf{v} + \mathbf{w} = \mathbf{w} + \mathbf{v}$$

and that the sum coincides with the diagonal of the parallelogram determined by **v** and **w** when these vectors are positioned so they have the same initial point.

The vector of length zero is called the ***zero vector*** and is denoted by **0**. We define

$$\mathbf{0} + \mathbf{v} = \mathbf{v} + \mathbf{0} = \mathbf{v}$$

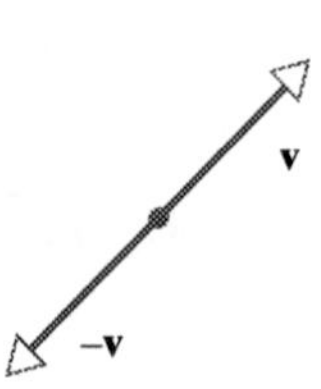

Figure 3.1.3 The negative of **v** has the same length as **v**, but is oppositely directed.

for every vector **v**. Since there is no natural direction for the zero vector, we shall agree that it can be assigned any direction that is convenient for the problem being considered. If **v** is any nonzero vector, then $-\mathbf{v}$, the ***negative*** of **v**, is defined to be the vector having the same magnitude as **v**, but oppositely directed (Figure 3.1.3). This vector has the property

$$\mathbf{v} + (-\mathbf{v}) = \mathbf{0}$$

(Why?) In addition, we define $-\mathbf{0} = \mathbf{0}$. Subtraction of vectors is defined as follows.

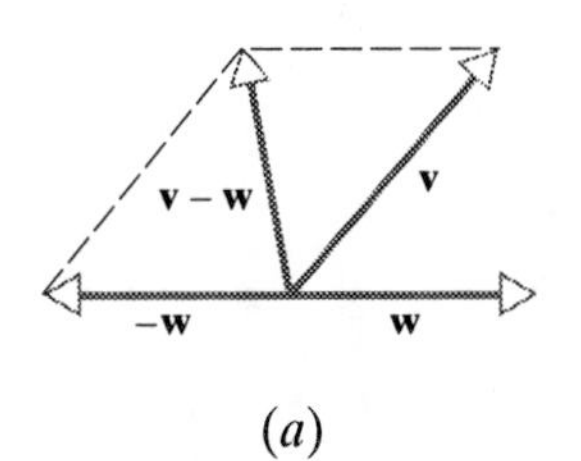

(a)

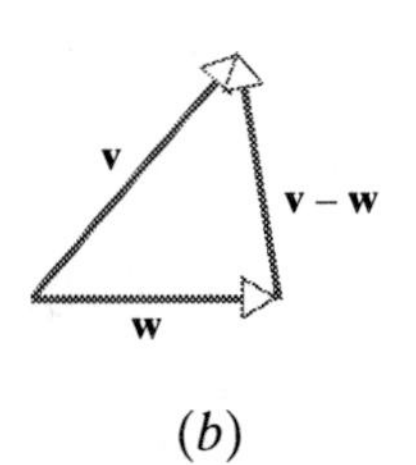

(b)

Figure 3.1.4

Definition

If $\mathbf{v}$ and $\mathbf{w}$ are any two vectors, then the ***difference*** of $\mathbf{w}$ from $\mathbf{v}$ is defined by

$$\mathbf{v} - \mathbf{w} = \mathbf{v} + (-\mathbf{w})$$

(Figure 3.1.4*a*).

To obtain the difference $\mathbf{v} - \mathbf{w}$ without constructing $-\mathbf{w}$, position $\mathbf{v}$ and $\mathbf{w}$ so their initial points coincide; the vector from the terminal point of $\mathbf{w}$ to the terminal point of $\mathbf{v}$ is then the vector $\mathbf{v} - \mathbf{w}$ (Figure 3.1.4*b*).

Definition

If $\mathbf{v}$ is a nonzero vector and k is a nonzero real number (scalar), then the ***product*** $k\mathbf{v}$ is defined to be the vector whose length is $|k|$ times the length of $\mathbf{v}$ and whose direction is the same as that of $\mathbf{v}$ if $k > 0$ and opposite to that of $\mathbf{v}$ if $k < 0$. We define $k\mathbf{v} = \mathbf{0}$ if $k = 0$ or $\mathbf{v} = \mathbf{0}$.

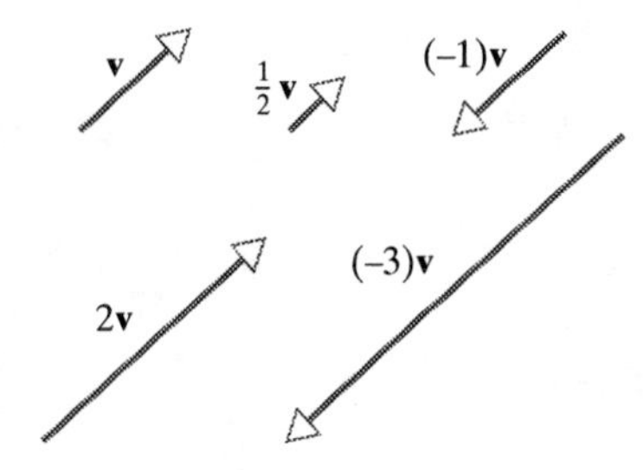

Figure 3.1.5

Figure 3.1.5 illustrates the relation between a vector $\mathbf{v}$ and the vectors $\frac{1}{2}\mathbf{v}$, $(-1)\mathbf{v}$, $2\mathbf{v}$, and $(-3)\mathbf{v}$. Note that the vector $(-1)\mathbf{v}$ has the same length as $\mathbf{v}$, but is oppositely directed. Thus, $(-1)\mathbf{v}$ is just the negative of $\mathbf{v}$; that is,

$$(-1)\mathbf{v} = -\mathbf{v}$$

A vector of the form $k\mathbf{v}$ is called a ***scalar multiple*** of $\mathbf{v}$. As evidenced by Figure 3.1.5, vectors that are scalar multiples of each other are parallel. Conversely, it can be shown that nonzero parallel vectors are scalar multiples of each other. We omit the proof.

Vectors in Coordinate Systems Problems involving vectors can often be simplified by introducing a rectangular coordinate system. For the moment we shall restrict the discussion to vectors in 2-space (the plane). Let $\mathbf{v}$ be any vector in the plane, and assume, as in Figure 3.1.6, that $\mathbf{v}$ has been positioned so its initial point is at the origin of a rectangular coordinate system. The coordinates (v_1, v_2) of the terminal point of $\mathbf{v}$ are called the ***components of*** $\mathbf{v}$, and we write

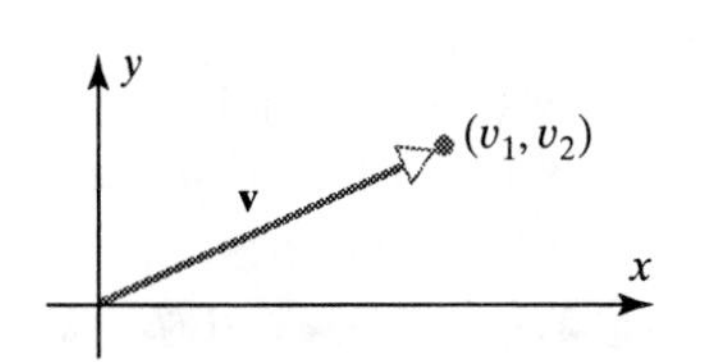

Figure 3.1.6
v_1 and v_2 are the components of $\mathbf{v}$.

$$\mathbf{v} = (v_1, v_2)$$

If equivalent vectors, $\mathbf{v}$ and $\mathbf{w}$, are located so their initial points fall at the origin, then it is obvious that their terminal points must coincide (since the vectors have the same length and direction); thus, the vectors have the same components. Conversely, vectors with the same components are equivalent since they have the same length and same direction. In summary, two vectors

$$\mathbf{v} = (v_1, v_2) \quad \text{and} \quad \mathbf{w} = (w_1, w_2)$$

are equivalent if and only if

$$v_1 = w_1 \quad \text{and} \quad v_2 = w_2$$

The operations of vector addition and multiplication by scalars are easy to carry out in terms of components. As illustrated in Figure 3.1.7, if

$$\mathbf{v} = (v_1, v_2) \quad \text{and} \quad \mathbf{w} = (w_1, w_2)$$

then

$$\mathbf{v} + \mathbf{w} = (v_1 + w_1, v_2 + w_2) \tag{1}$$

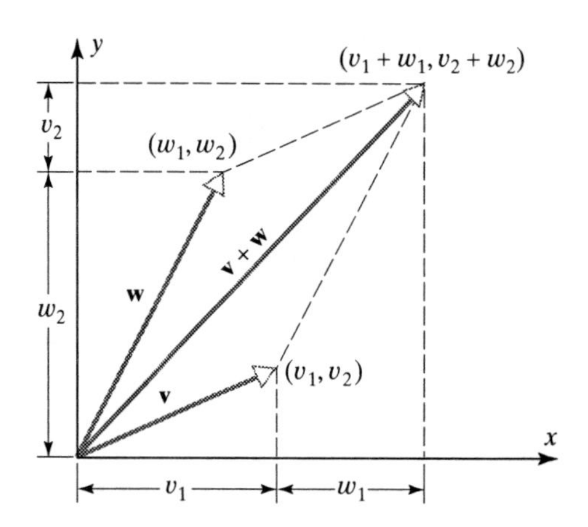

Figure 3.1.7

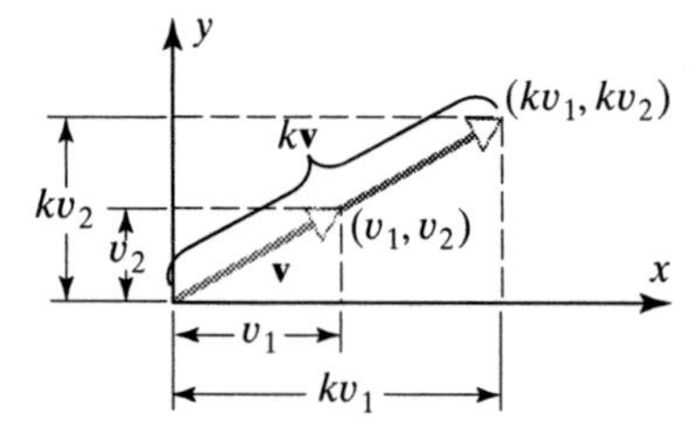

Figure 3.1.8

If $\mathbf{v} = (v_1, v_2)$ and k is any scalar, then by using a geometric argument involving similar triangles, it can be shown (Exercise 15) that

$$k\mathbf{v} = (kv_1, kv_2) \tag{2}$$

(Figure 3.1.8). Thus, for example, if $\mathbf{v} = (1, -2)$ and $\mathbf{w} = (7, 6)$, then

$$\mathbf{v} + \mathbf{w} = (1, -2) + (7, 6) = (1 + 7, -2 + 6) = (8, 4)$$

and

$$4\mathbf{v} = 4(1, -2) = (4(1), 4(-2)) = (4, -8)$$

Since $\mathbf{v} - \mathbf{w} = \mathbf{v} + (-1)\mathbf{w}$, it follows from Formulas (1) and (2) that

$$\mathbf{v} - \mathbf{w} = (v_1 - w_1, v_2 - w_2)$$

(Verify.)

Vectors in 3-Space Just as vectors in the plane can be described by pairs of real numbers, vectors in 3-space can be described by triples of real numbers by introducing a ***rectangular coordinate*** system. To construct such a coordinate system, select a point O, called the ***origin***, and choose three mutually perpendicular lines, called ***coordinate axes***, passing through the origin. Label these axes x, y, and z, and select a positive direction for each coordinate axis as well as a unit of length for measuring distances (Figure 3.1.9*a*). Each pair of coordinate axes determines a plane called a ***coordinate plane***. These are referred to as the ***xy-plane***, the ***xz-plane***, and the ***yz-plane***. To each point P in 3-space we assign a triple of numbers (x, y, z), called the ***coordinates of P***, as follows: Pass three planes through P parallel to the coordinate planes, and denote

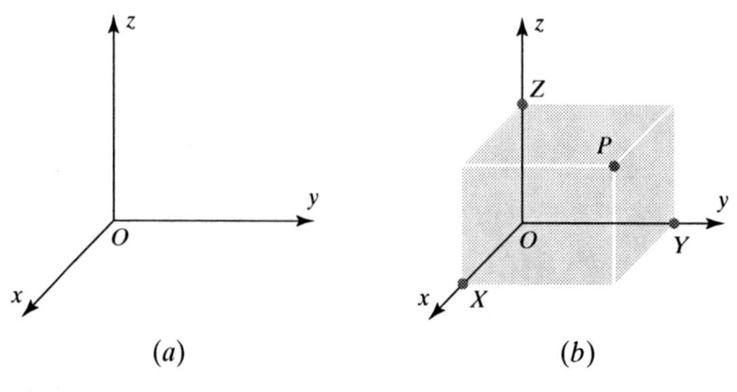

Figure 3.1.9

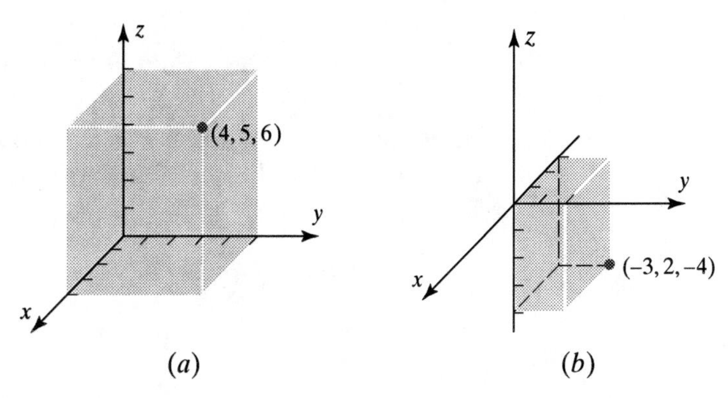

Figure 3.1.10

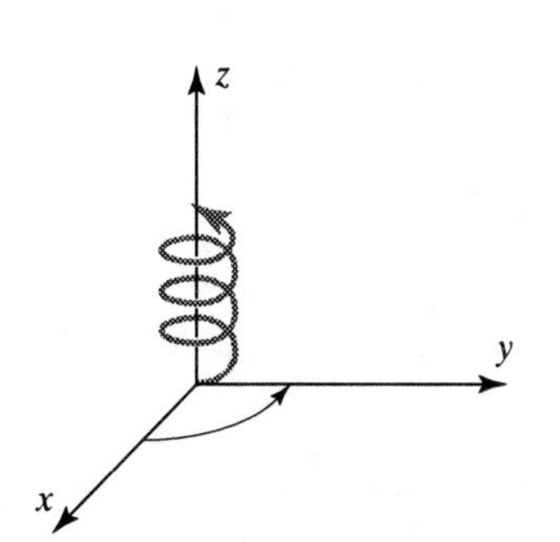

(a) Right-handed

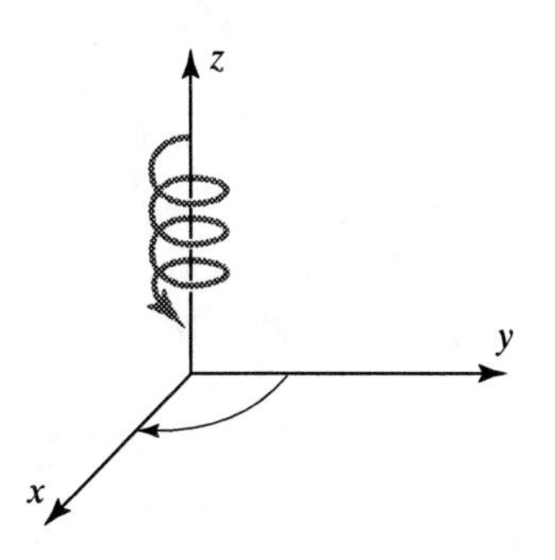

(b) Left-handed

Figure 3.1.11

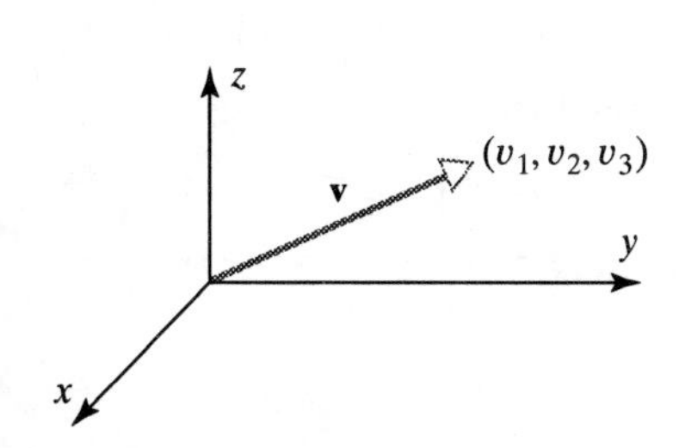

Figure 3.1.12

the points of intersections of these planes with the three coordinate axes by X, Y, and Z (Figure 3.1.9*b*).

The coordinates of P are defined to be the signed lengths

$$x = OX, \qquad y = OY, \qquad z = OZ$$

In Figure 3.1.10*a* we have constructed the point whose coordinates are (4, 5, 6) and in Figure 3.1.10*b* the point whose coordinates are (−3, 2, −4).

Rectangular coordinate systems in 3-space fall into two categories, ***left-handed*** and ***right-handed***. A right-handed system has the property that an ordinary screw pointed in the positive direction on the z-axis would be advanced if the positive x-axis is rotated 90° toward the positive y-axis (Figure 3.1.11*a*); the system is left-handed if the screw would be retracted (Figure 3.1.11*b*).

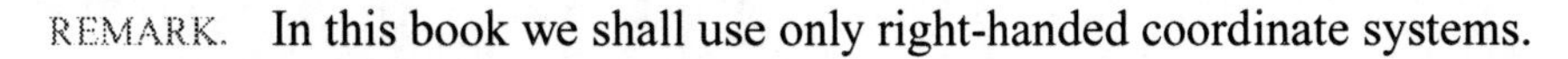

REMARK. In this book we shall use only right-handed coordinate systems.

If, as in Figure 3.1.12, a vector $\mathbf{v}$ in 3-space is positioned so its initial point is at the origin of a rectangular coordinate system, then the coordinates of the terminal point are called the ***components*** of $\mathbf{v}$, and we write

$$\mathbf{v} = (v_1, v_2, v_3)$$

If $\mathbf{v} = (v_1, v_2, v_3)$ and $\mathbf{w} = (w_1, w_2, w_3)$ are two vectors in 3-space, then arguments similar to those used for vectors in a plane can be used to establish the following results.

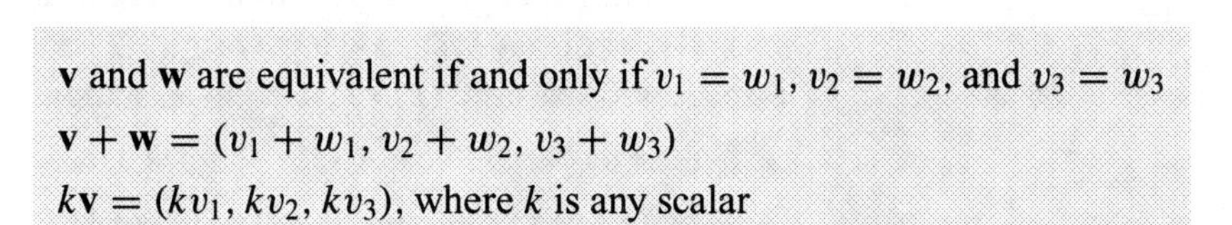

$\mathbf{v}$ and $\mathbf{w}$ are equivalent if and only if $v_1 = w_1$, $v_2 = w_2$, and $v_3 = w_3$

$\mathbf{v} + \mathbf{w} = (v_1 + w_1, v_2 + w_2, v_3 + w_3)$

$k\mathbf{v} = (kv_1, kv_2, kv_3)$, where k is any scalar

EXAMPLE 1 Vector Computations with Components

If $\mathbf{v} = (1, -3, 2)$ and $\mathbf{w} = (4, 2, 1)$, then

$$\mathbf{v} + \mathbf{w} = (5, -1, 3), \qquad 2\mathbf{v} = (2, -6, 4), \qquad -\mathbf{w} = (-4, -2, -1),$$
$$\mathbf{v} - \mathbf{w} = \mathbf{v} + (-\mathbf{w}) = (-3, -5, 1)$$ ◆

Sometimes a vector is positioned so that its initial point is not at the origin. If the vector $\overrightarrow{P_1P_2}$ has initial point $P_1(x_1, y_1, z_1)$ and terminal point $P_2(x_2, y_2, z_2)$, then

$$\overrightarrow{P_1P_2} = (x_2 - x_1, y_2 - y_1, z_2 - z_1)$$

That is, the components of $\overrightarrow{P_1P_2}$ are obtained by subtracting the coordinates of the initial point from the coordinates of the terminal point. This may be seen using Figure 3.1.13: The vector $\overrightarrow{P_1P_2}$ is the difference of vectors $\overrightarrow{OP_2}$ and $\overrightarrow{OP_1}$, so

$$\overrightarrow{P_1P_2} = \overrightarrow{OP_2} - \overrightarrow{OP_1} = (x_2, y_2, z_2) - (x_1, y_1, z_1) = (x_2 - x_1, y_2 - y_1, z_2 - z_1)$$

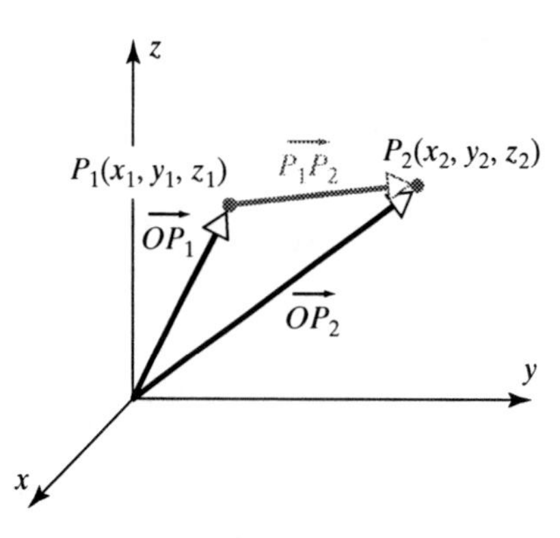

Figure 3.1.13

EXAMPLE 2 Finding the Components of a Vector

The components of the vector $\mathbf{v} = \overrightarrow{P_1P_2}$ with initial point $P_1(2, -1, 4)$ and terminal point $P_2(7, 5, -8)$ are

$$\mathbf{v} = (7 - 2, 5 - (-1), (-8) - 4) = (5, 6, -12)$$ ♦

In 2-space the vector with initial point $P_1(x_1, y_1)$ and terminal point $P_2(x_2, y_2)$ is

$$\overrightarrow{P_1P_2} = (x_2 - x_1, y_2 - y_1)$$

Translation of Axes The solutions to many problems can be simplified by translating the coordinate axes to obtain new axes parallel to the original ones.

In Figure 3.1.14a we have translated the axes of an xy-coordinate system to obtain an $x'y'$-coordinate system whose origin O' is at the point $(x, y) = (k, l)$. A point P in 2-space now has both (x, y) coordinates and (x', y') coordinates. To see how the two are related, consider the vector $\overrightarrow{O'P}$ (Figure 3.1.14b). In the xy-system its initial point is at (k, l) and its terminal point is at (x, y), so $\overrightarrow{O'P} = (x - k, y - l)$. In the $x'y'$-system its initial point is at $(0, 0)$ and its terminal point is at (x', y'), so $\overrightarrow{O'P} = (x', y')$. Therefore,

$$x' = x - k, \qquad y' = y - l$$

These formulas are called the ***translation equations***.

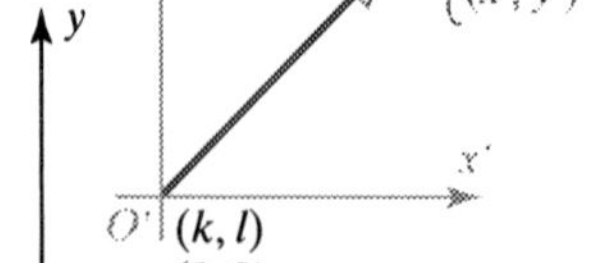

(a) (b)

Figure 3.1.14

EXAMPLE 3 Using the Translation Equations

Suppose that an xy-coordinate system is translated to obtain an $x'y'$-coordinate system whose origin has xy-coordinates $(k, l) = (4, 1)$.

(a) Find the $x'y'$-coordinates of the point with the xy-coordinates $P(2, 0)$.
(b) Find the xy-coordinates of the point with $x'y'$-coordinates $Q(-1, 5)$.

Solution (a). The translation equations are

$$x' = x - 4, \qquad y' = y - 1$$

so the $x'y'$-coordinates of $P(2, 0)$ are $x' = 2 - 4 = -2$ and $y' = 0 - 1 = -1$.

Solution (b). The translation equations in (a) can be rewritten as

$$x = x' + 4, \qquad y = y' + 1$$

so the xy-coordinates of Q are $x = -1 + 4 = 3$ and $y = 5 + 1 = 6$. ♦

In 3-space the translation equations are

$$x' = x - k, \qquad y' = y - l, \qquad z' = z - m$$

where (k, l, m) are the xyz-coordinates of the $x'y'z'$-origin.

Exercise Set 3.1

1. Draw a right-handed coordinate system and locate the points whose coordinates are

(a) $(3, 4, 5)$ (b) $(-3, 4, 5)$ (c) $(3, -4, 5)$ (d) $(3, 4, -5)$
(e) $(-3, -4, 5)$ (f) $(-3, 4, -5)$ (g) $(3, -4, -5)$ (h) $(-3, -4, -5)$
(i) $(-3, 0, 0)$ (j) $(3, 0, 3)$ (k) $(0, 0, -3)$ (l) $(0, 3, 0)$

2. Sketch the following vectors with the initial points located at the origin:

(a) $\mathbf{v}_1 = (3, 6)$ (b) $\mathbf{v}_2 = (-4, -8)$ (c) $\mathbf{v}_3 = (-4, -3)$ (d) $\mathbf{v}_4 = (5, -4)$ (e) $\mathbf{v}_5 = (3, 0)$
(f) $\mathbf{v}_6 = (0, -7)$ (g) $\mathbf{v}_7 = (3, 4, 5)$ (h) $\mathbf{v}_8 = (3, 3, 0)$ (i) $\mathbf{v}_9 = (0, 0, -3)$

3. Find the components of the vector having initial point P_1 and terminal point P_2.

(a) $P_1(4, 8)$, $P_2(3, 7)$ (b) $P_1(3, -5)$, $P_2(-4, -7)$ (c) $P_1(-5, 0)$, $P_2(-3, 1)$
(d) $P_1(0, 0)$, $P_2(a, b)$ (e) $P_1(3, -7, 2)$, $P_2(-2, 5, -4)$ (f) $P_1(-1, 0, 2)$, $P_2(0, -1, 0)$
(g) $P_1(a, b, c)$, $P_2(0, 0, 0)$ (h) $P_1(0, 0, 0)$, $P_2(a, b, c)$

4. Find a nonzero vector $\mathbf{u}$ with initial point $P(-1, 3, -5)$ such that

(a) $\mathbf{u}$ has the same direction as $\mathbf{v} = (6, 7, -3)$ (b) $\mathbf{u}$ is oppositely directed to $\mathbf{v} = (6, 7, -3)$

5. Find a nonzero vector $\mathbf{u}$ with terminal point $Q(3, 0, -5)$ such that

(a) $\mathbf{u}$ has the same direction as $\mathbf{v} = (4, -2, -1)$ (b) $\mathbf{u}$ is oppositely directed to $\mathbf{v} = (4, -2, -1)$

6. Let $\mathbf{u} = (-3, 1, 2)$, $\mathbf{v} = (4, 0, -8)$, and $\mathbf{w} = (6, -1, -4)$. Find the components of

(a) $\mathbf{v} - \mathbf{w}$ (b) $6\mathbf{u} + 2\mathbf{v}$ (c) $-\mathbf{v} + \mathbf{u}$ (d) $5(\mathbf{v} - 4\mathbf{u})$ (e) $-3(\mathbf{v} - 8\mathbf{w})$ (f) $(2\mathbf{u} - 7\mathbf{w}) - (8\mathbf{v} + \mathbf{u})$

7. Let $\mathbf{u}$, $\mathbf{v}$, and $\mathbf{w}$ be the vectors in Exercise 6. Find the components of the vector $\mathbf{x}$ that satisfies $2\mathbf{u} - \mathbf{v} + \mathbf{x} = 7\mathbf{x} + \mathbf{w}$.

8. Let $\mathbf{u}$, $\mathbf{v}$, and $\mathbf{w}$ be the vectors in Exercise 6. Find scalars c_1, c_2, and c_3 such that

$$c_1\mathbf{u} + c_2\mathbf{v} + c_3\mathbf{w} = (2, 0, 4)$$

9. Show that there do not exist scalars c_1, c_2, and c_3 such that

$$c_1(-2, 9, 6) + c_2(-3, 2, 1) + c_3(1, 7, 5) = (0, 5, 4)$$

10. Find all scalars c_1, c_2, and c_3 such that

$$c_1(1, 2, 0) + c_2(2, 1, 1) + c_3(0, 3, 1) = (0, 0, 0)$$

11. Let P be the point $(2, 3, -2)$ and Q the point $(7, -4, 1)$.

(a) Find the midpoint of the line segment connecting P and Q.
(b) Find the point on the line segment connecting P and Q that is $\frac{3}{4}$ of the way from P to Q.

12. Suppose an xy-coordinate system is translated to obtain an $x'y'$-coordinate system whose origin O' has xy-coordinates $(2, -3)$.

(a) Find the $x'y'$-coordinates of the point P whose xy-coordinates are $(7, 5)$.

(b) Find the xy-coordinates of the point Q whose $x'y'$-coordinates are $(-3, 6)$.

(c) Draw the xy and $x'y'$-coordinate axes and locate the points P and Q.

13. Suppose that an xyz-coordinate system is translated to obtain an $x'y'z'$-coordinate system. Let $\mathbf{v}$ be a vector whose components are $\mathbf{v} = (v_1, v_2, v_3)$ in the xyz-system. Show that $\mathbf{v}$ has the same components in the $x'y'z'$-system.

14. Find the components of $\mathbf{u}$, $\mathbf{v}$, $\mathbf{u} + \mathbf{v}$, and $\mathbf{u} - \mathbf{v}$ for the vectors shown in the accompanying figure.

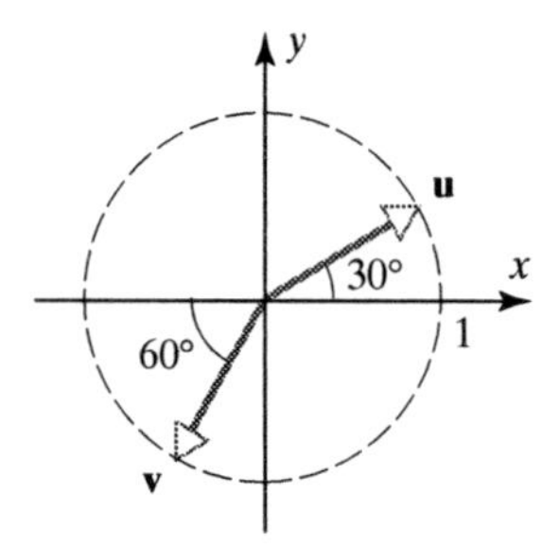

Figure Ex-14

15. Prove geometrically that if $\mathbf{v} = (v_1, v_2)$, then $k\mathbf{v} = (kv_1, kv_2)$. (Restrict the proof to the case $k > 0$ illustrated in Figure 3.1.8. The complete proof would involve various cases that depend on the sign of k and the quadrant in which the vector falls.)

Discussion and Discovery

16. Consider Figure 3.1.13. Discuss a geometric interpretation of the vector

$$\mathbf{u} = \overrightarrow{OP_1} + \tfrac{1}{2}(\overrightarrow{OP_2} - \overrightarrow{OP_1})$$

17. Draw a picture that shows four nonzero vectors whose sum is zero.

18. If you were given four nonzero vectors, how would you construct a fifth vector geometrically that is equal to the sum of the first four? Draw a picture to illustrate your method.

3.2 NORM OF A VECTOR; VECTOR ARITHMETIC

In this section we shall establish the basic rules of vector arithmetic.

Properties of Vector Operations The following theorem lists the most important properties of vectors in 2-space and 3-space.

Theorem 3.2.1 Properties of Vector Arithmetic

If $\mathbf{u}$, $\mathbf{v}$, and $\mathbf{w}$ are vectors in 2- or 3-space and k and l are scalars, then the following relationships hold.

(a) $\mathbf{u} + \mathbf{v} = \mathbf{v} + \mathbf{u}$

(b) $(\mathbf{u} + \mathbf{v}) + \mathbf{w} = \mathbf{u} + (\mathbf{v} + \mathbf{w})$

(c) $\mathbf{u} + \mathbf{0} = \mathbf{0} + \mathbf{u} = \mathbf{u}$

(d) $\mathbf{u} + (-\mathbf{u}) = \mathbf{0}$

(e) $k(l\mathbf{u}) = (kl)\mathbf{u}$

(f) $k(\mathbf{u} + \mathbf{v}) = k\mathbf{u} + k\mathbf{v}$

(g) $(k + l)\mathbf{u} = k\mathbf{u} + l\mathbf{u}$

(h) $1\mathbf{u} = \mathbf{u}$

Before discussing the proof, we note that we have developed two approaches to vectors: *geometric*, in which vectors are represented by arrows or directed line segments, and *analytic*, in which vectors are represented by pairs or triples of numbers called components. As a consequence, the equations in Theorem 3.2.1 can be proved either geometrically or analytically. To illustrate, we shall prove part (*b*) both ways. The remaining proofs are left as exercises.

Proof of part (b) (analytic). We shall give the proof for vectors in 3-space; the proof for 2-space is similar. If $\mathbf{u} = (u_1, u_2, u_3)$, $\mathbf{v} = (v_1, v_2, v_3)$, and $\mathbf{w} = (w_1, w_2, w_3)$, then

$$\begin{aligned}
(\mathbf{u}+\mathbf{v})+\mathbf{w} &= [(u_1, u_2, u_3) + (v_1, v_2, v_3)] + (w_1, w_2, w_3) \\
&= (u_1 + v_1, u_2 + v_2, u_3 + v_3) + (w_1, w_2, w_3) \\
&= ([u_1 + v_1] + w_1, [u_2 + v_2] + w_2, [u_3 + v_3] + w_3) \\
&= (u_1 + [v_1 + w_1], u_2 + [v_2 + w_2], u_3 + [v_3 + w_3]) \\
&= (u_1, u_2, u_3) + (v_1 + w_1, v_2 + w_2, v_3 + w_3) \\
&= \mathbf{u} + (\mathbf{v} + \mathbf{w})
\end{aligned}$$

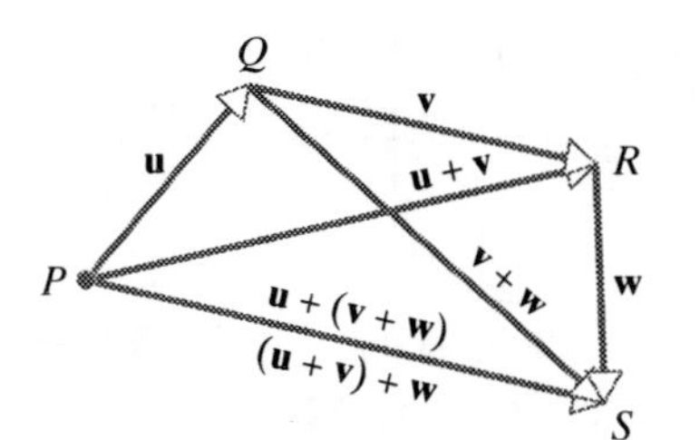

Figure 3.2.1
The vectors $\mathbf{u} + (\mathbf{v} + \mathbf{w})$ and $(\mathbf{u} + \mathbf{v}) + \mathbf{w}$ are equal.

Proof of part (b) (geometric). Let $\mathbf{u}$, $\mathbf{v}$, and $\mathbf{w}$ be represented by $\overrightarrow{PQ}$, $\overrightarrow{QR}$, and $\overrightarrow{RS}$ as shown in Figure 3.2.1. Then

$$\mathbf{v} + \mathbf{w} = \overrightarrow{QS} \quad \text{and} \quad \mathbf{u} + (\mathbf{v} + \mathbf{w}) = \overrightarrow{PS}$$

Also,

$$\mathbf{u} + \mathbf{v} = \overrightarrow{PR} \quad \text{and} \quad (\mathbf{u} + \mathbf{v}) + \mathbf{w} = \overrightarrow{PS}$$

Therefore,

$$\mathbf{u} + (\mathbf{v} + \mathbf{w}) = (\mathbf{u} + \mathbf{v}) + \mathbf{w}$$ ■

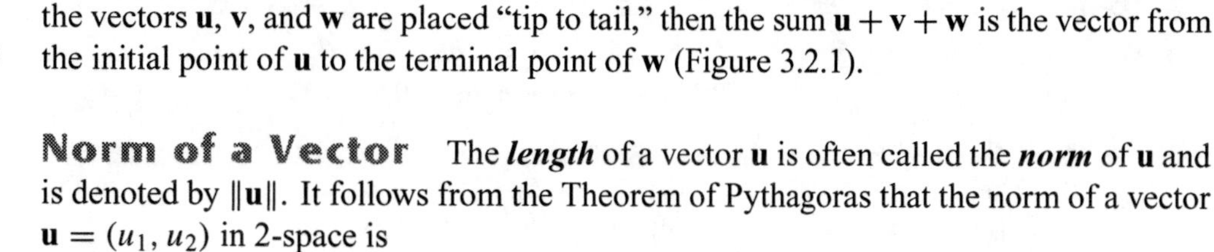

REMARK. In light of part (*b*) of this theorem, the symbol $\mathbf{u} + \mathbf{v} + \mathbf{w}$ is unambiguous since the same sum is obtained no matter where parentheses are inserted. Moreover, if the vectors $\mathbf{u}$, $\mathbf{v}$, and $\mathbf{w}$ are placed "tip to tail," then the sum $\mathbf{u} + \mathbf{v} + \mathbf{w}$ is the vector from the initial point of $\mathbf{u}$ to the terminal point of $\mathbf{w}$ (Figure 3.2.1).

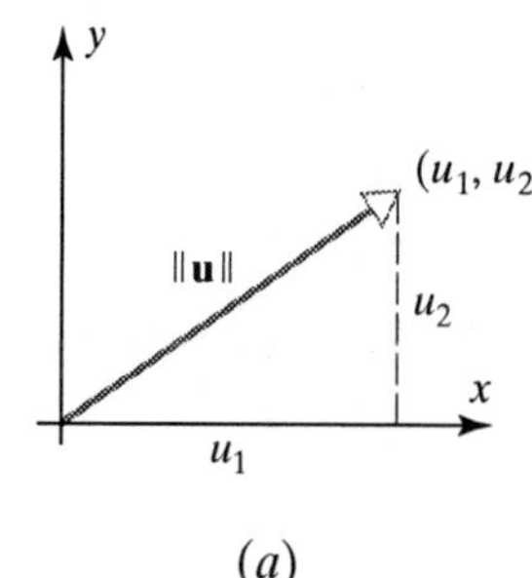

(*a*)

(*b*)

Figure 3.2.2

Norm of a Vector

The ***length*** of a vector $\mathbf{u}$ is often called the ***norm*** of $\mathbf{u}$ and is denoted by $\|\mathbf{u}\|$. It follows from the Theorem of Pythagoras that the norm of a vector $\mathbf{u} = (u_1, u_2)$ in 2-space is

$$\|\mathbf{u}\| = \sqrt{u_1^2 + u_2^2} \tag{1}$$

(Figure 3.2.2*a*). Let $\mathbf{u} = (u_1, u_2, u_3)$ be a vector in 3-space. Using Figure 3.2.2*b* and two applications of the Theorem of Pythagoras, we obtain

$$\|\mathbf{u}\|^2 = (OR)^2 + (RP)^2 = (OQ)^2 + (OS)^2 + (RP)^2 = u_1^2 + u_2^2 + u_3^2$$

Thus,

$$\|\mathbf{u}\| = \sqrt{u_1^2 + u_2^2 + u_3^2} \tag{2}$$

A vector of norm 1 is called a ***unit vector***.

If $P_1(x_1, y_1, z_1)$ and $P_2(x_2, y_2, z_2)$ are two points in 3-space, then the ***distance*** d between them is the norm of the vector $\overrightarrow{P_1P_2}$ (Figure 3.2.3). Since

$$\overrightarrow{P_1P_2} = (x_2 - x_1, y_2 - y_1, z_2 - z_1)$$

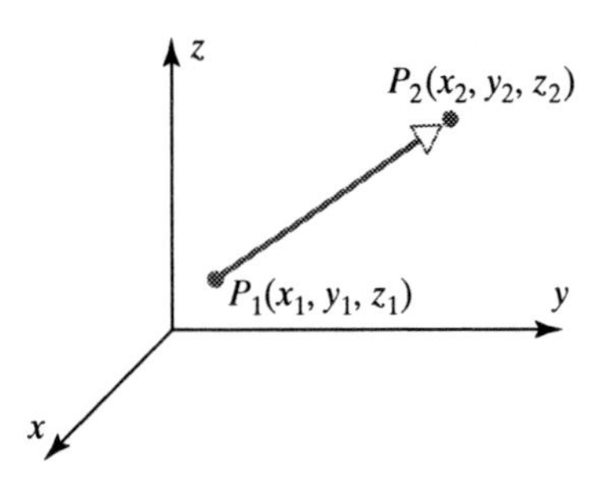

Figure 3.2.3 The distance between P_1 and P_2 is the norm of the vector $\overrightarrow{P_1P_2}$.

it follows from (2) that

$$d = \sqrt{(x_2 - x_1)^2 + (y_2 - y_1)^2 + (z_2 - z_1)^2} \tag{3}$$

Similarly, if $P_1(x_1, y_1)$ and $P_2(x_2, y_2)$ are points in 2-space, then the distance between them is given by

$$d = \sqrt{(x_2 - x_1)^2 + (y_2 - y_1)^2} \tag{4}$$

EXAMPLE 1 Finding Norm and Distance

The norm of the vector $\mathbf{u} = (-3, 2, 1)$ is

$$\|\mathbf{u}\| = \sqrt{(-3)^2 + (2)^2 + (1)^2} = \sqrt{14}$$

The distance d between the points $P_1(2, -1, -5)$ and $P_2(4, -3, 1)$ is

$$d = \sqrt{(4 - 2)^2 + (-3 + 1)^2 + (1 + 5)^2} = \sqrt{44} = 2\sqrt{11}$$ ♦

From the definition of the product $k\mathbf{u}$, the length of the vector $k\mathbf{u}$ is $|k|$ times the length of $\mathbf{u}$. Expressed as an equation, this statement says that

$$\|k\mathbf{u}\| = |k|\|\mathbf{u}\| \tag{5}$$

This useful formula is applicable in both 2-space and 3-space.

Exercise Set 3.2

1. Find the norm of $\mathbf{v}$.

(a) $\mathbf{v} = (4, -3)$ (b) $\mathbf{v} = (2, 3)$ (c) $\mathbf{v} = (-5, 0)$
(d) $\mathbf{v} = (2, 2, 2)$ (e) $\mathbf{v} = (-7, 2, -1)$ (f) $\mathbf{v} = (0, 6, 0)$

2. Find the distance between P_1 and P_2.

(a) $P_1(3, 4)$, $P_2(5, 7)$ (b) $P_1(-3, 6)$, $P_2(-1, -4)$
(c) $P_1(7, -5, 1)$, $P_2(-7, -2, -1)$ (d) $P_1(3, 3, 3)$, $P_2(6, 0, 3)$

3. Let $\mathbf{u} = (2, -2, 3)$, $\mathbf{v} = (1, -3, 4)$, $\mathbf{w} = (3, 6, -4)$. In each part evaluate the expression.

(a) $\|\mathbf{u} + \mathbf{v}\|$ (b) $\|\mathbf{u}\| + \|\mathbf{v}\|$ (c) $\|-2\mathbf{u}\| + 2\|\mathbf{u}\|$
(d) $\|3\mathbf{u} - 5\mathbf{v} + \mathbf{w}\|$ (e) $\dfrac{1}{\|\mathbf{w}\|}\mathbf{w}$ (f) $\left\|\dfrac{1}{\|\mathbf{w}\|}\mathbf{w}\right\|$

4. Let $\mathbf{v} = (-1, 2, 5)$. Find all scalars k such that $\|k\mathbf{v}\| = 4$.

5. Let $\mathbf{u} = (7, -3, 1)$, $\mathbf{v} = (9, 6, 6)$, $\mathbf{w} = (2, 1, -8)$, $k = -2$, and $l = 5$. Verify that these vectors and scalars satisfy the stated equalities from Theorem 3.2.1.

(a) part (b) (b) part (e) (c) part (f) (d) part (g)

6. (a) Show that if $\mathbf{v}$ is any nonzero vector, then $\dfrac{1}{\|\mathbf{v}\|}\mathbf{v}$ is a unit vector.

(b) Use the result in part (a) to find a unit vector that has the same direction as the vector $\mathbf{v} = (3, 4)$.

(c) Use the result in part (a) to find a unit vector that is oppositely directed to the vector $\mathbf{v} = (-2, 3, -6)$.

7. (a) Show that the components of the vector $\mathbf{v} = (v_1, v_2)$ in the accompanying figure are $v_1 = \|\mathbf{v}\| \cos\theta$ and $v_2 = \|\mathbf{v}\| \sin\theta$.
   (b) Let **u** and **v** be the vectors in the accompanying figure. Use the result in part (a) to find the components of $4\mathbf{u} - 5\mathbf{v}$.

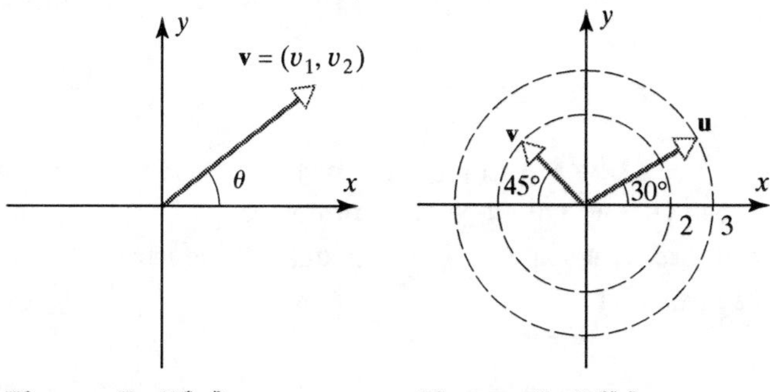

Figure Ex-7(a) **Figure Ex-7(b)**

8. Let $\mathbf{p}_0 = (x_0, y_0, z_0)$ and $\mathbf{p} = (x, y, z)$. Describe the set of all points (x, y, z) for which $\|\mathbf{p} - \mathbf{p}_0\| = 1$.

9. Prove geometrically that if **u** and **v** are vectors in 2- or 3-space, then $\|\mathbf{u} + \mathbf{v}\| \leq \|\mathbf{u}\| + \|\mathbf{v}\|$.

10. Prove parts (a), (c), and (e) of Theorem 3.2.1 analytically.

11. Prove parts (d), (g), and (h) of Theorem 3.2.1 analytically.

Discussion and Discovery

12. For the inequality stated in Exercise 9, is it possible to have $\|\mathbf{u} + \mathbf{v}\| = \|\mathbf{u}\| + \|\mathbf{v}\|$? Explain your reasoning.

13. (a) What relationship must hold for the point $\mathbf{p} = (a, b, c)$ to be equidistant from the origin and the xz-plane? Make sure that the relationship you state is valid for positive and negative values of a, b, and c.
    (b) What relationship must hold for the point $\mathbf{p} = (a, b, c)$ to be farther from the origin than from the xz-plane? Make sure that the relationship you state is valid for positive and negative values of a, b, and c.

14. (a) What does the inequality $\|\mathbf{x}\| < 1$ tell you about the point $\mathbf{x}$?
    (b) Write down an inequality that describes the set of points that lie outside the circle of radius 1, centered at the point $\mathbf{x}_0$.

15. The triangles in the accompanying figure should suggest a geometric proof of Theorem 3.2.1(f) for the case where $k > 0$. Give the proof.

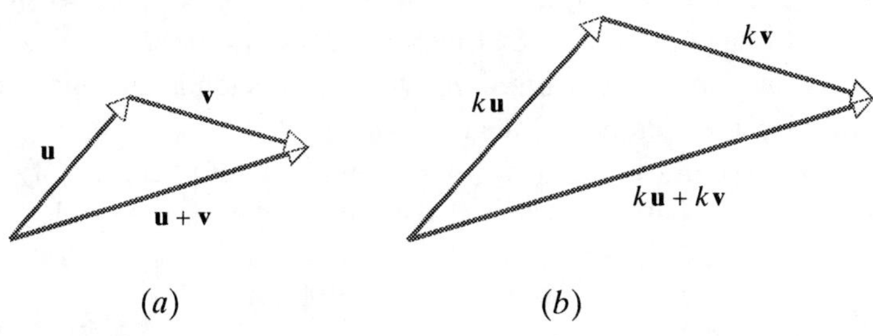

Figure Ex-15

3.3 DOT PRODUCT; PROJECTIONS

In this section we shall discuss an important way of multiplying vectors in 2-space or 3-space. We shall then give some applications of this multiplication to geometry.

Dot Product of Vectors Let $\mathbf{u}$ and $\mathbf{v}$ be two nonzero vectors in 2-space or 3-space, and assume these vectors have been positioned so their initial points coincide. By the ***angle between* u *and* v**, we shall mean the angle θ determined by $\mathbf{u}$ and $\mathbf{v}$ that satisfies $0 \le \theta \le \pi$ (Figure 3.3.1).

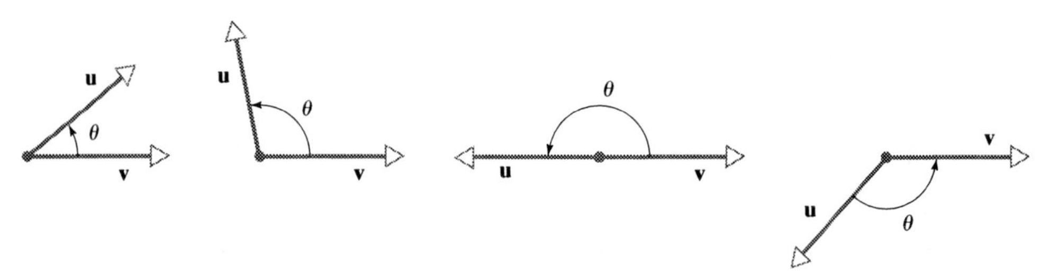

Figure 3.3.1 The angle θ between $\mathbf{u}$ and $\mathbf{v}$ satisfies $0 \le \theta \le \pi$.

Definition

If $\mathbf{u}$ and $\mathbf{v}$ are vectors in 2-space or 3-space and θ is the angle between $\mathbf{u}$ and $\mathbf{v}$, then the ***dot product*** or ***Euclidean inner product*** $\mathbf{u} \cdot \mathbf{v}$ is defined by

$$\mathbf{u} \cdot \mathbf{v} = \begin{cases} \|\mathbf{u}\|\,\|\mathbf{v}\|\cos\theta & \text{if } \mathbf{u} \neq \mathbf{0} \text{ and } \mathbf{v} \neq \mathbf{0} \\ 0 & \text{if } \mathbf{u} = \mathbf{0} \text{ or } \mathbf{v} = \mathbf{0} \end{cases} \tag{1}$$

EXAMPLE 1 Dot Product

As shown in Figure 3.3.2, the angle between the vectors $\mathbf{u} = (0, 0, 1)$ and $\mathbf{v} = (0, 2, 2)$ is 45°. Thus,

$$\mathbf{u} \cdot \mathbf{v} = \|\mathbf{u}\|\,\|\mathbf{v}\|\cos\theta = (\sqrt{0^2 + 0^2 + 1^2})(\sqrt{0^2 + 2^2 + 2^2})\left(\frac{1}{\sqrt{2}}\right) = 2$$ ♦

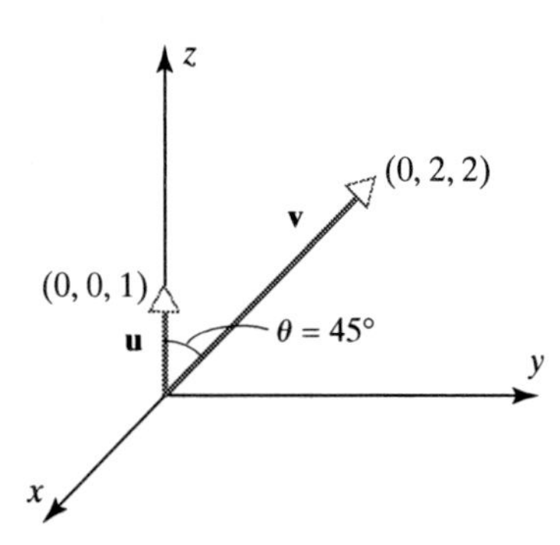

Figure 3.3.2

Component Form of the Dot Product For purposes of computation, it is desirable to have a formula that expresses the dot product of two vectors in terms of the components of the vectors. We will derive such a formula for vectors in 3-space; the derivation for vectors in 2-space is similar.

Let $\mathbf{u} = (u_1, u_2, u_3)$ and $\mathbf{v} = (v_1, v_2, v_3)$ be two nonzero vectors. If, as shown in Figure 3.3.3, θ is the angle between $\mathbf{u}$ and $\mathbf{v}$, then the law of cosines yields

$$\|\overrightarrow{PQ}\|^2 = \|\mathbf{u}\|^2 + \|\mathbf{v}\|^2 - 2\|\mathbf{u}\|\,\|\mathbf{v}\|\cos\theta \tag{2}$$

Since $\overrightarrow{PQ} = \mathbf{v} - \mathbf{u}$, we can rewrite (2) as

$$\|\mathbf{u}\|\,\|\mathbf{v}\|\cos\theta = \tfrac{1}{2}(\|\mathbf{u}\|^2 + \|\mathbf{v}\|^2 - \|\mathbf{v} - \mathbf{u}\|^2)$$

or

$$\mathbf{u} \cdot \mathbf{v} = \tfrac{1}{2}(\|\mathbf{u}\|^2 + \|\mathbf{v}\|^2 - \|\mathbf{v} - \mathbf{u}\|^2)$$

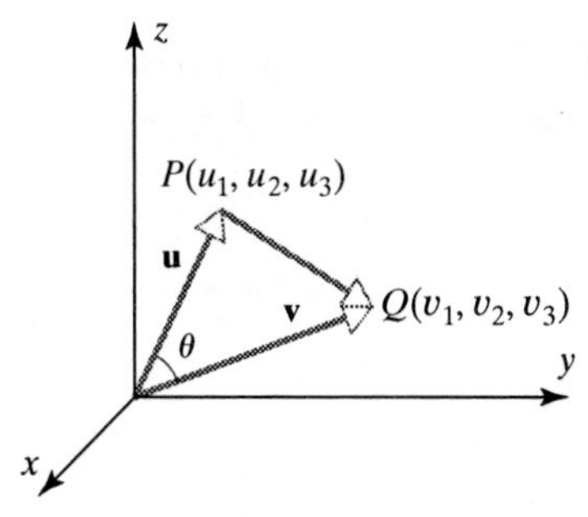

Figure 3.3.3

Substituting

$$\|\mathbf{u}\|^2 = u_1^2 + u_2^2 + u_3^2, \qquad \|\mathbf{v}\|^2 = v_1^2 + v_2^2 + v_3^2,$$

and

$$\|\mathbf{v} - \mathbf{u}\|^2 = (v_1 - u_1)^2 + (v_2 - u_2)^2 + (v_3 - u_3)^2$$

we obtain after simplifying

$$\mathbf{u} \cdot \mathbf{v} = u_1 v_1 + u_2 v_2 + u_3 v_3 \tag{3}$$

Although we derived this formula under the assumption that **u** and **v** are nonzero, the formula is also valid if $\mathbf{u} = \mathbf{0}$ or $\mathbf{v} = \mathbf{0}$ (verify).

If $\mathbf{u} = (u_1, u_2)$ and $\mathbf{v} = (v_1, v_2)$ are two vectors in 2-space, then the formula corresponding to (3) is

$$\mathbf{u} \cdot \mathbf{v} = u_1 v_1 + u_2 v_2 \tag{4}$$

Finding the Angle Between Vectors If **u** and **v** are nonzero vectors, then Formula (1) can be written as

$$\cos\theta = \frac{\mathbf{u} \cdot \mathbf{v}}{\|\mathbf{u}\|\|\mathbf{v}\|} \tag{5}$$

EXAMPLE 2 Dot Product Using (3)

Consider the vectors $\mathbf{u} = (2, -1, 1)$ and $\mathbf{v} = (1, 1, 2)$. Find $\mathbf{u} \cdot \mathbf{v}$ and determine the angle θ between **u** and **v**.

Solution.

$$\mathbf{u} \cdot \mathbf{v} = u_1 v_1 + u_2 v_2 + u_3 v_3 = (2)(1) + (-1)(1) + (1)(2) = 3$$

For the given vectors we have $\|\mathbf{u}\| = \|\mathbf{v}\| = \sqrt{6}$, so that from (5)

$$\cos\theta = \frac{\mathbf{u} \cdot \mathbf{v}}{\|\mathbf{u}\|\|\mathbf{v}\|} = \frac{3}{\sqrt{6}\sqrt{6}} = \frac{1}{2}$$

Thus, $\theta = 60°$. ♦

EXAMPLE 3 A Geometric Problem

Find the angle between a diagonal of a cube and one of its edges.

Solution.

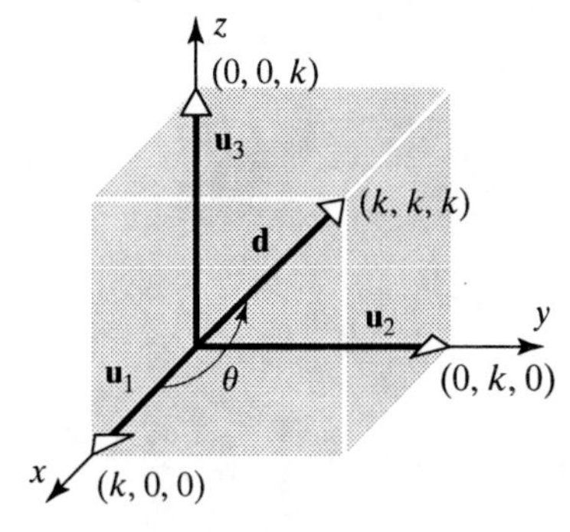

Figure 3.3.4

Let k be the length of an edge and introduce a coordinate system as shown in Figure 3.3.4. If we let $\mathbf{u}_1 = (k, 0, 0)$, $\mathbf{u}_2 = (0, k, 0)$, and $\mathbf{u}_3 = (0, 0, k)$, then the vector

$$\mathbf{d} = (k, k, k) = \mathbf{u}_1 + \mathbf{u}_2 + \mathbf{u}_3$$

is a diagonal of the cube. The angle θ between **d** and the edge $\mathbf{u}_1$ satisfies

$$\cos\theta = \frac{\mathbf{u}_1 \cdot \mathbf{d}}{\|\mathbf{u}_1\|\|\mathbf{d}\|} = \frac{k^2}{(k)(\sqrt{3k^2})} = \frac{1}{\sqrt{3}}$$

Thus,

$$\theta = \cos^{-1}\left(\frac{1}{\sqrt{3}}\right) \approx 54.74°$$

♦

The following theorem shows how the dot product can be used to obtain information about the angle between two vectors; it also establishes an important relationship between the norm and the dot product.

Theorem 3.3.1

Let $\mathbf{u}$ *and* $\mathbf{v}$ *be vectors in 2- or 3-space.*

(*a*) $\mathbf{v}\cdot\mathbf{v} = \|\mathbf{v}\|^2$; *that is,* $\|\mathbf{v}\| = (\mathbf{v}\cdot\mathbf{v})^{1/2}$

(*b*) *If the vectors* $\mathbf{u}$ *and* $\mathbf{v}$ *are nonzero and* θ *is the angle between them, then*

$$\begin{array}{lll} \theta \text{ is acute} & \text{if and only if} & \mathbf{u}\cdot\mathbf{v} > 0 \\ \theta \text{ is obtuse} & \text{if and only if} & \mathbf{u}\cdot\mathbf{v} < 0 \\ \theta = \pi/2 & \text{if and only if} & \mathbf{u}\cdot\mathbf{v} = 0 \end{array}$$

Proof (*a*). Since the angle θ between $\mathbf{v}$ and $\mathbf{v}$ is 0, we have

$$\mathbf{v}\cdot\mathbf{v} = \|\mathbf{v}\|\,\|\mathbf{v}\|\cos\theta = \|\mathbf{v}\|^2\cos 0 = \|\mathbf{v}\|^2$$

Proof (*b*). Since θ satisfies $0 \le \theta \le \pi$, it follows that: θ is acute if and only if $\cos\theta > 0$; θ is obtuse if and only if $\cos\theta < 0$; and $\theta = \pi/2$ if and only if $\cos\theta = 0$. But $\cos\theta$ has the same sign as $\mathbf{u}\cdot\mathbf{v}$ since $\mathbf{u}\cdot\mathbf{v} = \|\mathbf{u}\|\,\|\mathbf{v}\|\cos\theta$, $\|\mathbf{u}\| > 0$, and $\|\mathbf{v}\| > 0$. Thus, the result follows. ■

EXAMPLE 4 Finding Dot Products from Components

If $\mathbf{u} = (1, -2, 3)$, $\mathbf{v} = (-3, 4, 2)$, and $\mathbf{w} = (3, 6, 3)$, then

$$\begin{aligned} \mathbf{u}\cdot\mathbf{v} &= (1)(-3) + (-2)(4) + (3)(2) = -5 \\ \mathbf{v}\cdot\mathbf{w} &= (-3)(3) + (4)(6) + (2)(3) = 21 \\ \mathbf{u}\cdot\mathbf{w} &= (1)(3) + (-2)(6) + (3)(3) = 0 \end{aligned}$$

Therefore, $\mathbf{u}$ and $\mathbf{v}$ make an obtuse angle, $\mathbf{v}$ and $\mathbf{w}$ make an acute angle, and $\mathbf{u}$ and $\mathbf{w}$ are perpendicular. ♦

Orthogonal Vectors Perpendicular vectors are also called ***orthogonal*** vectors. In light of Theorem 3.3.1*b*, two *nonzero* vectors are orthogonal if and only if their dot product is zero. If we agree to consider $\mathbf{u}$ and $\mathbf{v}$ to be perpendicular when either or both of these vectors is $\mathbf{0}$, then we can state without exception that *two vectors* $\mathbf{u}$ *and* $\mathbf{v}$ *are orthogonal* (*perpendicular*) *if and only if* $\mathbf{u}\cdot\mathbf{v} = 0$. To indicate that $\mathbf{u}$ and $\mathbf{v}$ are orthogonal vectors we write $\mathbf{u} \perp \mathbf{v}$.

EXAMPLE 5 A Vector Perpendicular to a Line

Show that in 2-space the nonzero vector $\mathbf{n} = (a, b)$ is perpendicular to the line $ax + by + c = 0$.

Solution.

Let $P_1(x_1, y_1)$ and $P_2(x_2, y_2)$ be distinct points on the line, so that

$$\begin{aligned} ax_1 + by_1 + c &= 0 \\ ax_2 + by_2 + c &= 0 \end{aligned} \tag{6}$$

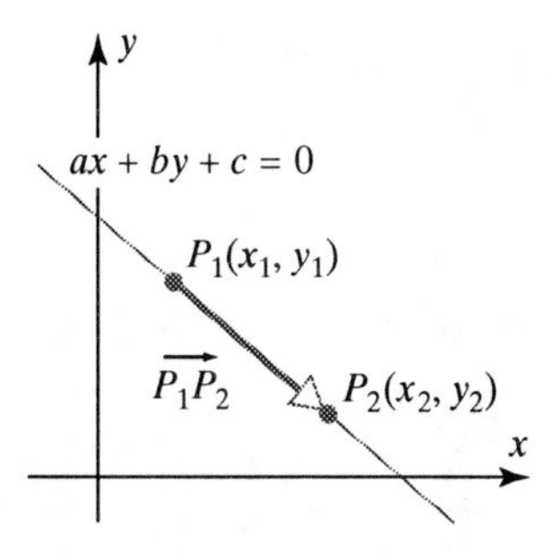

Figure 3.3.5

Since the vector $\overrightarrow{P_1P_2} = (x_2 - x_1, y_2 - y_1)$ runs along the line (Figure 3.3.5), we need only show that $\mathbf{n}$ and $\overrightarrow{P_1P_2}$ are perpendicular. But on subtracting the equations in (6) we obtain

$$a(x_2 - x_1) + b(y_2 - y_1) = 0$$

which can be expressed in the form

$$(a, b) \cdot (x_2 - x_1, y_2 - y_1) = 0 \quad \text{or} \quad \mathbf{n} \cdot \overrightarrow{P_1P_2} = 0$$

Thus, $\mathbf{n}$ and $\overrightarrow{P_1P_2}$ are perpendicular. ♦

The following theorem lists the most important properties of the dot product. They are useful in calculations involving vectors.

Theorem 3.3.2 Properties of the Dot Product

If $\mathbf{u}$, $\mathbf{v}$, *and* $\mathbf{w}$ *are vectors in 2- or 3-space and* k *is a scalar, then*:

(*a*) $\mathbf{u} \cdot \mathbf{v} = \mathbf{v} \cdot \mathbf{u}$
(*b*) $\mathbf{u} \cdot (\mathbf{v} + \mathbf{w}) = \mathbf{u} \cdot \mathbf{v} + \mathbf{u} \cdot \mathbf{w}$
(*c*) $k(\mathbf{u} \cdot \mathbf{v}) = (k\mathbf{u}) \cdot \mathbf{v} = \mathbf{u} \cdot (k\mathbf{v})$
(*d*) $\mathbf{v} \cdot \mathbf{v} > 0$ *if* $\mathbf{v} \neq \mathbf{0}$, *and* $\mathbf{v} \cdot \mathbf{v} = 0$ *if* $\mathbf{v} = \mathbf{0}$

Proof. We shall prove (*c*) for vectors in 3-space and leave the remaining proofs as exercises. Let $\mathbf{u} = (u_1, u_2, u_3)$ and $\mathbf{v} = (v_1, v_2, v_3)$; then

$$\begin{aligned} k(\mathbf{u} \cdot \mathbf{v}) &= k(u_1v_1 + u_2v_2 + u_3v_3) \\ &= (ku_1)v_1 + (ku_2)v_2 + (ku_3)v_3 \\ &= (k\mathbf{u}) \cdot \mathbf{v} \end{aligned}$$

Similarly,

$$k(\mathbf{u} \cdot \mathbf{v}) = \mathbf{u} \cdot (k\mathbf{v})$$ ■

An Orthogonal Projection In many applications it is of interest to "decompose" a vector $\mathbf{u}$ into a sum of two terms, one parallel to a specified nonzero vector $\mathbf{a}$ and the other perpendicular to $\mathbf{a}$. If $\mathbf{u}$ and $\mathbf{a}$ are positioned so their initial points coincide at a point Q, we can decompose the vector $\mathbf{u}$ as follows (Figure 3.3.6): Drop a perpendicular from the tip of $\mathbf{u}$ to the line through $\mathbf{a}$, and construct the vector $\mathbf{w}_1$ from Q to the foot of this perpendicular. Next form the difference

$$\mathbf{w}_2 = \mathbf{u} - \mathbf{w}_1$$

As indicated in Figure 3.3.6, the vector $\mathbf{w}_1$ is parallel to $\mathbf{a}$, the vector $\mathbf{w}_2$ is perpendicular to $\mathbf{a}$, and

$$\mathbf{w}_1 + \mathbf{w}_2 = \mathbf{w}_1 + (\mathbf{u} - \mathbf{w}_1) = \mathbf{u}$$

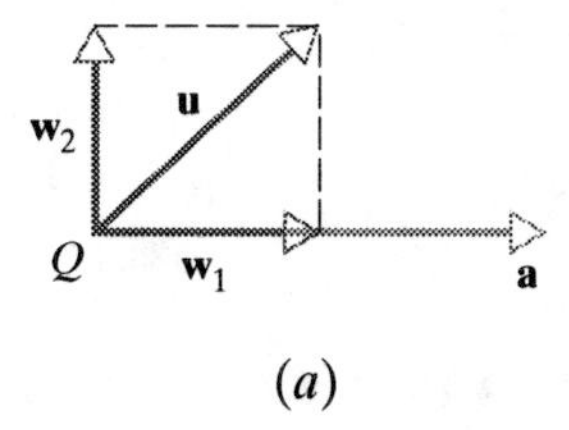

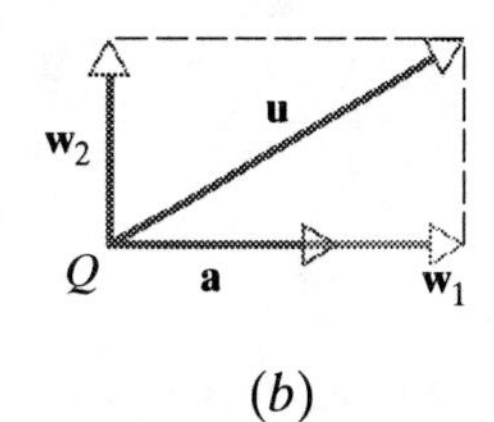

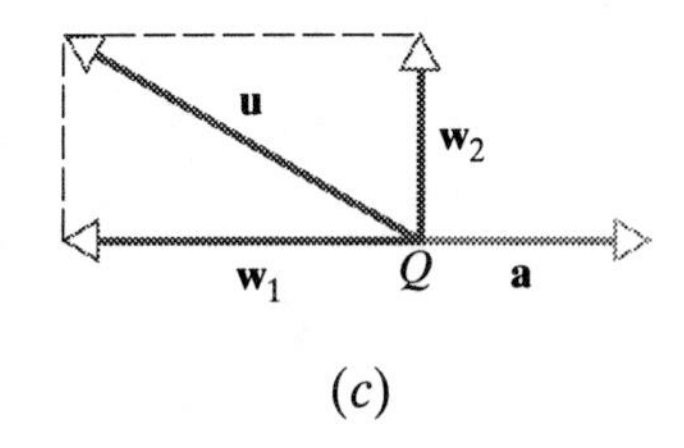

Figure 3.3.6 The vector $\mathbf{u}$ is the sum of $\mathbf{w}_1$ and $\mathbf{w}_2$, where $\mathbf{w}_1$ is parallel to $\mathbf{a}$ and $\mathbf{w}_2$ is perpendicular to $\mathbf{a}$.

The vector $\mathbf{w}_1$ is called the ***orthogonal projection of* u *on* a** or sometimes the ***vector component of* u *along* a**. It is denoted by

$$\operatorname{proj}_{\mathbf{a}} \mathbf{u} \tag{7}$$

The vector $\mathbf{w}_2$ is called the ***vector component of* u *orthogonal to* a**. Since we have $\mathbf{w}_2 = \mathbf{u} - \mathbf{w}_1$, this vector can be written in notation (7) as

$$\mathbf{w}_2 = \mathbf{u} - \operatorname{proj}_{\mathbf{a}} \mathbf{u}$$

The following theorem gives formulas for calculating $\operatorname{proj}_{\mathbf{a}} \mathbf{u}$ and $\mathbf{u} - \operatorname{proj}_{\mathbf{a}} \mathbf{u}$.

Theorem 3.3.3

If **u** *and* **a** *are vectors in 2-space or 3-space and if* $\mathbf{a} \neq \mathbf{0}$, *then*

$$\operatorname{proj}_{\mathbf{a}} \mathbf{u} = \frac{\mathbf{u} \cdot \mathbf{a}}{\|\mathbf{a}\|^2}\mathbf{a} \quad \text{(vector component of } \mathbf{u} \text{ along } \mathbf{a}\text{)}$$

$$\mathbf{u} - \operatorname{proj}_{\mathbf{a}} \mathbf{u} = \mathbf{u} - \frac{\mathbf{u} \cdot \mathbf{a}}{\|\mathbf{a}\|^2}\mathbf{a} \quad \text{(vector component of } \mathbf{u} \text{ orthogonal to } \mathbf{a}\text{)}$$

Proof. Let $\mathbf{w}_1 = \operatorname{proj}_{\mathbf{a}} \mathbf{u}$ and $\mathbf{w}_2 = \mathbf{u} - \operatorname{proj}_{\mathbf{a}} \mathbf{u}$. Since $\mathbf{w}_1$ is parallel to **a**, it must be a scalar multiple of **a**, so it can be written in the form $\mathbf{w}_1 = k\mathbf{a}$. Thus

$$\mathbf{u} = \mathbf{w}_1 + \mathbf{w}_2 = k\mathbf{a} + \mathbf{w}_2 \tag{8}$$

Taking the dot product of both sides of (8) with **a** and using Theorems 3.3.1*a* and 3.3.2 yields

$$\mathbf{u} \cdot \mathbf{a} = (k\mathbf{a} + \mathbf{w}_2) \cdot \mathbf{a} = k\|\mathbf{a}\|^2 + \mathbf{w}_2 \cdot \mathbf{a} \tag{9}$$

But $\mathbf{w}_2 \cdot \mathbf{a} = 0$ since $\mathbf{w}_2$ is perpendicular to **a**; so (9) yields

$$k = \frac{\mathbf{u} \cdot \mathbf{a}}{\|\mathbf{a}\|^2}$$

Since $\operatorname{proj}_{\mathbf{a}} \mathbf{u} = \mathbf{w}_1 = k\mathbf{a}$, we obtain

$$\operatorname{proj}_{\mathbf{a}} \mathbf{u} = \frac{\mathbf{u} \cdot \mathbf{a}}{\|\mathbf{a}\|^2}\mathbf{a}$$

■

EXAMPLE 6 Vector Component of u Along a

Let $\mathbf{u} = (2, -1, 3)$ and $\mathbf{a} = (4, -1, 2)$. Find the vector component of **u** along **a** and the vector component of **u** orthogonal to **a**.

Solution.

$$\mathbf{u} \cdot \mathbf{a} = (2)(4) + (-1)(-1) + (3)(2) = 15$$
$$\|\mathbf{a}\|^2 = 4^2 + (-1)^2 + 2^2 = 21$$

Thus, the vector component of **u** along **a** is

$$\operatorname{proj}_{\mathbf{a}} \mathbf{u} = \frac{\mathbf{u} \cdot \mathbf{a}}{\|\mathbf{a}\|^2}\mathbf{a} = \tfrac{15}{21}(4, -1, 2) = \left(\tfrac{20}{7}, -\tfrac{5}{7}, \tfrac{10}{7}\right)$$

and the vector component of **u** orthogonal to **a** is

$$\mathbf{u} - \text{proj}_{\mathbf{a}}\, \mathbf{u} = (2, -1, 3) - \left(\tfrac{20}{7}, -\tfrac{5}{7}, \tfrac{10}{7}\right) = \left(-\tfrac{6}{7}, -\tfrac{2}{7}, \tfrac{11}{7}\right)$$

As a check, the reader may wish to verify that the vectors $\mathbf{u} - \text{proj}_{\mathbf{a}}\, \mathbf{u}$ and **a** are perpendicular by showing that their dot product is zero. ◆

A formula for the length of the vector component of **u** along **a** can be obtained by writing

$$\|\text{proj}_{\mathbf{a}}\, \mathbf{u}\| = \left\| \frac{\mathbf{u} \cdot \mathbf{a}}{\|\mathbf{a}\|^2}\mathbf{a} \right\|$$

$$= \left| \frac{\mathbf{u} \cdot \mathbf{a}}{\|\mathbf{a}\|^2} \right| \|\mathbf{a}\| \quad \longleftarrow \text{Formula (5) of Section 3.2}$$

$$= \frac{|\mathbf{u} \cdot \mathbf{a}|}{\|\mathbf{a}\|^2} \|\mathbf{a}\| \quad \longleftarrow \text{Since } \|\mathbf{a}\|^2 > 0$$

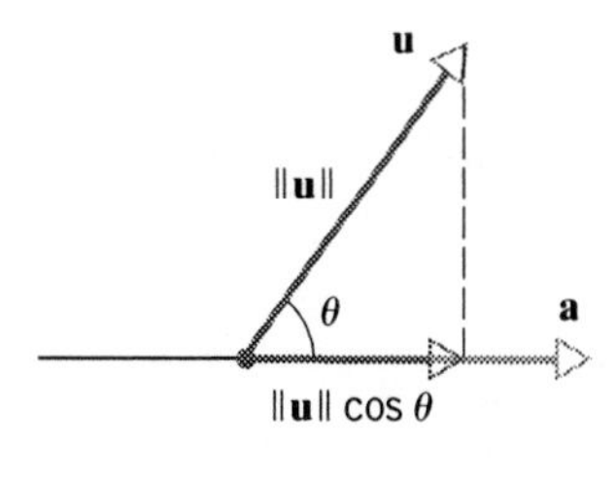

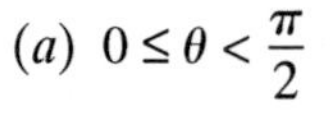

(a) $0 \le \theta < \frac{\pi}{2}$

which yields

$$\|\text{proj}_{\mathbf{a}}\, \mathbf{u}\| = \frac{|\mathbf{u} \cdot \mathbf{a}|}{\|\mathbf{a}\|} \tag{10}$$

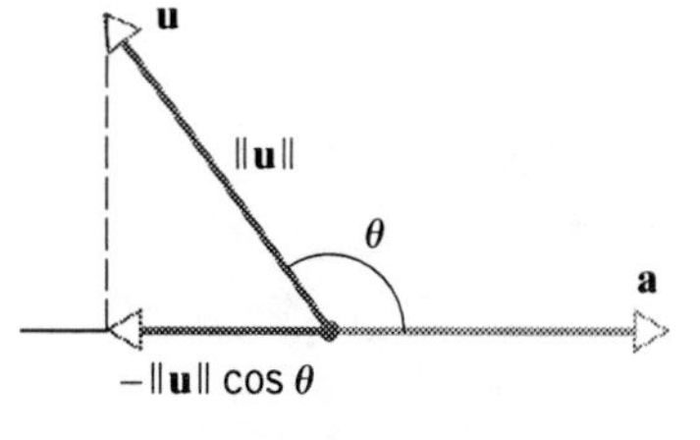

(b) $\frac{\pi}{2} < \theta \le \pi$

Figure 3.3.7

If θ denotes the angle between **u** and **a**, then $\mathbf{u} \cdot \mathbf{a} = \|\mathbf{u}\|\|\mathbf{a}\| \cos\theta$, so that (10) can also be written as

$$\|\text{proj}_{\mathbf{a}}\, \mathbf{u}\| = \|\mathbf{u}\|\, |\cos\theta| \tag{11}$$

(Verify.) A geometric interpretation of this result is given in Figure 3.3.7.

As an example, we will use vector methods to derive a formula for the distance from a point in the plane to a line.

EXAMPLE 7 Distance Between a Point and a Line

Find a formula for the distance D between point $P_0(x_0, y_0)$ and the line $ax + by + c = 0$.

Solution.

Let $Q(x_1, y_1)$ be any point on the line and position the vector $\mathbf{n} = (a, b)$ so that its initial point is at Q.

By virtue of Example 5, the vector **n** is perpendicular to the line (Figure 3.3.8). As indicated in the figure, the distance D is equal to the length of the orthogonal projection of $\overrightarrow{QP_0}$ on **n**; thus, from (10),

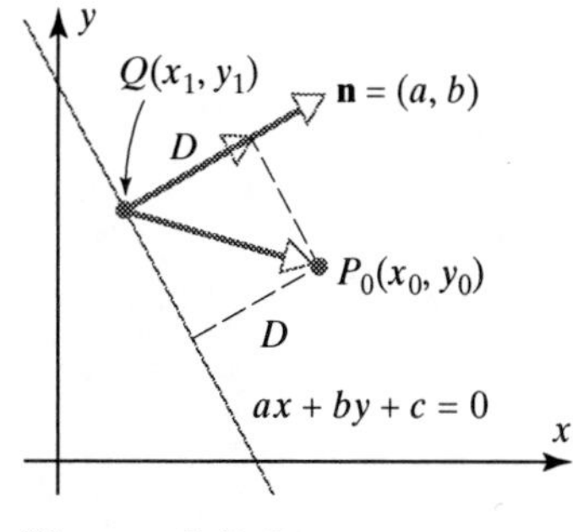

Figure 3.3.8

$$D = \|\text{proj}_{\mathbf{n}}\, \overrightarrow{QP_0}\| = \frac{|\overrightarrow{QP_0} \cdot \mathbf{n}|}{\|\mathbf{n}\|}$$

But

$$\overrightarrow{QP_0} = (x_0 - x_1, y_0 - y_1)$$

$$\overrightarrow{QP_0} \cdot \mathbf{n} = a(x_0 - x_1) + b(y_0 - y_1)$$

$$\|\mathbf{n}\| = \sqrt{a^2 + b^2}$$

so that

$$D = \frac{|a(x_0 - x_1) + b(y_0 - y_1)|}{\sqrt{a^2 + b^2}} \tag{12}$$

Since the point $Q(x_1, y_1)$ lies on the line, its coordinates satisfy the equation of the line, so

$$ax_1 + by_1 + c = 0 \quad \text{or} \quad c = -ax_1 - by_1$$

Substituting this expression in (12) yields the formula

$$D = \frac{|ax_0 + by_0 + c|}{\sqrt{a^2 + b^2}} \tag{13}$$

♦

EXAMPLE 8 Using the Distance Formula

It follows from Formula (13) that the distance D from the point $(1, -2)$ to the line $3x + 4y - 6 = 0$ is

$$D = \frac{|(3)(1) + 4(-2) - 6|}{\sqrt{3^2 + 4^2}} = \frac{|-11|}{\sqrt{25}} = \frac{11}{5}$$

♦

Exercise Set 3.3

1. Find $\mathbf{u} \cdot \mathbf{v}$.

(a) $\mathbf{u} = (2, 3),\ \mathbf{v} = (5, -7)$ (b) $\mathbf{u} = (-6, -2),\ \mathbf{v} = (4, 0)$

(c) $\mathbf{u} = (1, -5, 4),\ \mathbf{v} = (3, 3, 3)$ (d) $\mathbf{u} = (-2, 2, 3),\ \mathbf{v} = (1, 7, -4)$

2. In each part of Exercise 1, find the cosine of the angle θ between $\mathbf{u}$ and $\mathbf{v}$.

3. Determine whether $\mathbf{u}$ and $\mathbf{v}$ make an acute angle, make an obtuse angle, or are orthogonal.

(a) $\mathbf{u} = (6, 1, 4),\ \mathbf{v} = (2, 0, -3)$ (b) $\mathbf{u} = (0, 0, -1),\ \mathbf{v} = (1, 1, 1)$

(c) $\mathbf{u} = (-6, 0, 4),\ \mathbf{v} = (3, 1, 6)$ (d) $\mathbf{u} = (2, 4, -8),\ \mathbf{v} = (5, 3, 7)$

4. Find the orthogonal projection of $\mathbf{u}$ on $\mathbf{a}$.

(a) $\mathbf{u} = (6, 2),\ \mathbf{a} = (3, -9)$ (b) $\mathbf{u} = (-1, -2),\ \mathbf{a} = (-2, 3)$

(c) $\mathbf{u} = (3, 1, -7),\ \mathbf{a} = (1, 0, 5)$ (d) $\mathbf{u} = (1, 0, 0),\ \mathbf{a} = (4, 3, 8)$

5. In each part of Exercise 4, find the vector component of $\mathbf{u}$ orthogonal to $\mathbf{a}$.

6. In each part find $\|\text{proj}_{\mathbf{a}}\, \mathbf{u}\|$.

(a) $\mathbf{u} = (1, -2),\ \mathbf{a} = (-4, -3)$ (b) $\mathbf{u} = (5, 6),\ \mathbf{a} = (2, -1)$

(c) $\mathbf{u} = (3, 0, 4),\ \mathbf{a} = (2, 3, 3)$ (d) $\mathbf{u} = (3, -2, 6),\ \mathbf{a} = (1, 2, -7)$

7. Let $\mathbf{u} = (5, -2, 1)$, $\mathbf{v} = (1, 6, 3)$, and $k = -4$. Verify Theorem 3.3.2 for these quantities.

8. (a) Show that $\mathbf{v} = (a, b)$ and $\mathbf{w} = (-b, a)$ are orthogonal vectors.

(b) Use the result in part (a) to find two vectors that are orthogonal to $\mathbf{v} = (2, -3)$.

(c) Find two unit vectors that are orthogonal to $(-3, 4)$.

9. Let $\mathbf{u} = (3, 4)$, $\mathbf{v} = (5, -1)$, and $\mathbf{w} = (7, 1)$. Evaluate the expressions.

(a) $\mathbf{u} \cdot (7\mathbf{v} + \mathbf{w})$ (b) $\|(\mathbf{u} \cdot \mathbf{w})\mathbf{w}\|$ (c) $\|\mathbf{u}\|(\mathbf{v} \cdot \mathbf{w})$ (d) $(\|\mathbf{u}\|\mathbf{v}) \cdot \mathbf{w}$

10. Find five different nonzero vectors that are orthogonal to $\mathbf{u} = (5, -2, 3)$.

11. Use vectors to find the cosines of the interior angles of the triangle with vertices $(0, -1)$, $(1, -2)$, and $(4, 1)$.

12. Show that $A(3, 0, 2)$, $B(4, 3, 0)$, and $C(8, 1, -1)$ are vertices of a right triangle. At which vertex is the right angle?

13. Find a unit vector that is orthogonal to both $\mathbf{u} = (1, 0, 1)$ and $\mathbf{v} = (0, 1, 1)$.

14. Let $\mathbf{p} = (2, k)$ and $\mathbf{q} = (3, 5)$. Find k such that

(a) $\mathbf{p}$ and $\mathbf{q}$ are parallel
(b) $\mathbf{p}$ and $\mathbf{q}$ are orthogonal
(c) the angle between $\mathbf{p}$ and $\mathbf{q}$ is $\pi/3$
(d) the angle between $\mathbf{p}$ and $\mathbf{q}$ is $\pi/4$

15. Use Formula (13) to calculate the distance between the point and the line.

(a) $4x + 3y + 4 = 0$; $(-3, 1)$
(b) $y = -4x + 2$; $(2, -5)$
(c) $3x + y = 5$; $(1, 8)$

16. Establish the identity $\|\mathbf{u} + \mathbf{v}\|^2 + \|\mathbf{u} - \mathbf{v}\|^2 = 2\|\mathbf{u}\|^2 + 2\|\mathbf{v}\|^2$.

17. Establish the identity $\mathbf{u} \cdot \mathbf{v} = \frac{1}{4}\|\mathbf{u} + \mathbf{v}\|^2 - \frac{1}{4}\|\mathbf{u} - \mathbf{v}\|^2$.

18. Find the angle between a diagonal of a cube and one of its faces.

19. Let $\mathbf{i}$, $\mathbf{j}$, and $\mathbf{k}$ be unit vectors along the positive x, y, and z axes of a rectangular coordinate system in 3-space. If $\mathbf{v} = (a, b, c)$ is a nonzero vector, then the angles α, β, and γ between $\mathbf{v}$ and the vectors $\mathbf{i}$, $\mathbf{j}$, and $\mathbf{k}$, respectively, are called the ***direction angles*** of $\mathbf{v}$ (see accompanying figure), and the numbers $\cos\alpha$, $\cos\beta$, and $\cos\gamma$ are called the ***direction cosines*** of $\mathbf{v}$.

(a) Show that $\cos\alpha = a/\|\mathbf{v}\|$.
(b) Find $\cos\beta$ and $\cos\gamma$.
(c) Show that $\mathbf{v}/\|\mathbf{v}\| = (\cos\alpha, \cos\beta, \cos\gamma)$.
(d) Show that $\cos^2\alpha + \cos^2\beta + \cos^2\gamma = 1$.

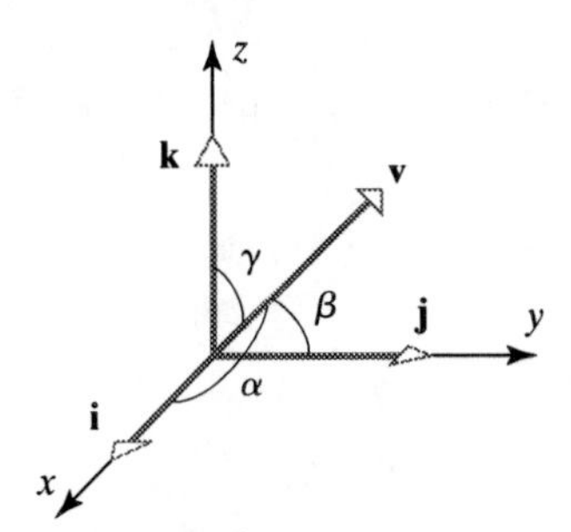

Figure Ex-19

20. Use the result in Exercise 19 to estimate, to the nearest degree, the angles that a diagonal of a box with dimensions 10 cm × 15 cm × 25 cm makes with edges of the box. [*Note.* A calculator is needed.]

21. Referring to Exercise 19, show that two nonzero vectors, $\mathbf{v}_1$ and $\mathbf{v}_2$, in 3-space are perpendicular if and only if their direction cosines satisfy

$$\cos\alpha_1 \cos\alpha_2 + \cos\beta_1 \cos\beta_2 + \cos\gamma_1 \cos\gamma_2 = 0$$

22. Show that if $\mathbf{v}$ is orthogonal to both $\mathbf{w}_1$ and $\mathbf{w}_2$, then $\mathbf{v}$ is orthogonal to $k_1\mathbf{w}_1 + k_2\mathbf{w}_2$ for all scalars k_1 and k_2.

23. Let $\mathbf{u}$ and $\mathbf{v}$ be nonzero vectors in 2- or 3-space, and let $k = \|\mathbf{u}\|$ and $l = \|\mathbf{v}\|$. Show that the vector $\mathbf{w} = l\mathbf{u} + k\mathbf{v}$ bisects the angle between $\mathbf{u}$ and $\mathbf{v}$.

Discussion and Discovery

24. In each part, something is wrong with the expression. What?

(a) $\mathbf{u} \cdot (\mathbf{v} \cdot \mathbf{w})$ (b) $(\mathbf{u} \cdot \mathbf{v}) + \mathbf{w}$ (c) $\|\mathbf{u} \cdot \mathbf{v}\|$ (d) $k \cdot (\mathbf{u} + \mathbf{v})$

25. Is it possible to have $\text{proj}_{\mathbf{a}}\, \mathbf{u} = \text{proj}_{\mathbf{u}}\, \mathbf{a}$? Explain your reasoning.

26. If $\mathbf{u} \neq \mathbf{0}$, is it valid to cancel $\mathbf{u}$ from both sides of the equation $\mathbf{u} \cdot \mathbf{v} = \mathbf{u} \cdot \mathbf{w}$ and conclude that $\mathbf{v} = \mathbf{w}$? Explain your reasoning.

27. Suppose that $\mathbf{u}$, $\mathbf{v}$, and $\mathbf{w}$ are mutually orthogonal nonzero vectors in 3-space, and suppose that you know the dot products of these vectors with a vector $\mathbf{r}$ in 3-space. Find an expression for $\mathbf{r}$ in terms of $\mathbf{u}$, $\mathbf{v}$, $\mathbf{w}$, and the dot products. [***Hint.*** Look for an expression of the form $\mathbf{r} = c_1\mathbf{u} + c_2\mathbf{v} + c_3\mathbf{w}$.]

28. Suppose that $\mathbf{u}$ and $\mathbf{v}$ are orthogonal vectors in 2-space or 3-space. What famous theorem is described by the equation $\|\mathbf{u}+\mathbf{v}\|^2 = \|\mathbf{u}\|^2 + \|\mathbf{v}\|^2$? Draw a picture to support your answer.

3.4 CROSS PRODUCT

In many applications of vectors to problems in geometry, physics, and engineering, it is of interest to construct a vector in 3-space that is perpendicular to two given vectors. In this section we shall show how to do this.

Cross Product of Vectors Recall from Section 3.3 that the dot product of two vectors in 2-space or 3-space produces a scalar. We will now define a type of vector multiplication that produces a vector as the product, but which is applicable only in 3-space.

Definition

If $\mathbf{u} = (u_1, u_2, u_3)$ and $\mathbf{v} = (v_1, v_2, v_3)$ are vectors in 3-space, then the ***cross product*** $\mathbf{u} \times \mathbf{v}$ is the vector defined by

$$\mathbf{u} \times \mathbf{v} = (u_2v_3 - u_3v_2,\ u_3v_1 - u_1v_3,\ u_1v_2 - u_2v_1) \tag{1a}$$

or in determinant notation

$$\mathbf{u} \times \mathbf{v} = \left(\begin{vmatrix} u_2 & u_3 \\ v_2 & v_3 \end{vmatrix}, -\begin{vmatrix} u_1 & u_3 \\ v_1 & v_3 \end{vmatrix}, \begin{vmatrix} u_1 & u_2 \\ v_1 & v_2 \end{vmatrix} \right) \tag{1b}$$

REMARK. Instead of memorizing (1b), you can obtain the components of $\mathbf{u} \times \mathbf{v}$ as follows:

- Form the 2×3 matrix $\begin{bmatrix} u_1 & u_2 & u_3 \\ v_1 & v_2 & v_3 \end{bmatrix}$ whose first row contains the components of $\mathbf{u}$ and whose second row contains the components of $\mathbf{v}$.
- To find the first component of $\mathbf{u} \times \mathbf{v}$, delete the first column and take the determinant; to find the second component, delete the second column and take the negative of the determinant; and to find the third component, delete the third column and take the determinant.

EXAMPLE 1 Calculating a Cross Product

Find $\mathbf{u} \times \mathbf{v}$, where $\mathbf{u} = (1, 2, -2)$ and $\mathbf{v} = (3, 0, 1)$.

Solution.

From either (1) or the mnemonic in the preceding remark, we have

$$\mathbf{u} \times \mathbf{v} = \left(\begin{vmatrix} 2 & -2 \\ 0 & 1 \end{vmatrix}, -\begin{vmatrix} 1 & -2 \\ 3 & 1 \end{vmatrix}, \begin{vmatrix} 1 & 2 \\ 3 & 0 \end{vmatrix} \right)$$

$$= (2, -7, -6)$$

♦

Joseph Louis Lagrange (1736–1813) was a French-Italian mathematician and astronomer. Although his father wanted him to become a lawyer, Lagrange was attracted to mathematics and astronomy after reading a memoir by the astronomer Halley. At age 16 he began to study mathematics on his own and by age 19 was appointed to a professorship at the Royal Artillery School in Turin. The following year he solved some famous problems using new methods that eventually blossomed into a branch of mathematics called the *calculus of variations*. These methods and Lagrange's applications of them to problems in celestial mechanics were so monumental that by age 25 he was regarded by many of his contemporaries as the greatest living mathematician. One of Lagrange's most famous works is a memoir, *Mécanique Analytique*, in which he reduced the theory of mechanics to a few general formulas from which all other necessary equations could be derived.

Napoleon was a great admirer of Lagrange and showered him with many honors. In spite of his fame, Lagrange was a shy and modest man. On his death, he was buried with honor in the Pantheon.

There is an important difference between the dot product and cross product of two vectors—the dot product is a scalar and the cross product is a vector. The following theorem gives some important relationships between the dot product and cross product and also shows that $\mathbf{u} \times \mathbf{v}$ is orthogonal to both $\mathbf{u}$ and $\mathbf{v}$.

Theorem 3.4.1 Relationships Involving Cross Product and Dot Product

If $\mathbf{u}$, $\mathbf{v}$, *and* $\mathbf{w}$ *are vectors in 3-space, then*:

(*a*) $\mathbf{u} \cdot (\mathbf{u} \times \mathbf{v}) = 0$ **($\mathbf{u} \times \mathbf{v}$ *is orthogonal to* $\mathbf{u}$)**

(*b*) $\mathbf{v} \cdot (\mathbf{u} \times \mathbf{v}) = 0$ **($\mathbf{u} \times \mathbf{v}$ *is orthogonal to* $\mathbf{v}$)**

(*c*) $\|\mathbf{u} \times \mathbf{v}\|^2 = \|\mathbf{u}\|^2\|\mathbf{v}\|^2 - (\mathbf{u} \cdot \mathbf{v})^2$ ***(Lagrange's identity)***

(*d*) $\mathbf{u} \times (\mathbf{v} \times \mathbf{w}) = (\mathbf{u} \cdot \mathbf{w})\mathbf{v} - (\mathbf{u} \cdot \mathbf{v})\mathbf{w}$ ***(relationship between cross and dot products)***

(*e*) $(\mathbf{u} \times \mathbf{v}) \times \mathbf{w} = (\mathbf{u} \cdot \mathbf{w})\mathbf{v} - (\mathbf{v} \cdot \mathbf{w})\mathbf{u}$ ***(relationship between cross and dot products)***

Proof (*a*). Let $\mathbf{u} = (u_1, u_2, u_3)$ and $\mathbf{v} = (v_1, v_2, v_3)$. Then

$$\begin{aligned}\mathbf{u} \cdot (\mathbf{u} \times \mathbf{v}) &= (u_1, u_2, u_3) \cdot (u_2v_3 - u_3v_2, u_3v_1 - u_1v_3, u_1v_2 - u_2v_1) \\ &= u_1(u_2v_3 - u_3v_2) + u_2(u_3v_1 - u_1v_3) + u_3(u_1v_2 - u_2v_1) = 0\end{aligned}$$

Proof (*b*). Similar to (*a*).

Proof (*c*). Since

$$\|\mathbf{u} \times \mathbf{v}\|^2 = (u_2v_3 - u_3v_2)^2 + (u_3v_1 - u_1v_3)^2 + (u_1v_2 - u_2v_1)^2 \tag{2}$$

and

$$\|\mathbf{u}\|^2\|\mathbf{v}\|^2 - (\mathbf{u} \cdot \mathbf{v})^2 = (u_1^2 + u_2^2 + u_3^2)(v_1^2 + v_2^2 + v_3^2) - (u_1v_1 + u_2v_2 + u_3v_3)^2 \tag{3}$$

the proof can be completed by "multiplying out" the right sides of (2) and (3) and verifying their equality.

Proof (*d*) *and* (*e*). See Exercises 26 and 27. ■

EXAMPLE 2 u × v Is Perpendicular to u and to v

Consider the vectors

$$\mathbf{u} = (1, 2, -2) \quad \text{and} \quad \mathbf{v} = (3, 0, 1)$$

In Example 1 we showed that

$$\mathbf{u} \times \mathbf{v} = (2, -7, -6)$$

Since

$$\mathbf{u} \cdot (\mathbf{u} \times \mathbf{v}) = (1)(2) + (2)(-7) + (-2)(-6) = 0$$

and

$$\mathbf{v} \cdot (\mathbf{u} \times \mathbf{v}) = (3)(2) + (0)(-7) + (1)(-6) = 0$$

$\mathbf{u} \times \mathbf{v}$ is orthogonal to both $\mathbf{u}$ and $\mathbf{v}$ as guaranteed by Theorem 3.4.1. ♦

The main arithmetic properties of the cross product are listed in the next theorem.

Theorem 3.4.2 Properties of Cross Product

If $\mathbf{u}$, $\mathbf{v}$, *and* $\mathbf{w}$ *are any vectors in 3-space and* k *is any scalar, then*:

(*a*) $\mathbf{u} \times \mathbf{v} = -(\mathbf{v} \times \mathbf{u})$
(*b*) $\mathbf{u} \times (\mathbf{v} + \mathbf{w}) = (\mathbf{u} \times \mathbf{v}) + (\mathbf{u} \times \mathbf{w})$
(*c*) $(\mathbf{u} + \mathbf{v}) \times \mathbf{w} = (\mathbf{u} \times \mathbf{w}) + (\mathbf{v} \times \mathbf{w})$
(*d*) $k(\mathbf{u} \times \mathbf{v}) = (k\mathbf{u}) \times \mathbf{v} = \mathbf{u} \times (k\mathbf{v})$
(*e*) $\mathbf{u} \times \mathbf{0} = \mathbf{0} \times \mathbf{u} = \mathbf{0}$
(*f*) $\mathbf{u} \times \mathbf{u} = \mathbf{0}$

The proofs follow immediately from Formula (1b) and properties of determinants; for example, (*a*) can be proved as follows:

Proof (*a*). Interchanging $\mathbf{u}$ and $\mathbf{v}$ in (1b) interchanges the rows of the three determinants on the right side of (1b) and hence changes the sign of each component in the cross product. Thus, $\mathbf{u} \times \mathbf{v} = -(\mathbf{v} \times \mathbf{u})$. ■

The proofs of the remaining parts are left as exercises.

EXAMPLE 3 Standard Unit Vectors

Consider the vectors

$$\mathbf{i} = (1, 0, 0), \quad \mathbf{j} = (0, 1, 0), \quad \mathbf{k} = (0, 0, 1)$$

These vectors each have length 1 and lie along the coordinate axes (Figure 3.4.1). They are called the ***standard unit vectors*** in 3-space. Every vector $\mathbf{v} = (v_1, v_2, v_3)$ in 3-space is expressible in terms of $\mathbf{i}$, $\mathbf{j}$, and $\mathbf{k}$ since we can write

$$\mathbf{v} = (v_1, v_2, v_3) = v_1(1, 0, 0) + v_2(0, 1, 0) + v_3(0, 0, 1) = v_1\mathbf{i} + v_2\mathbf{j} + v_3\mathbf{k}$$

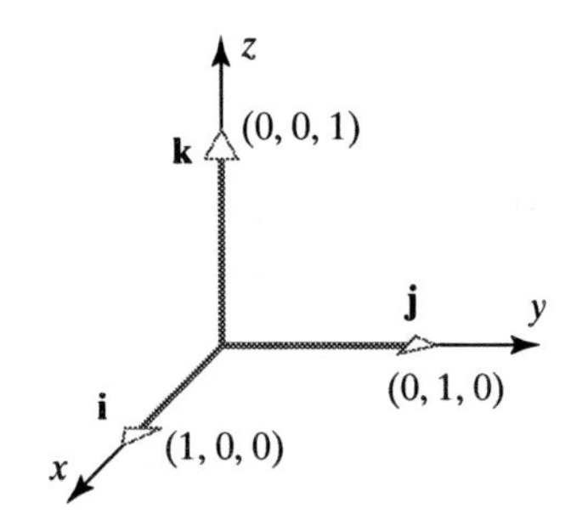

Figure 3.4.1
The standard unit vectors.

For example,

$$(2, -3, 4) = 2\mathbf{i} - 3\mathbf{j} + 4\mathbf{k}$$

From (1b) we obtain

$$\mathbf{i} \times \mathbf{j} = \left(\begin{vmatrix} 0 & 0 \\ 1 & 0 \end{vmatrix}, -\begin{vmatrix} 1 & 0 \\ 0 & 0 \end{vmatrix}, \begin{vmatrix} 1 & 0 \\ 0 & 1 \end{vmatrix} \right) = (0, 0, 1) = \mathbf{k}$$ ◆

The reader should have no trouble obtaining the following results:

$$\begin{array}{lll} \mathbf{i} \times \mathbf{i} = \mathbf{0} & \mathbf{j} \times \mathbf{j} = \mathbf{0} & \mathbf{k} \times \mathbf{k} = \mathbf{0} \\ \mathbf{i} \times \mathbf{j} = \mathbf{k} & \mathbf{j} \times \mathbf{k} = \mathbf{i} & \mathbf{k} \times \mathbf{i} = \mathbf{j} \\ \mathbf{j} \times \mathbf{i} = -\mathbf{k} & \mathbf{k} \times \mathbf{j} = -\mathbf{i} & \mathbf{i} \times \mathbf{k} = -\mathbf{j} \end{array}$$

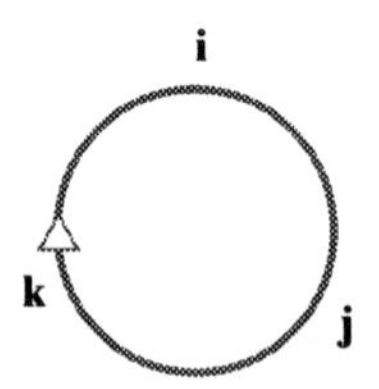

Figure 3.4.2

Figure 3.4.2 is helpful for remembering these results. Referring to this diagram, the cross product of two consecutive vectors going clockwise is the next vector around, and the cross product of two consecutive vectors going counterclockwise is the negative of the next vector around.

Determinant Form of Cross Product It is also worth noting that a cross product can be represented symbolically in the form of a 3×3 determinant:

$$\mathbf{u} \times \mathbf{v} = \begin{vmatrix} \mathbf{i} & \mathbf{j} & \mathbf{k} \\ u_1 & u_2 & u_3 \\ v_1 & v_2 & v_3 \end{vmatrix} = \begin{vmatrix} u_2 & u_3 \\ v_2 & v_3 \end{vmatrix} \mathbf{i} - \begin{vmatrix} u_1 & u_3 \\ v_1 & v_3 \end{vmatrix} \mathbf{j} + \begin{vmatrix} u_1 & u_2 \\ v_1 & v_2 \end{vmatrix} \mathbf{k} \tag{4}$$

For example, if $\mathbf{u} = (1, 2, -2)$ and $\mathbf{v} = (3, 0, 1)$, then

$$\mathbf{u} \times \mathbf{v} = \begin{vmatrix} \mathbf{i} & \mathbf{j} & \mathbf{k} \\ 1 & 2 & -2 \\ 3 & 0 & 1 \end{vmatrix} = 2\mathbf{i} - 7\mathbf{j} - 6\mathbf{k}$$

which agrees with the result obtained in Example 1.

Warning. It is not true in general that $\mathbf{u} \times (\mathbf{v} \times \mathbf{w}) = (\mathbf{u} \times \mathbf{v}) \times \mathbf{w}$. For example,

$$\mathbf{i} \times (\mathbf{j} \times \mathbf{j}) = \mathbf{i} \times \mathbf{0} = \mathbf{0}$$

and

$$(\mathbf{i} \times \mathbf{j}) \times \mathbf{j} = \mathbf{k} \times \mathbf{j} = -\mathbf{i}$$

so that

$$\mathbf{i} \times (\mathbf{j} \times \mathbf{j}) \neq (\mathbf{i} \times \mathbf{j}) \times \mathbf{j}$$

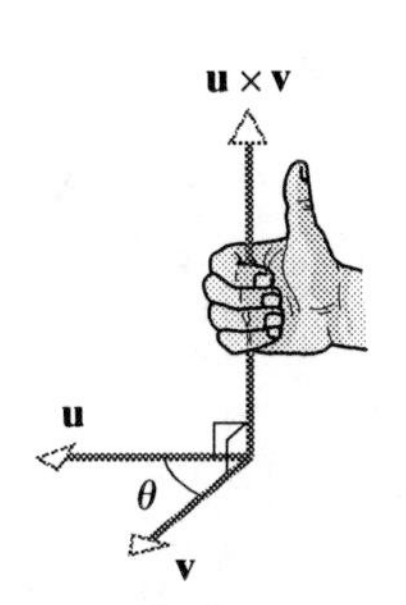

Figure 3.4.3

We know from Theorem 3.4.1 that $\mathbf{u} \times \mathbf{v}$ is orthogonal to both $\mathbf{u}$ and $\mathbf{v}$. If $\mathbf{u}$ and $\mathbf{v}$ are nonzero vectors, it can be shown that the direction of $\mathbf{u} \times \mathbf{v}$ can be determined using the following "right-hand rule"† (Figure 3.4.3): Let θ be the angle between $\mathbf{u}$ and $\mathbf{v}$, and suppose $\mathbf{u}$ is rotated through the angle θ until it coincides with $\mathbf{v}$. If the fingers of the right hand are cupped so they point in the direction of rotation, then the thumb indicates (roughly) the direction of $\mathbf{u} \times \mathbf{v}$.

The reader may find it instructive to practice this rule with the products

$$\mathbf{i} \times \mathbf{j} = \mathbf{k}, \qquad \mathbf{j} \times \mathbf{k} = \mathbf{i}, \qquad \mathbf{k} \times \mathbf{i} = \mathbf{j}$$

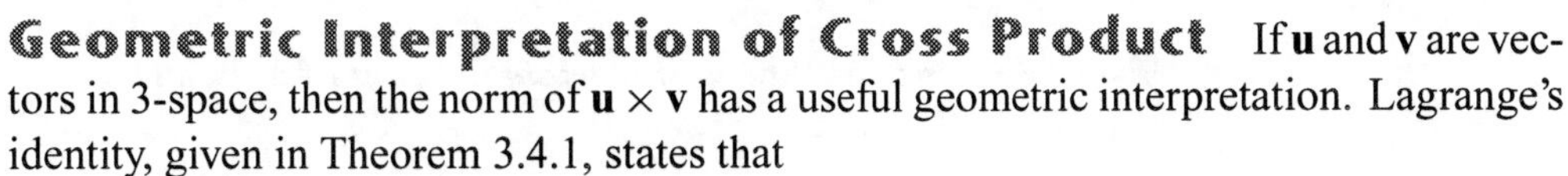

Geometric Interpretation of Cross Product If $\mathbf{u}$ and $\mathbf{v}$ are vectors in 3-space, then the norm of $\mathbf{u} \times \mathbf{v}$ has a useful geometric interpretation. Lagrange's identity, given in Theorem 3.4.1, states that

$$\|\mathbf{u} \times \mathbf{v}\|^2 = \|\mathbf{u}\|^2\|\mathbf{v}\|^2 - (\mathbf{u} \cdot \mathbf{v})^2 \tag{5}$$

If θ denotes the angle between $\mathbf{u}$ and $\mathbf{v}$, then $\mathbf{u} \cdot \mathbf{v} = \|\mathbf{u}\|\|\mathbf{v}\| \cos\theta$, so that (5) can be rewritten as

$$\begin{aligned} \|\mathbf{u} \times \mathbf{v}\|^2 &= \|\mathbf{u}\|^2\|\mathbf{v}\|^2 - \|\mathbf{u}\|^2\|\mathbf{v}\|^2 \cos^2\theta \\ &= \|\mathbf{u}\|^2\|\mathbf{v}\|^2(1 - \cos^2\theta) \\ &= \|\mathbf{u}\|^2\|\mathbf{v}\|^2 \sin^2\theta \end{aligned}$$

Since $0 \leq \theta \leq \pi$, it follows that $\sin\theta \geq 0$, so this can be rewritten as

$$\|\mathbf{u} \times \mathbf{v}\| = \|\mathbf{u}\|\|\mathbf{v}\| \sin\theta \tag{6}$$

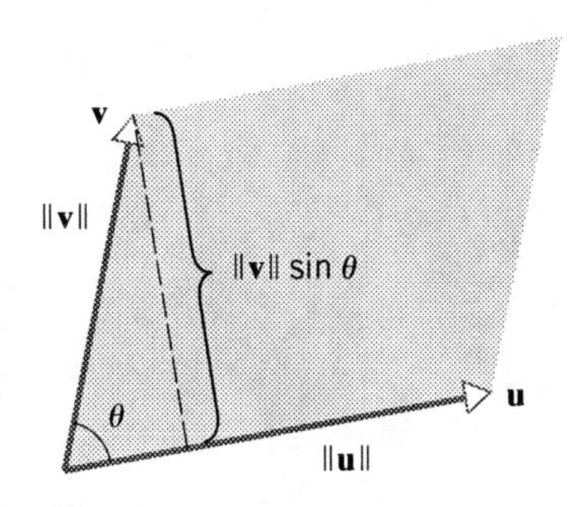

Figure 3.4.4

But $\|\mathbf{v}\| \sin\theta$ is the altitude of the parallelogram determined by $\mathbf{u}$ and $\mathbf{v}$ (Figure 3.4.4). Thus, from (6), the area A of this parallelogram is given by

$$A = \text{(base)(altitude)} = \|\mathbf{u}\|\|\mathbf{v}\| \sin\theta = \|\mathbf{u} \times \mathbf{v}\|$$

---

† Recall that we agreed to consider only right-handed coordinate systems in this text. Had we used left-handed systems instead, a "left-hand rule" would apply here.

This result is even correct if **u** and **v** are collinear, since the parallelogram determined by **u** and **v** has zero area and from (6) we have $\mathbf{u} \times \mathbf{v} = \mathbf{0}$ because $\theta = 0$ in this case. Thus we have the following theorem.

Theorem 3.4.3 Area of a Parallelogram

If **u** *and* **v** *are vectors in 3-space, then* $\|\mathbf{u} \times \mathbf{v}\|$ *is equal to the area of the parallelogram determined by* **u** *and* **v**.

EXAMPLE 4 Area of a Triangle

Find the area of the triangle determined by the points $P_1(2, 2, 0)$, $P_2(-1, 0, 2)$, and $P_3(0, 4, 3)$.

Solution.

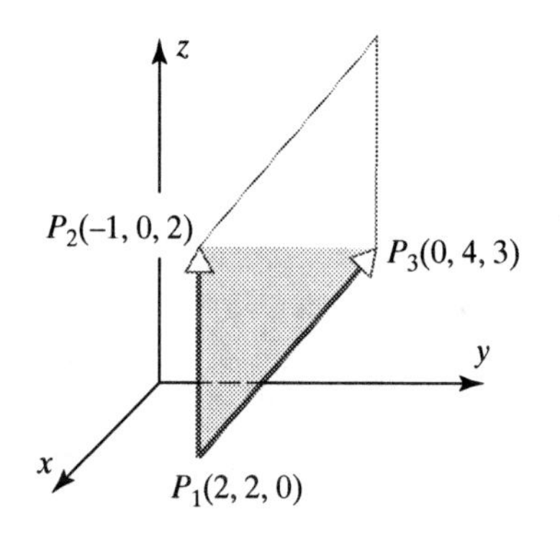

Figure 3.4.5

The area A of the triangle is $\frac{1}{2}$ the area of the parallelogram determined by the vectors $\overrightarrow{P_1P_2}$ and $\overrightarrow{P_1P_3}$ (Figure 3.4.5). Using the method discussed in Example 2 of Section 3.1, $\overrightarrow{P_1P_2} = (-3, -2, 2)$ and $\overrightarrow{P_1P_3} = (-2, 2, 3)$. It follows that

$$\overrightarrow{P_1P_2} \times \overrightarrow{P_1P_3} = (-10, 5, -10)$$

and consequently

$$A = \tfrac{1}{2}\|\overrightarrow{P_1P_2} \times \overrightarrow{P_1P_3}\| = \tfrac{1}{2}(15) = \tfrac{15}{2}$$ ♦

Definition

If **u**, **v**, and **w** are vectors in 3-space, then

$$\mathbf{u} \cdot (\mathbf{v} \times \mathbf{w})$$

is called the ***scalar triple product*** of **u**, **v**, and **w**.

The scalar triple product of $\mathbf{u} = (u_1, u_2, u_3)$, $\mathbf{v} = (v_1, v_2, v_3)$, and $\mathbf{w} = (w_1, w_2, w_3)$ can be calculated from the formula

$$\mathbf{u} \cdot (\mathbf{v} \times \mathbf{w}) = \begin{vmatrix} u_1 & u_2 & u_3 \\ v_1 & v_2 & v_3 \\ w_1 & w_2 & w_3 \end{vmatrix} \tag{7}$$

This follows from Formula (4) since

$$\begin{aligned} \mathbf{u} \cdot (\mathbf{v} \times \mathbf{w}) &= \mathbf{u} \cdot \left(\begin{vmatrix} v_2 & v_3 \\ w_2 & w_3 \end{vmatrix} \mathbf{i} - \begin{vmatrix} v_1 & v_3 \\ w_1 & w_3 \end{vmatrix} \mathbf{j} + \begin{vmatrix} v_1 & v_2 \\ w_1 & w_2 \end{vmatrix} \mathbf{k} \right) \\ &= \begin{vmatrix} v_2 & v_3 \\ w_2 & w_3 \end{vmatrix} u_1 - \begin{vmatrix} v_1 & v_3 \\ w_1 & w_3 \end{vmatrix} u_2 + \begin{vmatrix} v_1 & v_2 \\ w_1 & w_2 \end{vmatrix} u_3 \\ &= \begin{vmatrix} u_1 & u_2 & u_3 \\ v_1 & v_2 & v_3 \\ w_1 & w_2 & w_3 \end{vmatrix} \end{aligned}$$

EXAMPLE 5 Calculating a Scalar Triple Product

Calculate the scalar triple product $\mathbf{u} \cdot (\mathbf{v} \times \mathbf{w})$ of the vectors

$$\mathbf{u} = 3\mathbf{i} - 2\mathbf{j} - 5\mathbf{k}, \qquad \mathbf{v} = \mathbf{i} + 4\mathbf{j} - 4\mathbf{k}, \qquad \mathbf{w} = 3\mathbf{j} + 2\mathbf{k}$$

Solution.

From (7)

$$\begin{aligned}\mathbf{u} \cdot (\mathbf{v} \times \mathbf{w}) &= \begin{vmatrix} 3 & -2 & -5 \\ 1 & 4 & -4 \\ 0 & 3 & 2 \end{vmatrix} \\ &= 3\begin{vmatrix} 4 & -4 \\ 3 & 2 \end{vmatrix} - (-2)\begin{vmatrix} 1 & -4 \\ 0 & 2 \end{vmatrix} + (-5)\begin{vmatrix} 1 & 4 \\ 0 & 3 \end{vmatrix} \\ &= 60 + 4 - 15 = 49\end{aligned}$$

♦

REMARK. The symbol $(\mathbf{u} \cdot \mathbf{v}) \times \mathbf{w}$ makes no sense since we cannot form the cross product of a scalar and a vector. Thus, no ambiguity arises if we write $\mathbf{u} \cdot \mathbf{v} \times \mathbf{w}$ rather than $\mathbf{u} \cdot (\mathbf{v} \times \mathbf{w})$. However, for clarity we shall usually keep the parentheses.

It follows from (7) that

$$\mathbf{u} \cdot (\mathbf{v} \times \mathbf{w}) = \mathbf{w} \cdot (\mathbf{u} \times \mathbf{v}) = \mathbf{v} \cdot (\mathbf{w} \times \mathbf{u})$$

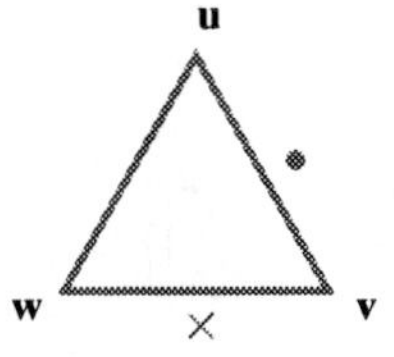

Figure 3.4.6

since the 3×3 determinants that represent these products can be obtained from one another by *two* row interchanges. (Verify.) These relationships can be remembered by moving the vectors **u**, **v**, and **w** clockwise around the vertices of the triangle in Figure 3.4.6.

Geometric Interpretation of Determinants The next theorem provides a useful geometric interpretation of 2×2 and 3×3 determinants.

Theorem 3.4.4

(*a*) *The absolute value of the determinant*

$$\det \begin{bmatrix} u_1 & u_2 \\ v_1 & v_2 \end{bmatrix}$$

is equal to the area of the parallelogram in 2-space determined by the vectors $\mathbf{u} = (u_1, u_2)$ *and* $\mathbf{v} = (v_1, v_2)$. *(See Figure 3.4.7a.)*

(*b*) *The absolute value of the determinant*

$$\det \begin{bmatrix} u_1 & u_2 & u_3 \\ v_1 & v_2 & v_3 \\ w_1 & w_2 & w_3 \end{bmatrix}$$

is equal to the volume of the parallelepiped in 3-space determined by the vectors $\mathbf{u} = (u_1, u_2, u_3)$, $\mathbf{v} = (v_1, v_2, v_3)$, *and* $\mathbf{w} = (w_1, w_2, w_3)$. *(See Figure 3.4.7b.)*

Proof (a). The key to the proof is to use Theorem 3.4.3. However, that theorem applies to vectors in 3-space, whereas $\mathbf{u} = (u_1, u_2)$ and $\mathbf{v} = (v_1, v_2)$ are vectors in 2-space. To circumvent this "dimension problem," we shall view **u** and **v** as vectors in the xy-plane

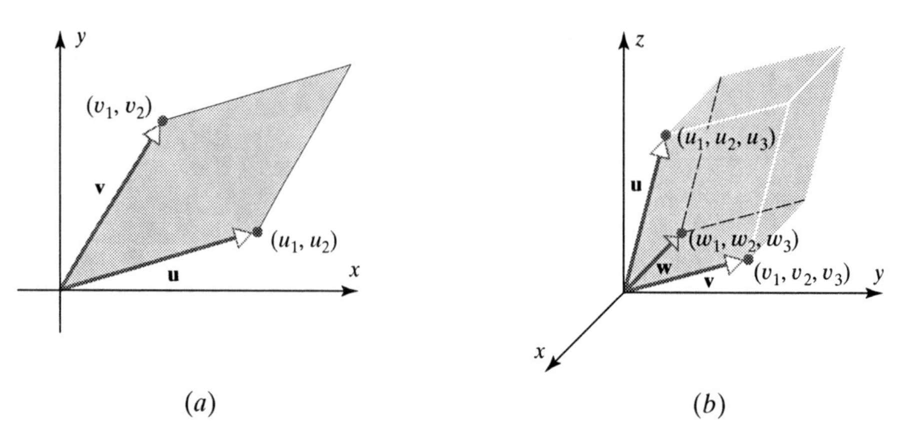

Figure 3.4.7

of an xyz-coordinate system (Figure 3.4.8a), in which case these vectors are expressed as $\mathbf{u} = (u_1, u_2, 0)$ and $\mathbf{v} = (v_1, v_2, 0)$. Thus,

$$\mathbf{u} \times \mathbf{v} = \begin{vmatrix} \mathbf{i} & \mathbf{j} & \mathbf{k} \\ u_1 & u_2 & 0 \\ v_1 & v_2 & 0 \end{vmatrix} = \begin{vmatrix} u_1 & u_2 \\ v_1 & v_2 \end{vmatrix} \mathbf{k} = \det \begin{bmatrix} u_1 & u_2 \\ v_1 & v_2 \end{bmatrix} \mathbf{k}$$

It now follows from Theorem 3.4.3 and the fact that $\|\mathbf{k}\| = 1$ that the area A of the parallelogram determined by $\mathbf{u}$ and $\mathbf{v}$ is

$$A = \|\mathbf{u} \times \mathbf{v}\| = \left\| \det \begin{bmatrix} u_1 & u_2 \\ v_1 & v_2 \end{bmatrix} \mathbf{k} \right\| = \left| \det \begin{bmatrix} u_1 & u_2 \\ v_1 & v_2 \end{bmatrix} \right| \|\mathbf{k}\| = \left| \det \begin{bmatrix} u_1 & u_2 \\ v_1 & v_2 \end{bmatrix} \right|$$

which completes the proof.

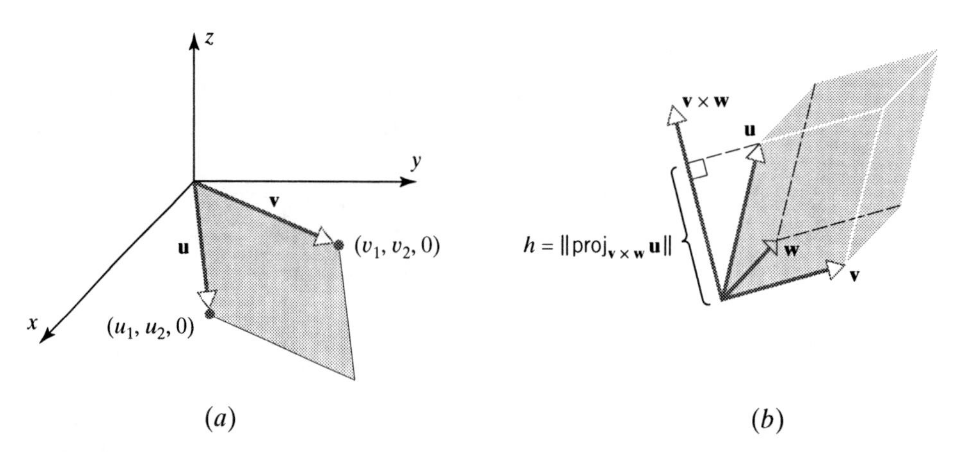

Figure 3.4.8

Proof (b). As shown in Figure 3.4.8b, take the base of the parallelepiped determined by $\mathbf{u}$, $\mathbf{v}$, and $\mathbf{w}$ to be the parallelogram determined by $\mathbf{v}$ and $\mathbf{w}$. It follows from Theorem 3.4.3 that the area of the base is $\|\mathbf{v} \times \mathbf{w}\|$ and, as illustrated in Figure 3.4.8b, the height h of the parallelepiped is the length of the orthogonal projection of $\mathbf{u}$ on $\mathbf{v} \times \mathbf{w}$. Therefore, by Formula (10) of Section 3.3,

$$h = \|\text{proj}_{\mathbf{v} \times \mathbf{w}} \mathbf{u}\| = \frac{|\mathbf{u} \cdot (\mathbf{v} \times \mathbf{w})|}{\|\mathbf{v} \times \mathbf{w}\|}$$

It follows that the volume V of the parallelepiped is

$$V = (\text{area of base}) \cdot \text{height} = \|\mathbf{v} \times \mathbf{w}\| \frac{|\mathbf{u} \cdot (\mathbf{v} \times \mathbf{w})|}{\|\mathbf{v} \times \mathbf{w}\|} = |\mathbf{u} \cdot (\mathbf{v} \times \mathbf{w})|$$

so from (7)

$$V = \left| \det \begin{bmatrix} u_1 & u_2 & u_3 \\ v_1 & v_2 & v_3 \\ w_1 & w_2 & w_3 \end{bmatrix} \right|$$

which completes the proof. ■

REMARK. If V denotes the volume of the parallelepiped determined by vectors **u**, **v**, and **w**, then it follows from Theorem 3.4.4 and Formula (7) that

$$V = \begin{bmatrix} \text{volume of parallelepiped} \\ \text{determined by } \mathbf{u}, \mathbf{v}, \text{ and } \mathbf{w} \end{bmatrix} = |\mathbf{u} \cdot (\mathbf{v} \times \mathbf{w})| \tag{8}$$

From this and Theorem 3.3.1*b*, we can conclude that

$$\mathbf{u} \cdot (\mathbf{v} \times \mathbf{w}) = \pm V$$

where the $+$ or $-$ results depending on whether **u** makes an acute or obtuse angle with $\mathbf{v} \times \mathbf{w}$.

Formula (8) leads to a useful test for ascertaining whether three given vectors lie in the same plane. Since three vectors not in the same plane determine a parallelepiped of positive volume, it follows from (8) that $|\mathbf{u} \cdot (\mathbf{v} \times \mathbf{w})| = 0$ if and only if the vectors **u**, **v**, and **w** lie in the same plane. Thus, we have the following result.

Theorem 3.4.5

If the vectors $\mathbf{u} = (u_1, u_2, u_3)$, $\mathbf{v} = (v_1, v_2, v_3)$, *and* $\mathbf{w} = (w_1, w_2, w_3)$ *have the same initial point, then they lie in the same plane if and only if*

$$\mathbf{u} \cdot (\mathbf{v} \times \mathbf{w}) = \begin{vmatrix} u_1 & u_2 & u_3 \\ v_1 & v_2 & v_3 \\ w_1 & w_2 & w_3 \end{vmatrix} = 0$$

Independence of Cross Product and Coordinates Initially, we defined a vector to be a directed line segment or arrow in 2-space or 3-space; coordinate systems and components were introduced later in order to simplify computations with vectors. Thus, a vector has a "mathematical existence" regardless of whether a coordinate system has been introduced. Further, the components of a vector are not determined by the vector alone; they depend as well on the coordinate system chosen. For example, in Figure 3.4.9 we have indicated a fixed vector **v** in the plane and two

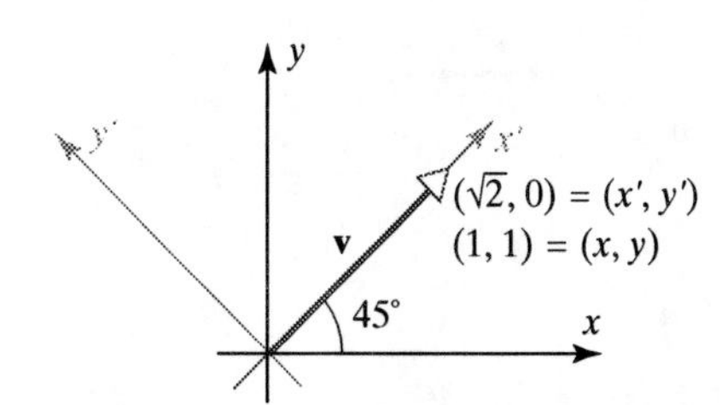

Figure 3.4.9

different coordinate systems. In the xy-coordinate system the components of $\mathbf{v}$ are $(1, 1)$, and in the $x'y'$-system they are $(\sqrt{2}, 0)$.

This raises an important question about our definition of cross product. Since we defined the cross product $\mathbf{u} \times \mathbf{v}$ in terms of the components of $\mathbf{u}$ and $\mathbf{v}$, and since these components depend on the coordinate system chosen, it seems possible that two *fixed* vectors $\mathbf{u}$ and $\mathbf{v}$ might have different cross products in different coordinate systems. Fortunately, this is not the case. To see that this is so, we need only recall that

- $\mathbf{u} \times \mathbf{v}$ is perpendicular to both $\mathbf{u}$ and $\mathbf{v}$.
- The orientation of $\mathbf{u} \times \mathbf{v}$ is determined by the right-hand rule.
- $\|\mathbf{u} \times \mathbf{v}\| = \|\mathbf{u}\|\|\mathbf{v}\| \sin\theta$.

These three properties completely determine the vector $\mathbf{u} \times \mathbf{v}$: the first and second properties determine the direction, and the third property determines the length. Since these properties of $\mathbf{u} \times \mathbf{v}$ depend only on the lengths and relative positions of $\mathbf{u}$ and $\mathbf{v}$ and not on the particular right-hand coordinate system being used, the vector $\mathbf{u} \times \mathbf{v}$ will remain unchanged if a different right-hand coordinate system is introduced. Thus, we say that the definition of $\mathbf{u} \times \mathbf{v}$ is ***coordinate free***. This result is of importance to physicists and engineers who often work with many coordinate systems in the same problem.

EXAMPLE 6 $\mathbf{u} \times \mathbf{v}$ Is Independent of the Coordinate System

Consider two perpendicular vectors $\mathbf{u}$ and $\mathbf{v}$, each of length 1 (Figure 3.4.10*a*). If we introduce an xyz-coordinate system as shown in Figure 3.4.10*b*, then

$$\mathbf{u} = (1, 0, 0) = \mathbf{i} \quad \text{and} \quad \mathbf{v} = (0, 1, 0) = \mathbf{j}$$

so that

$$\mathbf{u} \times \mathbf{v} = \mathbf{i} \times \mathbf{j} = \mathbf{k} = (0, 0, 1)$$

However, if we introduce an $x'y'z'$-coordinate system as shown in Figure 3.4.10*c*, then

$$\mathbf{u} = (0, 0, 1) = \mathbf{k} \quad \text{and} \quad \mathbf{v} = (1, 0, 0) = \mathbf{i}$$

so that

$$\mathbf{u} \times \mathbf{v} = \mathbf{k} \times \mathbf{i} = \mathbf{j} = (0, 1, 0)$$

But it is clear from Figures 3.4.10*b* and 3.4.10*c* that the vector $(0, 0, 1)$ in the xyz-system is the same as the vector $(0, 1, 0)$ in the $x'y'z'$-system. Thus, we obtain the same vector $\mathbf{u} \times \mathbf{v}$ whether we compute with coordinates from the xyz-system or with coordinates from the $x'y'z'$-system. ♦

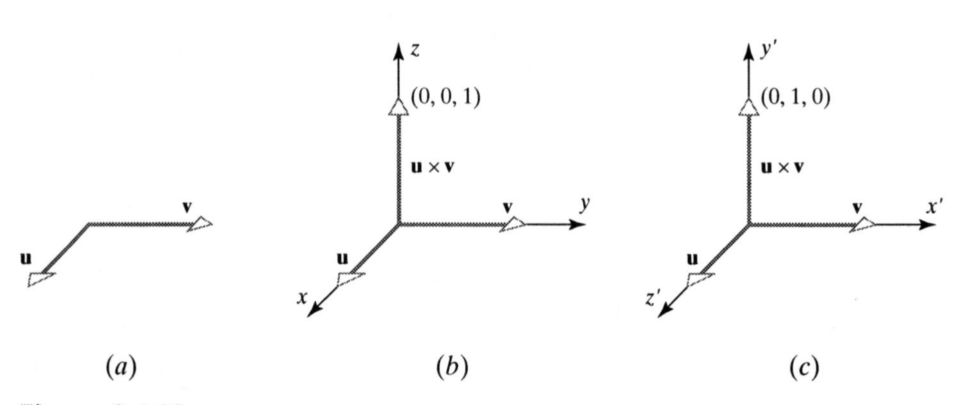

Figure 3.4.10

Exercise Set 3.4

1. Let $\mathbf{u} = (3, 2, -1)$, $\mathbf{v} = (0, 2, -3)$, and $\mathbf{w} = (2, 6, 7)$. Compute

(a) $\mathbf{v} \times \mathbf{w}$ (b) $\mathbf{u} \times (\mathbf{v} \times \mathbf{w})$ (c) $(\mathbf{u} \times \mathbf{v}) \times \mathbf{w}$
(d) $(\mathbf{u} \times \mathbf{v}) \times (\mathbf{v} \times \mathbf{w})$ (e) $\mathbf{u} \times (\mathbf{v} - 2\mathbf{w})$ (f) $(\mathbf{u} \times \mathbf{v}) - 2\mathbf{w}$

2. Find a vector that is orthogonal to both $\mathbf{u}$ and $\mathbf{v}$.

(a) $\mathbf{u} = (-6, 4, 2)$, $\mathbf{v} = (3, 1, 5)$ (b) $\mathbf{u} = (-2, 1, 5)$, $\mathbf{v} = (3, 0, -3)$

3. Find the area of the parallelogram determined by $\mathbf{u}$ and $\mathbf{v}$.

(a) $\mathbf{u} = (1, -1, 2)$, $\mathbf{v} = (0, 3, 1)$ (b) $\mathbf{u} = (2, 3, 0)$, $\mathbf{v} = (-1, 2, -2)$
(c) $\mathbf{u} = (3, -1, 4)$, $\mathbf{v} = (6, -2, 8)$

4. Find the area of the triangle having vertices P, Q, and R.

(a) $P(2, 6, -1)$, $Q(1, 1, 1)$, $R(4, 6, 2)$ (b) $P(1, -1, 2)$, $Q(0, 3, 4)$, $R(6, 1, 8)$

5. Verify parts (*a*), (*b*), and (*c*) of Theorem 3.4.1 for the vectors $\mathbf{u} = (4, 2, 1)$ and $\mathbf{v} = (-3, 2, 7)$.

6. Verify parts (*a*), (*b*), and (*c*) of Theorem 3.4.2 for $\mathbf{u} = (5, -1, 2)$, $\mathbf{v} = (6, 0, -2)$, and $\mathbf{w} = (1, 2, -1)$.

7. Find a vector $\mathbf{v}$ that is orthogonal to the vector $\mathbf{u} = (2, -3, 5)$.

8. Find the scalar triple product $\mathbf{u} \cdot (\mathbf{v} \times \mathbf{w})$.

(a) $\mathbf{u} = (-1, 2, 4)$, $\mathbf{v} = (3, 4, -2)$, $\mathbf{w} = (-1, 2, 5)$
(b) $\mathbf{u} = (3, -1, 6)$, $\mathbf{v} = (2, 4, 3)$, $\mathbf{w} = (5, -1, 2)$

9. Suppose that $\mathbf{u} \cdot (\mathbf{v} \times \mathbf{w}) = 3$. Find

(a) $\mathbf{u} \cdot (\mathbf{w} \times \mathbf{v})$ (b) $(\mathbf{v} \times \mathbf{w}) \cdot \mathbf{u}$ (c) $\mathbf{w} \cdot (\mathbf{u} \times \mathbf{v})$ (d) $\mathbf{v} \cdot (\mathbf{u} \times \mathbf{w})$ (e) $(\mathbf{u} \times \mathbf{w}) \cdot \mathbf{v}$ (f) $\mathbf{v} \cdot (\mathbf{w} \times \mathbf{w})$

10. Find the volume of the parallelepiped with sides $\mathbf{u}$, $\mathbf{v}$, and $\mathbf{w}$.

(a) $\mathbf{u} = (2, -6, 2)$, $\mathbf{v} = (0, 4, -2)$, $\mathbf{w} = (2, 2, -4)$ (b) $\mathbf{u} = (3, 1, 2)$, $\mathbf{v} = (4, 5, 1)$, $\mathbf{w} = (1, 2, 4)$

11. Determine whether $\mathbf{u}$, $\mathbf{v}$, and $\mathbf{w}$ lie in the same plane when positioned so that their initial points coincide.

(a) $\mathbf{u} = (-1, -2, 1)$, $\mathbf{v} = (3, 0, -2)$, $\mathbf{w} = (5, -4, 0)$
(b) $\mathbf{u} = (5, -2, 1)$, $\mathbf{v} = (4, -1, 1)$, $\mathbf{w} = (1, -1, 0)$
(c) $\mathbf{u} = (4, -8, 1)$, $\mathbf{v} = (2, 1, -2)$, $\mathbf{w} = (3, -4, 12)$

12. Find all unit vectors parallel to the yz-plane that are perpendicular to the vector $(3, -1, 2)$.

13. Find all unit vectors in the plane determined by $\mathbf{u} = (3, 0, 1)$ and $\mathbf{v} = (1, -1, 1)$ that are perpendicular to the vector $\mathbf{w} = (1, 2, 0)$.

14. Let $\mathbf{a} = (a_1, a_2, a_3)$, $\mathbf{b} = (b_1, b_2, b_3)$, $\mathbf{c} = (c_1, c_2, c_3)$, and $\mathbf{d} = (d_1, d_2, d_3)$. Show that

$$(\mathbf{a} + \mathbf{d}) \cdot (\mathbf{b} \times \mathbf{c}) = \mathbf{a} \cdot (\mathbf{b} \times \mathbf{c}) + \mathbf{d} \cdot (\mathbf{b} \times \mathbf{c})$$

15. Simplify $(\mathbf{u} + \mathbf{v}) \times (\mathbf{u} - \mathbf{v})$.

16. Use the cross product to find the sine of the angle between the vectors $\mathbf{u} = (2, 3, -6)$ and $\mathbf{v} = (2, 3, 6)$.

17. (a) Find the area of the triangle having vertices $A(1, 0, 1)$, $B(0, 2, 3)$, and $C(2, 1, 0)$.
(b) Use the result of part (a) to find the length of the altitude from vertex C to side AB.

18. Show that if $\mathbf{u}$ is a vector from any point on a line to a point P not on the line, and $\mathbf{v}$ is a vector parallel to the line, then the distance between P and the line is given by $\|\mathbf{u} \times \mathbf{v}\| / \|\mathbf{v}\|$.

19. Use the result of Exercise 18 to find the distance between the point P and the line through the points A and B:

(a) $P(-3, 1, 2)$, $A(1, 1, 0)$, $B(-2, 3, -4)$ (b) $P(4, 3, 0)$, $A(2, 1, -3)$, $B(0, 2, -1)$

20. Prove: If θ is the angle between $\mathbf{u}$ and $\mathbf{v}$ and $\mathbf{u} \cdot \mathbf{v} \neq 0$, then $\tan\theta = \|\mathbf{u} \times \mathbf{v}\| / (\mathbf{u} \cdot \mathbf{v})$.

21. Consider the parallelepiped with sides $\mathbf{u} = (3, 2, 1)$, $\mathbf{v} = (1, 1, 2)$, and $\mathbf{w} = (1, 3, 3)$.

(a) Find the area of the face determined by **u** and **w**.

(b) Find the angle between **u** and the plane containing the face determined by **v** and **w**. [*Note.* The ***angle between a vector and a plane*** is defined to be the complement of the angle θ between the vector and that normal to the plane for which $0 \le \theta \le \pi/2$.]

22. Find a vector **n** perpendicular to the plane determined by the points $A(0, -2, 1)$, $B(1, -1, -2)$, and $C(-1, 1, 0)$. [See the note in Exercise 21.]

23. Let **m** and **n** be vectors whose components in the xyz-system of Figure 3.4.10 are $\mathbf{m} = (0, 0, 1)$ and $\mathbf{n} = (0, 1, 0)$.

(a) Find the components of **m** and **n** in the $x'y'z'$-system of Figure 3.4.10.

(b) Compute $\mathbf{m} \times \mathbf{n}$ using the components in the xyz-system.

(c) Compute $\mathbf{m} \times \mathbf{n}$ using the components in the $xy'z'$-system.

(d) Show that the vectors obtained in (b) and (c) are the same.

24. Prove the following identities.

(a) $(\mathbf{u} + k\mathbf{v}) \times \mathbf{v} = \mathbf{u} \times \mathbf{v}$ (b) $\mathbf{u} \cdot (\mathbf{v} \times \mathbf{z}) = -(\mathbf{u} \times \mathbf{z}) \cdot \mathbf{v}$

25. Let **u**, **v**, and **w** be nonzero vectors in 3-space with the same initial point, but such that no two of them are collinear. Show that

(a) $\mathbf{u} \times (\mathbf{v} \times \mathbf{w})$ lies in the plane determined by **v** and **w**

(b) $(\mathbf{u} \times \mathbf{v}) \times \mathbf{w}$ lies in the plane determined by **u** and **v**

26. Prove part (d) of Theorem 3.4.1. [***Hint.*** First prove the result in the case where $\mathbf{w} = \mathbf{i} = (1, 0, 0)$, then when $\mathbf{w} = \mathbf{j} = (0, 1, 0)$, and then when $\mathbf{w} = \mathbf{k} = (0, 0, 1)$. Finally prove it for an arbitrary vector $\mathbf{w} = (w_1, w_2, w_3)$ by writing $\mathbf{w} = w_1\mathbf{i} + w_2\mathbf{j} + w_3\mathbf{k}$.]

27. Prove part (e) of Theorem 3.4.1. [***Hint.*** Apply part (a) of Theorem 3.4.2 to the result in part (d) of Theorem 3.4.1.]

28. Let $\mathbf{u} = (1, 3, -1)$, $\mathbf{v} = (1, 1, 2)$, and $\mathbf{w} = (3, -1, 2)$. Calculate $\mathbf{u} \times (\mathbf{v} \times \mathbf{w})$ using the technique of Exercise 26; then check your result by calculating directly.

29. Prove: If **a**, **b**, **c**, and **d** lie in the same plane, then $(\mathbf{a} \times \mathbf{b}) \times (\mathbf{c} \times \mathbf{d}) = \mathbf{0}$.

30. It is a theorem of solid geometry that the volume of a tetrahedron is $\frac{1}{3}$(area of base) · (height). Use this result to prove that the volume of a tetrahedron whose sides are the vectors **a**, **b**, and **c** is $\frac{1}{6}|\mathbf{a} \cdot (\mathbf{b} \times \mathbf{c})|$ (see accompanying figure).

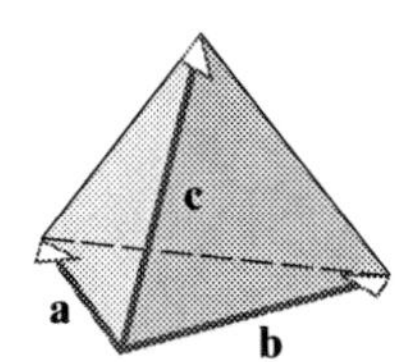

Figure Ex-30

31. Use the result of Exercise 30 to find the volume of the tetrahedron with vertices P, Q, R, S.

(a) $P(-1, 2, 0)$, $Q(2, 1, -3)$, $R(1, 0, 1)$, $S(3, -2, 3)$

(b) $P(0, 0, 0)$, $Q(1, 2, -1)$, $R(3, 4, 0)$, $S(-1, -3, 4)$

32. Prove part (b) of Theorem 3.4.2.

33. Prove parts (c) and (d) of Theorem 3.4.2.

34. Prove parts (e) and (f) of Theorem 3.4.2.

Discussion and Discovery

35. (a) Suppose that **u** and **v** are noncollinear vectors with their initial points at the origin in 3-space. Make a sketch that illustrates how $\mathbf{w} = \mathbf{v} \times (\mathbf{u} \times \mathbf{v})$ is oriented in relation to **u** and **v**.

(b) What can you say about the values of $\mathbf{u} \cdot \mathbf{w}$ and $\mathbf{v} \cdot \mathbf{w}$? Explain your reasoning.

36. If $\mathbf{u} \neq \mathbf{0}$, is it valid to cancel $\mathbf{u}$ from both sides of the equation $\mathbf{u} \times \mathbf{v} = \mathbf{u} \times \mathbf{w}$ and conclude that $\mathbf{v} = \mathbf{w}$? Explain your reasoning.

37. Something is wrong with one of the following expressions. Which one is it and what is wrong?

$$\mathbf{u} \cdot (\mathbf{v} \times \mathbf{w}), \qquad \mathbf{u} \times \mathbf{v} \times \mathbf{w}, \qquad \mathbf{u} \cdot \mathbf{v} \times \mathbf{w}$$

38. What can you say about the vectors $\mathbf{u}$ and $\mathbf{v}$ if $\mathbf{u} \times \mathbf{v} = \mathbf{0}$?

39. Give some examples of algebraic rules that hold for multiplication of real numbers but not for the cross product of vectors.

3.5 LINES AND PLANES IN 3-SPACE

In this section we shall use vectors to derive equations of lines and planes in 3-space. We shall then use these equations to solve some basic geometric problems.

Planes in 3-Space In analytic geometry a line in 2-space can be specified by giving its slope and one of its points. Similarly, one can specify a plane in 3-space by giving its inclination and specifying one of its points. A convenient method for describing the inclination of a plane is to specify a nonzero vector, called a ***normal***, that is perpendicular to the plane.

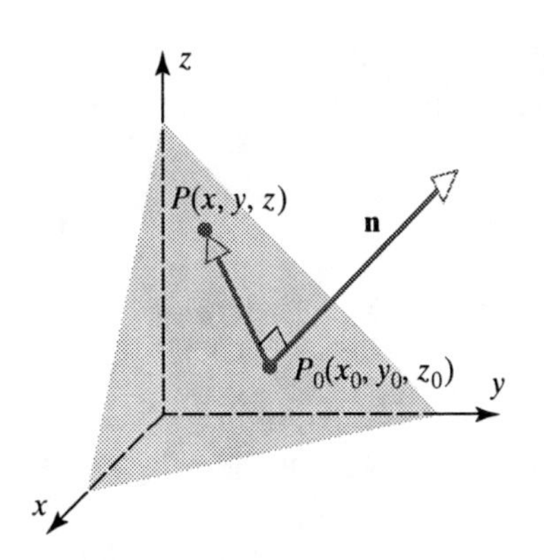

Figure 3.5.1
Plane with normal vector.

Suppose that we want to find the equation of the plane passing through the point $P_0(x_0, y_0, z_0)$ and having the nonzero vector $\mathbf{n} = (a, b, c)$ as a normal. It is evident from Figure 3.5.1 that the plane consists precisely of those points $P(x, y, z)$ for which the vector $\overrightarrow{P_0P}$ is orthogonal to $\mathbf{n}$, that is,

$$\mathbf{n} \cdot \overrightarrow{P_0P} = 0 \tag{1}$$

Since $\overrightarrow{P_0P} = (x - x_0, y - y_0, z - z_0)$, Equation (1) can be written as

$$a(x - x_0) + b(y - y_0) + c(z - z_0) = 0 \tag{2}$$

We call this the ***point-normal*** form of the equation of a plane.

EXAMPLE 1 Finding the Point-Normal Equation of a Plane

Find an equation of the plane passing through the point $(3, -1, 7)$ and perpendicular to the vector $\mathbf{n} = (4, 2, -5)$.

Solution.

From (2) a point-normal form is

$$4(x - 3) + 2(y + 1) - 5(z - 7) = 0$$

◆

By multiplying out and collecting terms, (2) can be rewritten in the form

$$ax + by + cz + d = 0$$

where a, b, c, and d are constants, and a, b, and c are not all zero. For example, the equation in Example 1 can be rewritten as

$$4x + 2y - 5z + 25 = 0$$

As the next theorem shows, planes in 3-space are represented by equations of the form $ax + by + cz + d = 0$.

Theorem 3.5.1

If a, b, c, and d are constants and a, b, and c are not all zero, then the graph of the equation

$$ax + by + cz + d = 0 \tag{3}$$

is a plane having the vector $\mathbf{n} = (a, b, c)$ as a normal.

Equation (3) is a linear equation in x, y, and z; it is called the ***general form*** of the equation of a plane.

Proof. By hypothesis, the coefficients a, b, and c are not all zero. Assume, for the moment, that $a \neq 0$. Then the equation $ax + by + cz + d = 0$ can be rewritten in the form $a(x + (d/a)) + by + cz = 0$. But this is a point-normal form of the plane passing through the point $(-d/a, 0, 0)$ and having $\mathbf{n} = (a, b, c)$ as a normal.

If $a = 0$, then either $b \neq 0$ or $c \neq 0$. A straightforward modification of the above argument will handle these other cases. ■

Just as the solutions of a system of linear equations

$$\begin{aligned} ax + by &= k_1 \\ cx + dy &= k_2 \end{aligned}$$

correspond to points of intersection of the lines $ax + by = k_1$ and $cx + dy = k_2$ in the xy-plane, so the solutions of a system

$$\begin{aligned} ax + by + cz &= k_1 \\ dx + ey + fz &= k_2 \\ gx + hy + iz &= k_3 \end{aligned} \tag{4}$$

correspond to the points of intersection of the three planes $ax + by + cz = k_1$, $dx + ey + fz = k_2$, and $gx + hy + iz = k_3$.

In Figure 3.5.2 we have illustrated the geometric possibilities that occur when (4) has zero, one, or infinitely many solutions.

EXAMPLE 2 Equation of a Plane Through Three Points

Find the equation of the plane passing through the points $P_1(1, 2, -1)$, $P_2(2, 3, 1)$, and $P_3(3, -1, 2)$.

Solution.

Since the three points lie in the plane, their coordinates must satisfy the general equation $ax + by + cz + d = 0$ of the plane. Thus,

$$\begin{aligned} a + 2b - c + d &= 0 \\ 2a + 3b + c + d &= 0 \\ 3a - b + 2c + d &= 0 \end{aligned}$$

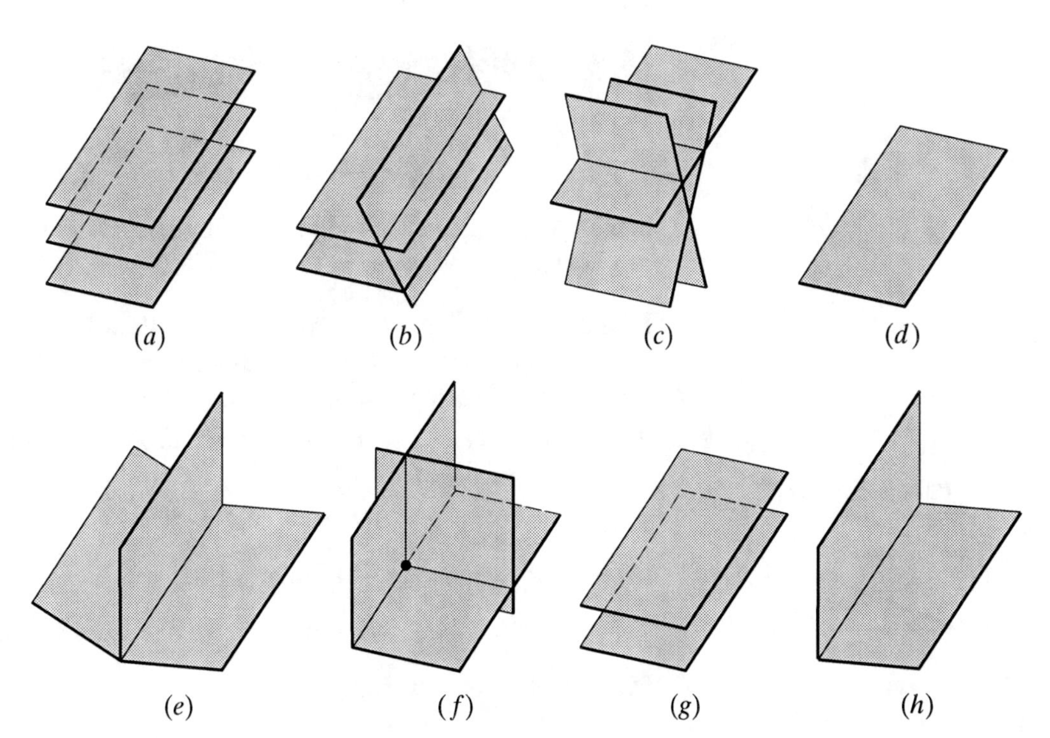

Figure 3.5.2 (*a*) No solutions (3 parallel planes). (*b*) No solutions (2 planes parallel). (*c*) No solutions (3 planes with no common intersection). (*d*) Infinitely many solutions (3 coincident planes). (*e*) Infinitely many solutions (3 planes intersecting in a line). (*f*) One solution (3 planes intersecting at a point). (*g*) No solutions (2 coincident planes parallel to a third plane). (*h*) Infinitely many solutions (2 coincident planes intersecting a third plane).

Solving this system gives $a = -\frac{9}{16}t$, $b = -\frac{1}{16}t$, $c = \frac{5}{16}t$, $d = t$. Letting $t = -16$, for example, yields the desired equation

$$9x + y - 5z - 16 = 0$$

We note that any other choice of t gives a multiple of this equation, so that any value of $t \neq 0$ would also give a valid equation of the plane.

Alternative Solution.

Since the points $P_1(1, 2, -1)$, $P_2(2, 3, 1)$, and $P_3(3, -1, 2)$ lie in the plane, the vectors $\overrightarrow{P_1P_2} = (1, 1, 2)$ and $\overrightarrow{P_1P_3} = (2, -3, 3)$ are parallel to the plane. Therefore, the equation $\overrightarrow{P_1P_2} \times \overrightarrow{P_1P_3} = (9, 1, -5)$ is normal to the plane, since it is perpendicular to both $\overrightarrow{P_1P_2}$ and $\overrightarrow{P_1P_3}$. From this and the fact that P_1 lies in the plane, a point-normal form for the equation of the plane is

$$9(x-1) + (y-2) - 5(z+1) = 0 \quad \text{or} \quad 9x + y - 5z - 16 = 0$$

◆

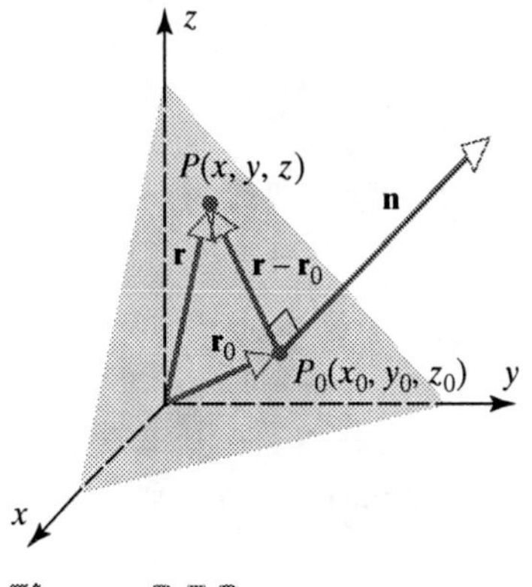

Figure 3.5.3

Vector Form of Equation of a Plane Vector notation provides a useful alternative way of writing the point-normal form of the equation of a plane: Referring to Figure 3.5.3, let $\mathbf{r} = (x, y, z)$ be the vector from the origin to the point $P(x, y, z)$, let $\mathbf{r}_0 = (x_0, y_0, z_0)$ be the vector from the origin to the point $P_0(x_0, y_0, z_0)$, and let $\mathbf{n} = (a, b, c)$ be a vector normal to the plane. Then $\overrightarrow{P_0P} = \mathbf{r} - \mathbf{r}_0$, so Formula (1) can be rewritten as

$$\mathbf{n} \cdot (\mathbf{r} - \mathbf{r}_0) = 0 \tag{5}$$

This is called the ***vector form of the equation of a plane***.

EXAMPLE 3 Vector Equation of a Plane Using [5]

The equation

$$(-1, 2, 5) \cdot (x - 6, y - 3, z + 4) = 0$$

is the vector equation of the plane that passes through the point $(6, 3, -4)$ and is perpendicular to the vector $\mathbf{n} = (-1, 2, 5)$. ♦

Lines in 3-Space We shall now show how to obtain equations for lines in 3-space. Suppose that l is the line in 3-space through the point $P_0(x_0, y_0, z_0)$ and parallel to the nonzero vector $\mathbf{v} = (a, b, c)$. It is clear (Figure 3.5.4) that l consists precisely of those points $P(x, y, z)$ for which the vector $\overrightarrow{P_0P}$ is parallel to $\mathbf{v}$, that is, for which there is a scalar t such that

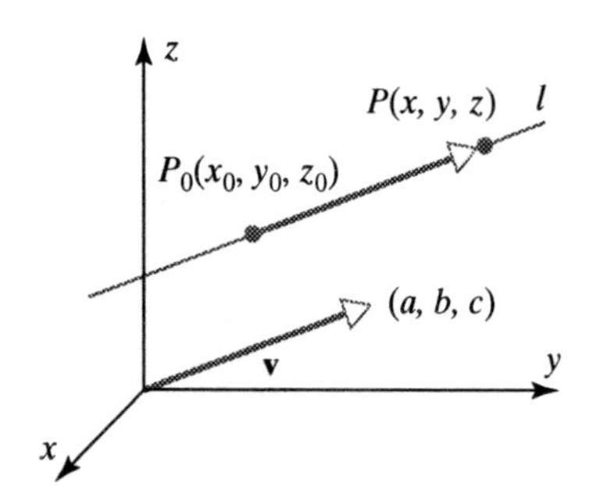

Figure 3.5.4
$\overrightarrow{P_0P}$ is parallel to $\mathbf{v}$.

$$\overrightarrow{P_0P} = t\mathbf{v} \tag{6}$$

In terms of components, (6) can be written as

$$(x - x_0, y - y_0, z - z_0) = (ta, tb, tc)$$

from which it follows that $x - x_0 = ta$, $y - y_0 = tb$, and $z - z_0 = tc$, so that

$$x = x_0 + ta, \qquad y = y_0 + tb, \qquad z = z_0 + tc$$

As the parameter t varies from $-\infty$ to $+\infty$, the point $P(x, y, z)$ traces out the line l. The equations

$$x = x_0 + ta, \quad y = y_0 + tb, \quad z = z_0 + tc \qquad (-\infty < t < +\infty) \tag{7}$$

are called ***parametric equations*** for l.

EXAMPLE 4 Parametric Equations of a Line

The line through the point $(1, 2, -3)$ and parallel to the vector $\mathbf{v} = (4, 5, -7)$ has parametric equations

$$x = 1 + 4t, \quad y = 2 + 5t, \quad z = -3 - 7t \qquad (-\infty < t < +\infty)$$ ♦

EXAMPLE 5 Intersection of a Line and the xy-Plane

(a) Find parametric equations for the line l passing through the points $P_1(2, 4, -1)$ and $P_2(5, 0, 7)$.
(b) Where does the line intersect the xy-plane?

Solution (a). Since the vector $\overrightarrow{P_1P_2} = (3, -4, 8)$ is parallel to l and $P_1(2, 4, -1)$ lies on l, the line l is given by

$$x = 2 + 3t, \quad y = 4 - 4t, \quad z = -1 + 8t \qquad (-\infty < t < +\infty)$$

Solution (b). The line intersects the xy-plane at the point where $z = -1 + 8t = 0$, that is, where $t = \frac{1}{8}$. Substituting this value of t in the parametric equations for l yields as the point of intersection

$$(x, y, z) = \left(\tfrac{19}{8}, \tfrac{7}{2}, 0\right)$$ ♦

EXAMPLE 6 Line of Intersection of Two Planes

Find parametric equations for the line of intersection of the planes

$$3x + 2y - 4z - 6 = 0 \quad \text{and} \quad x - 3y - 2z - 4 = 0$$

Solution.

The line of intersection consists of all points (x, y, z) that satisfy the two equations in the system

$$\begin{aligned} 3x + 2y - 4z &= 6 \\ x - 3y - 2z &= 4 \end{aligned}$$

Solving this system gives $x = \frac{26}{11} + \frac{16}{11}t$, $y = -\frac{6}{11} - \frac{2}{11}t$, $z = t$. Therefore, the line of intersection can be represented by the parametric equations

$$x = \tfrac{26}{11} + \tfrac{16}{11}t, \quad y = -\tfrac{6}{11} - \tfrac{2}{11}t, \quad z = t \qquad (-\infty < t < +\infty)$$

♦

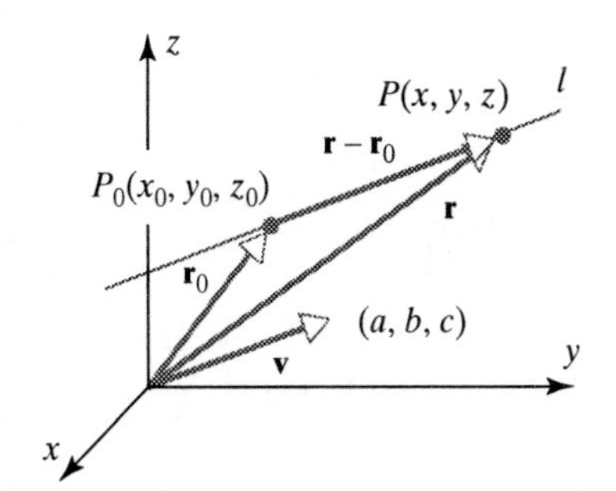

Figure 3.5.5
Vector interpretation of a line in 3-space.

Vector Form of Equation of a Line Vector notation provides a useful alternative way of writing the parametric equations of a line: Referring to Figure 3.5.5, let $\mathbf{r} = (x, y, z)$ be the vector from the origin to the point $P(x, y, z)$, let $\mathbf{r}_0 = (x_0, y_0, z_0)$ be the vector from the origin to the point $P_0(x_0, y_0, z_0)$, and let $\mathbf{v} = (a, b, c)$ be a vector parallel to the line. Then $\overrightarrow{P_0P} = \mathbf{r} - \mathbf{r}_0$, so Formula (6) can be rewritten as

$$\mathbf{r} - \mathbf{r}_0 = t\mathbf{v}$$

Taking into account the range of t-values, this can be rewritten as

$$\mathbf{r} = \mathbf{r}_0 + t\mathbf{v} \qquad (-\infty < t < +\infty) \tag{8}$$

This is called the ***vector form of the equation of a line*** in 3-space.

EXAMPLE 7 A Line Parallel to a Given Vector

The equation

$$(x, y, z) = (-2, 0, 3) + t(4, -7, 1) \qquad (-\infty < t < +\infty)$$

is the vector equation of the line through the point $(-2, 0, 3)$ that is parallel to the vector $\mathbf{v} = (4, -7, 1)$. ♦

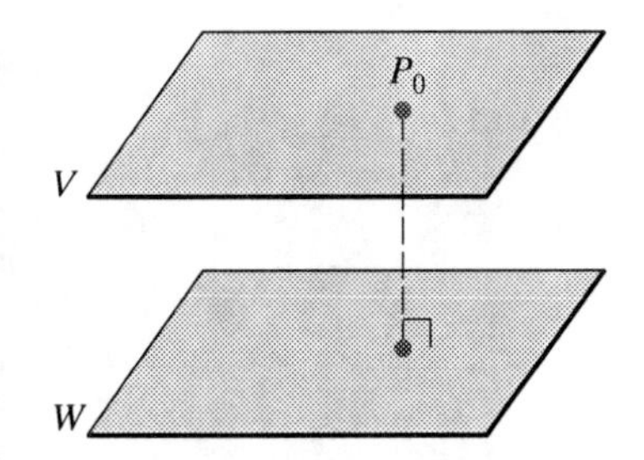

Figure 3.5.6
The distance between the parallel planes V and W is equal to the distance between P_0 and W.

Problems Involving Distance We conclude this section by discussing two basic "distance problems" in 3-space:

Problems:
(a) Find the distance between a point and a plane.
(b) Find the distance between two parallel planes.

The two problems are related. If we can find the distance between a point and a plane, then we can find the distance between parallel planes by computing the distance between either one of the planes and an arbitrary point P_0 in the other (Figure 3.5.6).

Theorem 3.5.2 Distance Between a Point and a Plane

The distance D between a point $P_0(x_0, y_0, z_0)$ and the plane $ax + by + cz + d = 0$ is

$$D = \frac{|ax_0 + by_0 + cz_0 + d|}{\sqrt{a^2 + b^2 + c^2}} \tag{9}$$

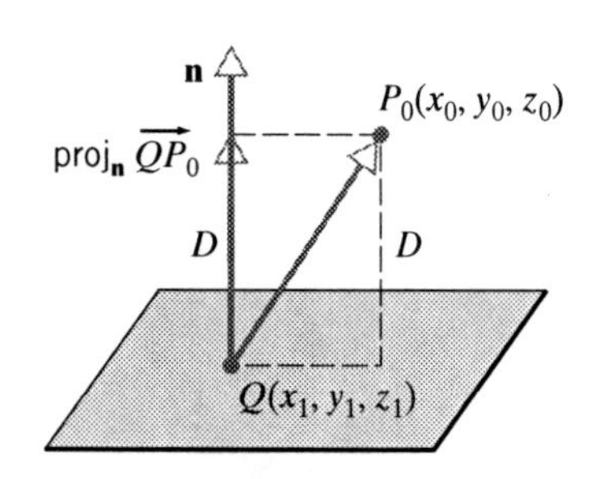

Figure 3.5.7
Distance from P_0 to plane.

Proof. Let $Q(x_1, y_1, z_1)$ be any point in the plane. Position the normal $\mathbf{n} = (a, b, c)$ so that its initial point is at Q. As illustrated in Figure 3.5.7, the distance D is equal to the length of the orthogonal projection of $\overrightarrow{QP_0}$ on $\mathbf{n}$. Thus, from (10) of Section 3.3,

$$D = \|\text{proj}_{\mathbf{n}} \overrightarrow{QP_0}\| = \frac{|\overrightarrow{QP_0} \cdot \mathbf{n}|}{\|\mathbf{n}\|}$$

But

$$\overrightarrow{QP_0} = (x_0 - x_1, y_0 - y_1, z_0 - z_1)$$

$$\overrightarrow{QP_0} \cdot \mathbf{n} = a(x_0 - x_1) + b(y_0 - y_1) + c(z_0 - z_1)$$

$$\|\mathbf{n}\| = \sqrt{a^2 + b^2 + c^2}$$

Thus,

$$D = \frac{|a(x_0 - x_1) + b(y_0 - y_1) + c(z_0 - z_1)|}{\sqrt{a^2 + b^2 + c^2}} \tag{10}$$

Since the point $Q(x_1, y_1, z_1)$ lies in the plane, its coordinates satisfy the equation of the plane; thus

$$ax_1 + by_1 + cz_1 + d = 0$$

or

$$d = -ax_1 - by_1 - cz_1$$

Substituting this expression in (10) yields (9). ■

REMARK. Note the similarity between (9) and the formula for the distance between a point and a line in 2-space [(13) of Section 3.3].

EXAMPLE 8 Distance Between a Point and a Plane

Find the distance D between the point $(1, -4, -3)$ and the plane $2x - 3y + 6z = -1$.

Solution.

To apply (9), we first rewrite the equation of the plane in the form

$$2x - 3y + 6z + 1 = 0$$

Then

$$D = \frac{|2(1) + (-3)(-4) + 6(-3) + 1|}{\sqrt{2^2 + (-3)^2 + 6^2}} = \frac{|-3|}{7} = \frac{3}{7}$$ ◆

Given two planes, either they intersect, in which case we can ask for their line of intersection, as in Example 6, or they are parallel, in which case we can ask for the distance between them. The following example illustrates the latter problem.

EXAMPLE 9 Distance Between Parallel Planes

The planes

$$x + 2y - 2z = 3 \quad \text{and} \quad 2x + 4y - 4z = 7$$

are parallel since their normals, $(1, 2, -2)$ and $(2, 4, -4)$, are parallel vectors. Find the distance between these planes.

Solution.

To find the distance D between the planes, we may select an arbitrary point in one of the planes and compute its distance to the other plane. By setting $y = z = 0$ in the equation $x + 2y - 2z = 3$, we obtain the point $P_0(3, 0, 0)$ in this plane. From (9), the distance between P_0 and the plane $2x + 4y - 4z = 7$ is

$$D = \frac{|2(3) + 4(0) + (-4)(0) - 7|}{\sqrt{2^2 + 4^2 + (-4)^2}} = \frac{1}{6}$$ ◆

Exercise Set 3.5

1. Find a point-normal form of the equation of the plane passing through P and having $\mathbf{n}$ as a normal.

(a) $P(-1, 3, -2)$; $\mathbf{n} = (-2, 1, -1)$ (b) $P(1, 1, 4)$; $\mathbf{n} = (1, 9, 8)$
(c) $P(2, 0, 0)$; $\mathbf{n} = (0, 0, 2)$ (d) $P(0, 0, 0)$; $\mathbf{n} = (1, 2, 3)$

2. Write the equations of the planes in Exercise 1 in general form.

3. Find a point-normal form.

(a) $-3x + 7y + 2z = 10$ (b) $x - 4z = 0$

4. Find an equation for the plane passing through the given points.

(a) $P(-4, -1, -1)$, $Q(-2, 0, 1)$, $R(-1, -2, -3)$ (b) $P(5, 4, 3)$, $Q(4, 3, 1)$, $R(1, 5, 4)$

5. Determine whether the planes are parallel.

(a) $4x - y + 2z = 5$ and $7x - 3y + 4z = 8$
(b) $x - 4y - 3z - 2 = 0$ and $3x - 12y - 9z - 7 = 0$
(c) $2y = 8x - 4z + 5$ and $x = \frac{1}{2}z + \frac{1}{4}y$

6. Determine whether the line and plane are parallel.

(a) $x = -5 - 4t$, $y = 1 - t$, $z = 3 + 2t$; $x + 2y + 3z - 9 = 0$
(b) $x = 3t$, $y = 1 + 2t$, $z = 2 - t$; $4x - y + 2z = 1$

7. Determine whether the planes are perpendicular.

(a) $3x - y + z - 4 = 0$, $x + 2z = -1$ (b) $x - 2y + 3z = 4$, $-2x + 5y + 4z = -1$

8. Determine whether the line and plane are perpendicular.

(a) $x = -2 - 4t$, $y = 3 - 2t$, $z = 1 + 2t$; $2x + y - z = 5$
(b) $x = 2 + t$, $y = 1 - t$, $z = 5 + 3t$; $6x + 6y - 7 = 0$

9. Find parametric equations for the line passing through P and parallel to $\mathbf{n}$.

(a) $P(3, -1, 2)$; $\mathbf{n} = (2, 1, 3)$ (b) $P(-2, 3, -3)$; $\mathbf{n} = (6, -6, -2)$
(c) $P(2, 2, 6)$; $\mathbf{n} = (0, 1, 0)$ (d) $P(0, 0, 0)$; $\mathbf{n} = (1, -2, 3)$

10. Find parametric equations for the line passing through the given points.

(a) $(5, -2, 4)$, $(7, 2, -4)$ (b) $(0, 0, 0)$, $(2, -1, -3)$

11. Find parametric equations for the line of intersection of the given planes.

(a) $7x - 2y + 3z = -2$ and $-3x + y + 2z + 5 = 0$ (b) $2x + 3y - 5z = 0$ and $y = 0$

12. Find the vector form of the equation of the plane that passes through P_0 and has normal $\mathbf{n}$.

(a) $P_0(-1, 2, 4)$; $\mathbf{n} = (-2, 4, 1)$ (b) $P_0(2, 0, -5)$; $\mathbf{n} = (-1, 4, 3)$

(c) $P_0(5, -2, 1)$; $\mathbf{n} = (-1, 0, 0)$ (d) $P_0(0, 0, 0)$; $\mathbf{n} = (a, b, c)$

13. Determine whether the planes are parallel.

(a) $(-1, 2, 4) \cdot (x - 5, y + 3, z - 7) = 0$; $(2, -4, -8) \cdot (x + 3, y + 5, z - 9) = 0$

(b) $(3, 0, -1) \cdot (x + 1, y - 2, z - 3) = 0$; $(-1, 0, 3) \cdot (x + 1, y - z, z - 3) = 0$

14. Determine whether the planes are perpendicular.

(a) $(-2, 1, 4) \cdot (x - 1, y, z + 3) = 0$; $(1, -2, 1) \cdot (x + 3, y - 5, z) = 0$

(b) $(3, 0, -2) \cdot (x + 4, y - 7, z + 1) = 0$; $(1, 1, 1) \cdot (x, y, z) = 0$

15. Find the vector form of the equation of the line through P_0 and parallel to $\mathbf{v}$.

(a) $P_0(-1, 2, 3)$; $\mathbf{v} = (7, -1, 5)$ (b) $P_0(2, 0, -1)$; $\mathbf{v} = (1, 1, 1)$

(c) $P_0(2, -4, 1)$; $\mathbf{v} = (0, 0, -2)$ (d) $P_0(0, 0, 0)$; $\mathbf{v} = (a, b, c)$

16. Show that the line

$$x = 0, \quad y = t, \quad z = t \qquad (-\infty < t < +\infty)$$

(a) lies in the plane $6x + 4y - 4z = 0$

(b) is parallel to and below the plane $5x - 3y + 3z = 1$

(c) is parallel to and above the plane $6x + 2y - 2z = 3$

17. Find an equation for the plane through $(-2, 1, 7)$ that is perpendicular to the line $x - 4 = 2t$, $y + 2 = 3t$, $z = -5t$.

18. Find an equation of

(a) the xy-plane (b) the xz-plane (c) the yz-plane

19. Find an equation of the plane that contains the point (x_0, y_0, z_0) and is

(a) parallel to the xy-plane (b) parallel to the yz-plane (c) parallel to the xz-plane

20. Find an equation for the plane that passes through the origin and is parallel to the plane $7x + 4y - 2z + 3 = 0$.

21. Find an equation for the plane that passes through the point $(3, -6, 7)$ and is parallel to the plane $5x - 2y + z - 5 = 0$.

22. Find the point of intersection of the line

$$x - 9 = -5t, \quad y + 1 = -t, \quad z - 3 = t \qquad (-\infty < t < +\infty)$$

and the plane $2x - 3y + 4z + 7 = 0$.

23. Find an equation for the plane that contains the line $x = -1 + 3t$, $y = 5 + 2t$, $z = 2 - t$ and is perpendicular to the plane $2x - 4y + 2z = 9$.

24. Find an equation for the plane that passes through $(2, 4, -1)$ and contains the line of intersection of the planes $x - y - 4z = 2$ and $-2x + y + 2z = 3$.

25. Show that the points $(-1, -2, -3)$, $(-2, 0, 1)$, $(-4, -1, -1)$, and $(2, 0, 1)$ lie in the same plane.

26. Find parametric equations for the line through $(-2, 5, 0)$ that is parallel to the planes $2x + y - 4z = 0$ and $-x + 2y + 3z + 1 = 0$.

27. Find an equation for the plane through $(-2, 1, 5)$ that is perpendicular to the planes $4x - 2y + 2z = -1$ and $3x + 3y - 6z = 5$.

28. Find an equation for the plane through $(2, -1, 4)$ that is perpendicular to the line of intersection of the planes $4x + 2y + 2z = -1$ and $3x + 6y + 3z = 7$.

29. Find an equation for the plane that is perpendicular to the plane $8x - 2y + 6z = 1$ and passes through the points $P_1(-1, 2, 5)$ and $P_2(2, 1, 4)$.

30. Show that the lines

$$x = 3 - 2t, \quad y = 4 + t, \quad z = 1 - t \qquad (-\infty < t < +\infty)$$

and

$$x = 5 + 2t, \quad y = 1 - t, \quad z = 7 + t \qquad (-\infty < t < +\infty)$$

are parallel, and find an equation for the plane they determine.

31. Find an equation for the plane that contains the point $(1, -1, 2)$ and the line $x = t$, $y = t + 1$, $z = -3 + 2t$.

32. Find an equation for the plane that contains the line $x = 1 + t$, $y = 3t$, $z = 2t$ and is parallel to the line of intersection of the planes $-x + 2y + z = 0$ and $x + z + 1 = 0$.

33. Find an equation for the plane, each of whose points is equidistant from $(-1, -4, -2)$ and $(0, -2, 2)$.

34. Show that the line

$$x - 5 = -t, \quad y + 3 = 2t, \quad z + 1 = -5t \qquad (-\infty < t < +\infty)$$

is parallel to the plane $-3x + y + z - 9 = 0$.

35. Show that the lines

$$x - 3 = 4t, \quad y - 4 = t, \quad z - 1 = 0 \qquad (-\infty < t < +\infty)$$

and

$$x + 1 = 12t, \quad y - 7 = 6t, \quad z - 5 = 3t \qquad (-\infty < t < +\infty)$$

intersect, and find the point of intersection.

36. Find an equation for the plane containing the lines in Exercise 35.

37. Find parametric equations for the line of intersection of the planes

(a) $-3x + 2y + z = -5$ and $7x + 3y - 2z = -2$
(b) $5x - 7y + 2z = 0$ and $y = 0$

38. Show that the plane whose intercepts with the coordinate axes are $x = a$, $y = b$, and $z = c$ has equation

$$\frac{x}{a} + \frac{y}{b} + \frac{z}{c} = 1$$

provided a, b, and c are nonzero.

39. Find the distance between the point and the plane.

(a) $(3, 1, -2)$; $x + 2y - 2z = 4$
(b) $(-1, 2, 1)$; $2x + 3y - 4z = 1$
(c) $(0, 3, -2)$; $x - y - z = 3$

40. Find the distance between the given parallel planes.

(a) $3x - 4y + z = 1$ and $6x - 8y + 2z = 3$
(b) $-4x + y - 3z = 0$ and $8x - 2y + 6z = 0$
(c) $2x - y + z = 1$ and $2x - y + z = -1$

41. Show that if a, b, and c are all nonzero, then the line

$$x = x_0 + at, \quad y = y_0 + bt, \quad z = z_0 + ct \qquad (-\infty < t < +\infty)$$

consists of all points (x, y, z) that satisfy

$$\frac{x - x_0}{a} = \frac{y - y_0}{b} = \frac{z - z_0}{c}$$

These are called ***symmetric equations*** for the line.

42. Find symmetric equations for the lines in parts (a) and (b) of Exercise 9. [***Note.*** See Exercise 41 for terminology.]

43. In each part find equations for two planes whose intersection is the given line.

(a) $x = 7 - 4t, \quad y = -5 - 2t, \quad z = 5 + t \qquad (-\infty < t < +\infty)$
(b) $x = 4t, \quad y = 2t, \quad z = 7t \qquad (-\infty < t < +\infty)$
[***Hint.*** Each equality in the symmetric equations of a line represents a plane containing the line. See Exercise 41 for terminology.]

10

Complex Vector Spaces

Chapter Contents

INTRODUCTION: Up to now we have considered only vector spaces for which the scalars are real numbers. However, for many important applications of vectors it is desirable to allow the scalars to be complex numbers. One advantage of allowing complex scalars is that all matrices with scalar entries have eigenvalues, which is not true if only real scalars are allowed.

In the first three sections of this chapter we will review some of the basic properties of complex numbers, and in subsequent sections we will discuss vector spaces in which scalars can be complex numbers.

10.1 COMPLEX NUMBERS

In this section we shall review the definition of a complex number and discuss the operations of addition, subtraction, and multiplication of such numbers. We will also consider matrices with complex entries and explain how addition and subtraction of complex numbers can be viewed as operations on vectors.

Complex Numbers Since $x^2 \geq 0$ for every real number x, the equation $x^2 = -1$ has no real solutions. To deal with this problem, mathematicians of the eighteenth century introduced the "imaginary" number,

$$i = \sqrt{-1}$$

which they assumed had the property

$$i^2 = (\sqrt{-1})^2 = -1$$

but which otherwise could be treated like an ordinary number. Expressions of the form

$$a + bi \tag{1}$$

where a and b are real numbers were called "complex numbers," and these were manipulated according to the standard rules of arithmetic with the added property that $i^2 = -1$.

By the beginning of the nineteenth century it was recognized that a complex number (1) could be regarded as an alternative symbol for the ordered pair

$$(a, b)$$

of real numbers, and that operations of addition, subtraction, multiplication, and division could be defined on these ordered pairs so that the familiar laws of arithmetic hold and $i^2 = -1$. This is the approach we will follow.

Definition

A ***complex number*** is an ordered pair of real numbers, denoted either by (a, b) or $a + bi$, where $i^2 = -1$.

EXAMPLE 1 Two Notations for a Complex Number

Some examples of complex numbers in both notations are as follows:

Ordered Pair	Equivalent Notation
(3, 4)	$3 + 4i$
(−1, 2)	$-1 + 2i$
(0, 1)	$0 + i$
(2, 0)	$2 + 0i$
(4, −2)	$4 + (-2)i$

For simplicity, the last three complex numbers would usually be abbreviated as

$$0 + i = i, \quad 2 + 0i = 2, \quad 4 + (-2)i = 4 - 2i$$

♦

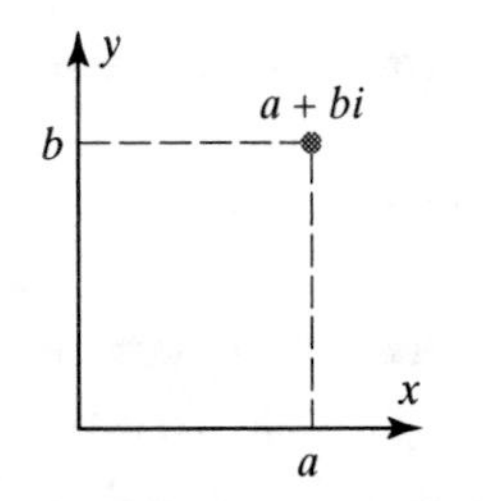

(a) Complex number as a point

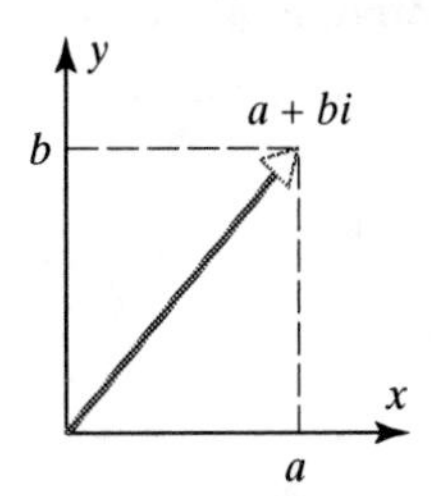

(b) Complex number as a vector

Figure 10.1.1

Geometrically, a complex number can be viewed either as a point or a vector in the xy-plane (Figure 10.1.1).

EXAMPLE 2 Complex Numbers as Points and as Vectors

Some complex numbers are shown as points in Figure 10.1.2*a* and as vectors in Figure 10.1.2*b*.

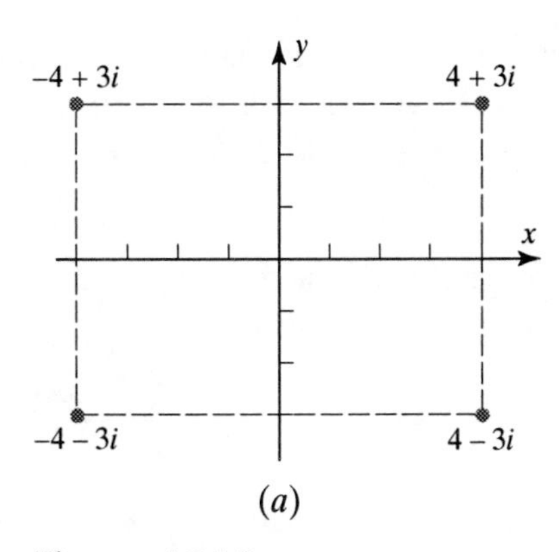

(a)

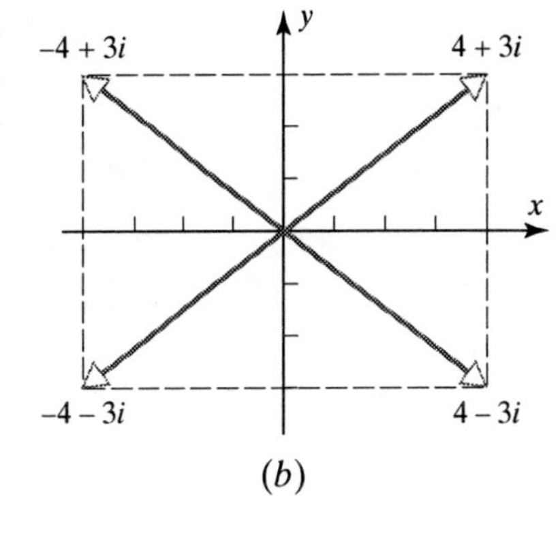

(b)

Figure 10.1.2 ◆

The Complex Plane Sometimes it is convenient to use a single letter, such as z, to denote a complex number. Thus, we might write

$$z = a + bi$$

The real number a is called the ***real part of** z* and the real number b the ***imaginary part of** z*. These numbers are denoted by $\text{Re}(z)$ and $\text{Im}(z)$, respectively. Thus,

$$\text{Re}(4 - 3i) = 4 \quad \text{and} \quad \text{Im}(4 - 3i) = -3$$

When complex numbers are represented geometrically in an xy-coordinate system, the x-axis is called the ***real axis***, the y-axis the ***imaginary axis***, and the plane is called the ***complex plane*** (Figure 10.1.3).

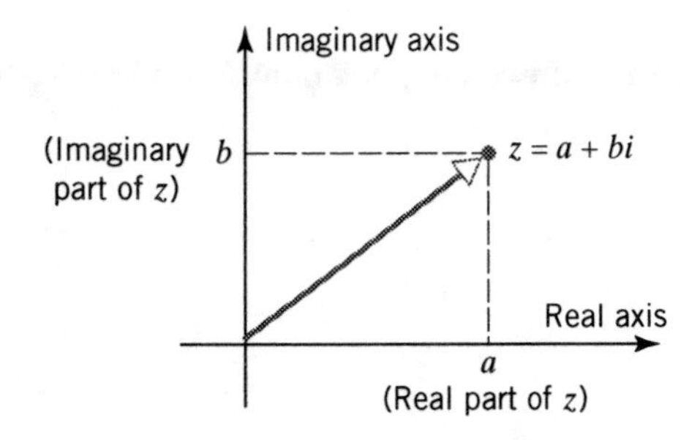

Figure 10.1.3 Complex plane.

Operations on Complex Numbers Just as two vectors in R^2 are defined to be equal if they have the same components, so we define two complex numbers to be equal if their real parts are equal and their imaginary parts are equal:

Definition

Two complex numbers, $a + bi$ and $c + di$, are defined to be ***equal***, written

$$a + bi = c + di$$

if $a = c$ and $b = d$.

If $b = 0$, then the complex number $a + bi$ reduces to $a + 0i$, which we write simply as a. Thus, for any real number a,

$$a = a + 0i$$

so that the real numbers can be regarded as complex numbers with an imaginary part of zero. Geometrically, the real numbers correspond to points on the real axis. If we have $a = 0$, then $a + bi$ reduces to $0 + bi$, which we usually write as bi. These complex numbers, which correspond to points on the imaginary axis, are called ***pure imaginary numbers***.

Just as vectors in R^2 are added by adding corresponding components, so complex numbers are added by adding their real parts and adding their imaginary parts:

$$(a + bi) + (c + di) = (a + c) + (b + d)i \tag{2}$$

The operations of subtraction and multiplication by a *real* number are also similar to the corresponding vector operations in R^2:

$$(a + bi) - (c + di) = (a - c) + (b - d)i \tag{3}$$

$$k(a + bi) = (ka) + (kb)i, \qquad k \text{ real} \tag{4}$$

Because the operations of addition, subtraction, and multiplication of a complex number by a real number parallel the corresponding operations for vectors in R^2, the familiar geometric interpretations of these operations hold for complex numbers (see Figure 10.1.4).

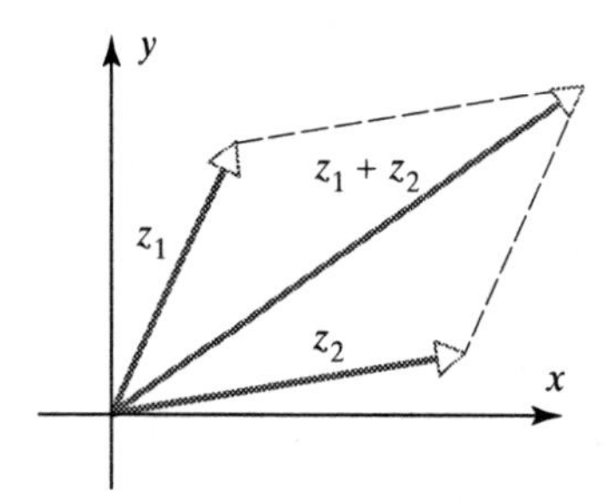

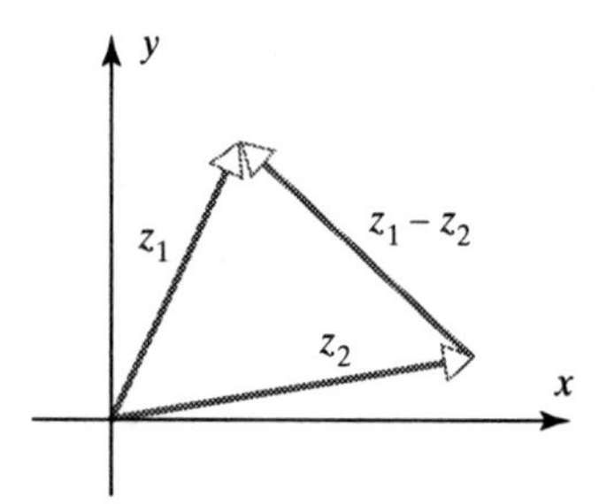

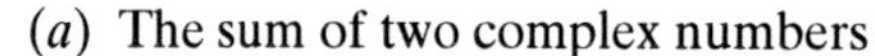

(a) The sum of two complex numbers

(b) The difference of two complex numbers

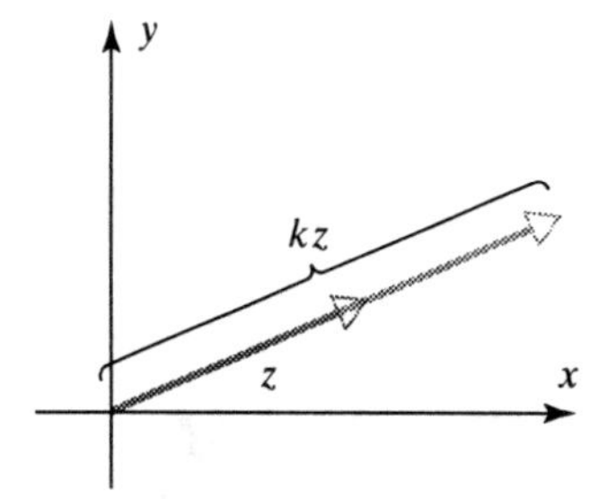

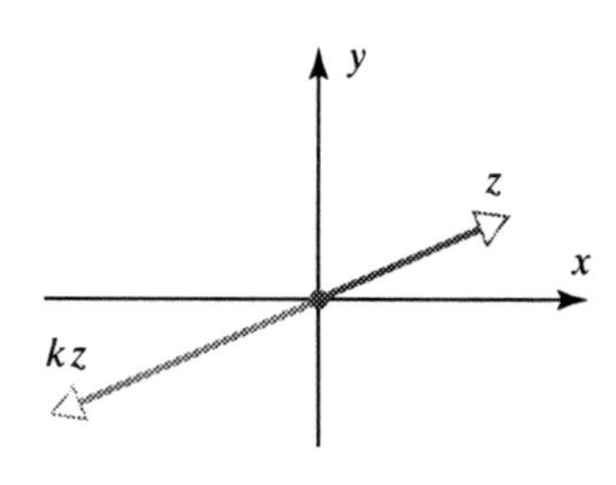

(c) The product of a complex number z and a positive real number k

(d) The product of a complex number z and a negative real number k

Figure 10.1.4

It follows from (4) that $(-1)z + z = 0$ (verify), so we denote $(-1)z$ as $-z$ and call it the ***negative of*** z.

EXAMPLE 3 Adding, Subtracting, and Multiplying by Real Numbers

If $z_1 = 4 - 5i$ and $z_2 = -1 + 6i$, find $z_1 + z_2$, $z_1 - z_2$, $3z_1$, and $-z_2$.

Solution.

$$\begin{aligned} z_1 + z_2 &= (4 - 5i) + (-1 + 6i) = (4 - 1) + (-5 + 6)i = 3 + i \\ z_1 - z_2 &= (4 - 5i) - (-1 + 6i) = (4 + 1) + (-5 - 6)i = 5 - 11i \\ 3z_1 &= 3(4 - 5i) = 12 - 15i \\ -z_2 &= (-1)z_2 = (-1)(-1 + 6i) = 1 - 6i \end{aligned}$$ ♦

So far, there has been a parallel between complex numbers and vectors in R^2. However, we now define multiplication of complex numbers, an operation with no vector analog in R^2. To motivate the definition, we expand the product

$$(a + bi)(c + di)$$

following the usual rules of algebra, but treating i^2 as -1. This yields

$$\begin{aligned} (a + bi)(c + di) &= ac + bdi^2 + adi + bci \\ &= (ac - bd) + (ad + bc)i \end{aligned}$$

which suggests the following *definition*:

$$(a + bi)(c + di) = (ac - bd) + (ad + bc)i \tag{5}$$

EXAMPLE 4 Multiplying Complex Numbers

$$\begin{aligned} (3 + 2i)(4 + 5i) &= (3 \cdot 4 - 2 \cdot 5) + (3 \cdot 5 + 2 \cdot 4)i \\ &= 2 + 23i \\ (4 - i)(2 - 3i) &= [4 \cdot 2 - (-1)(-3)] + [(4)(-3) + (-1)(2)]i \\ &= 5 - 14i \\ i^2 = (0 + i)(0 + i) &= (0 \cdot 0 - 1 \cdot 1) + (0 \cdot 1 + 1 \cdot 0)i = -1 \end{aligned}$$ ♦

We leave it as an exercise to verify the following rules of complex arithmetic:

$$\begin{aligned} z_1 + z_2 &= z_2 + z_1 \\ z_1 z_2 &= z_2 z_1 \\ z_1 + (z_2 + z_3) &= (z_1 + z_2) + z_3 \\ z_1(z_2 z_3) &= (z_1 z_2) z_3 \\ z_1(z_2 + z_3) &= z_1 z_2 + z_1 z_3 \\ 0 + z &= z \\ z + (-z) &= 0 \\ 1 \cdot z &= z \end{aligned}$$

These rules make it possible to multiply complex numbers without using Formula (5) directly. Following the procedure used to motivate this formula, we can simply multiply each term of $a + bi$ by each term of $c + di$, set $i^2 = -1$, and simplify.

EXAMPLE 5 Multiplication of Complex Numbers

$$(3+2i)(4+i) = 12+3i+8i+2i^2 = 12+11i-2 = 10+11i$$
$$(5-\tfrac{1}{2}i)(2+3i) = 10+15i-i-\tfrac{3}{2}i^2 = 10+14i+\tfrac{3}{2} = \tfrac{23}{2}+14i$$
$$i(1+i)(1-2i) = i(1-2i+i-2i^2) = i(3-i) = 3i-i^2 = 1+3i$$ ♦

REMARK. Unlike the real numbers, there is no size ordering for the complex numbers. Thus, the order symbols $<$, $\leq$, $>$, and $\geq$ are not used with complex numbers.

Now that we have defined addition, subtraction, and multiplication of complex numbers, it is possible to add, subtract, and multiply matrices with complex entries and multiply a matrix by a complex number. Without going into detail, we note that the matrix operations and terminology discussed in Chapter 1 carry over without change to matrices with complex entries.

EXAMPLE 6 Matrices with Complex Entries

If

$$A = \begin{bmatrix} 1 & -i \\ 1+i & 4-i \end{bmatrix} \quad \text{and} \quad B = \begin{bmatrix} i & 1-i \\ 2-3i & 4 \end{bmatrix}$$

then

$$A+B = \begin{bmatrix} 1+i & 1-2i \\ 3-2i & 8-i \end{bmatrix}, \qquad A-B = \begin{bmatrix} 1-i & -1 \\ -1+4i & -i \end{bmatrix}$$

$$iA = \begin{bmatrix} i & -i^2 \\ i+i^2 & 4i-i^2 \end{bmatrix} = \begin{bmatrix} i & 1 \\ -1+i & 1+4i \end{bmatrix}$$

$$AB = \begin{bmatrix} 1 & -i \\ 1+i & 4-i \end{bmatrix}\begin{bmatrix} i & 1-i \\ 2-3i & 4 \end{bmatrix}$$
$$= \begin{bmatrix} 1\cdot i+(-i)\cdot(2-3i) & 1\cdot(1-i)+(-i)\cdot 4 \\ (1+i)\cdot i+(4-i)\cdot(2-3i) & (1+i)\cdot(1-i)+(4-i)\cdot 4 \end{bmatrix}$$
$$= \begin{bmatrix} -3-i & 1-5i \\ 4-13i & 18-4i \end{bmatrix}$$ ♦

Exercise Set 10.1

1. In each part plot the point and sketch the vector that corresponds to the given complex number.

   (a) $2+3i$ (b) -4 (c) $-3-2i$ (d) $-5i$

2. Express each complex number in Exercise 1 as an ordered pair of real numbers.

3. In each part use the given information to find the real numbers x and y.

   (a) $x-iy=-2+3i$ (b) $(x+y)+(x-y)i=3+i$

4. Given that $z_1=1-2i$ and $z_2=4+5i$, find

   (a) z_1+z_2 (b) z_1-z_2 (c) $4z_1$ (d) $-z_2$ (e) $3z_1+4z_2$ (f) $\frac{1}{2}z_1-\frac{3}{2}z_2$

5. In each part solve for z.

   (a) $z+(1-i)=3+2i$ (b) $-5z=5+10i$ (c) $(i-z)+(2z-3i)=-2+7i$

6. In each part sketch the vectors z_1, z_2, $z_1 + z_2$, and $z_1 - z_2$.

(a) $z_1 = 3 + i,\ z_2 = 1 + 4i$ (b) $z_1 = -2 + 2i,\ z_2 = 4 + 5i$

7. In each part sketch the vectors z and kz.

(a) $z = 1 + i,\ k = 2$ (b) $z = -3 - 4i,\ k = -2$ (c) $z = 4 + 6i,\ k = \frac{1}{2}$

8. In each part find real numbers k_1 and k_2 that satisfy the equation.

(a) $k_1 i + k_2(1 + i) = 3 - 2i$ (b) $k_1(2 + 3i) + k_2(1 - 4i) = 7 + 5i$

9. In each part find $z_1 z_2$, z_1^2, and z_2^2.

(a) $z_1 = 3i,\ z_2 = 1 - i$ (b) $z_1 = 4 + 6i,\ z_2 = 2 - 3i$ (c) $z_1 = \frac{1}{3}(2 + 4i),\ z_2 = \frac{1}{2}(1 - 5i)$

10. Given that $z_1 = 2 - 5i$ and $z_2 = -1 - i$, find

(a) $z_1 - z_1 z_2$ (b) $(z_1 + 3z_2)^2$ (c) $[z_1 + (1 + z_2)]^2$ (d) $iz_2 - z_1^2$

In Exercises 11–18 perform the calculations and express the result in the form $a + bi$.

11. $(1 + 2i)(4 - 6i)^2$

12. $(2 - i)(3 + i)(4 - 2i)$

13. $(1 - 3i)^3$

14. $i(1 + 7i) - 3i(4 + 2i)$

15. $[(2 + i)(\frac{1}{2} + \frac{3}{4}i)]^2$

16. $(\sqrt{2} + i) - i\sqrt{2}(1 + \sqrt{2}i)$

17. $(1 + i + i^2 + i^3)^{100}$

18. $(3 - 2i)^2 - (3 + 2i)^2$

19. Let

$$A = \begin{bmatrix} 1 & i \\ -i & 3 \end{bmatrix}, \quad B = \begin{bmatrix} 2 & 2 + i \\ 3 - i & 4 \end{bmatrix}$$

Find

(a) $A + 3iB$ (b) BA (c) AB (d) $B^2 - A^2$

20. Let

$$A = \begin{bmatrix} 3 + 2i & 0 \\ -i & 2 \\ 1 + i & 1 - i \end{bmatrix}, \quad B = \begin{bmatrix} -i & 2 \\ 0 & i \end{bmatrix}, \quad C = \begin{bmatrix} -1 - i & 0 & -i \\ 3 & 2i & -5 \end{bmatrix}$$

Find

(a) $A(BC)$ (b) $(BC)A$ (c) $(CA)B^2$ (d) $(1 + i)(AB) + (3 - 4i)A$

21. Show that

(a) $\text{Im}(iz) = \text{Re}(z)$ (b) $\text{Re}(iz) = -\text{Im}(z)$

22. In each part solve the equation by the quadratic formula and check your results by substituting the solutions into the given equation.

(a) $z^2 + 2z + 2 = 0$ (b) $z^2 - z + 1 = 0$

23. (a) Show that if n is a positive integer, then the only possible values for i^n are $1, -1, i$, and $-i$.

(b) Find i^{2509}. [***Hint.*** The value of i^n can be determined from the remainder when n is divided by 4.]

24. Prove: If $z_1 z_2 = 0$, then $z_1 = 0$ or $z_2 = 0$.

25. Use the result of Exercise 24 to prove: If $zz_1 = zz_2$ and $z \neq 0$, then $z_1 = z_2$.

26. Prove that for all complex numbers z_1, z_2, and z_3

(a) $z_1 + z_2 = z_2 + z_1$ (b) $z_1 + (z_2 + z_3) = (z_1 + z_2) + z_3$

27. Prove that for all complex numbers z_1, z_2, and z_3

(a) $z_1 z_2 = z_2 z_1$ (b) $z_1(z_2 z_3) = (z_1 z_2) z_3$

28. Prove that $z_1(z_2 + z_3) = z_1 z_2 + z_1 z_3$ for all complex numbers z_1, z_2, and z_3.

29. In quantum mechanics the ***Dirac matrices*** are

$$\beta = \begin{bmatrix} 1 & 0 & 0 & 0 \\ 0 & 1 & 0 & 0 \\ 0 & 0 & -1 & 0 \\ 0 & 0 & 0 & -1 \end{bmatrix}, \quad \alpha_x = \begin{bmatrix} 0 & 0 & 0 & 1 \\ 0 & 0 & 1 & 0 \\ 0 & 1 & 0 & 0 \\ 1 & 0 & 0 & 0 \end{bmatrix},$$

$$\alpha_y = \begin{bmatrix} 0 & 0 & 0 & -i \\ 0 & 0 & i & 0 \\ 0 & -i & 0 & 0 \\ i & 0 & 0 & 0 \end{bmatrix}, \quad \alpha_z = \begin{bmatrix} 0 & 0 & 1 & 0 \\ 0 & 0 & 0 & -1 \\ 1 & 0 & 0 & 0 \\ 0 & -1 & 0 & 0 \end{bmatrix}$$

(a) Prove that $\beta^2 = \alpha_x^2 = \alpha_y^2 = \alpha_z^2 = I$.

(b) Two matrices A and B are called ***anticommutative*** if $AB = -BA$. Prove that any two distinct Dirac matrices are anticommutative.

10.2 DIVISION OF COMPLEX NUMBERS

In the last section we defined multiplication of complex numbers. In this section we shall define division of complex numbers as the inverse of multiplication.

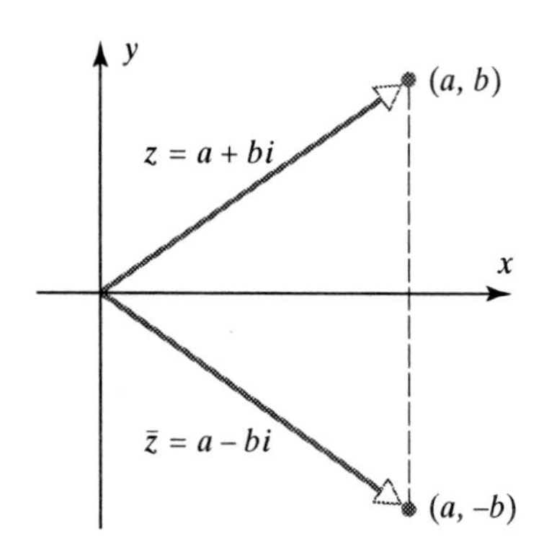

Figure 10.2.1
The conjugate of a complex number.

We begin with some preliminary ideas.

Complex Conjugates If $z = a + bi$ is any complex number, then the ***complex conjugate*** of z (also called the ***conjugate*** of z) is denoted by the symbol $\bar{z}$ (read "z bar") and is defined by

$$\bar{z} = a - bi$$

In words, $\bar{z}$ is obtained by reversing the sign of the imaginary part of z. Geometrically, $\bar{z}$ is the reflection of z about the real axis (Figure 10.2.1).

EXAMPLE 1 Examples of Conjugates

$$\begin{array}{ll} z = 3 + 2i & \bar{z} = 3 - 2i \\ z = -4 - 2i & \bar{z} = -4 + 2i \\ z = i & \bar{z} = -i \\ z = 4 & z = 4 \end{array}$$

◆

REMARK. The last line in Example 1 illustrates the fact that a real number is the same as its conjugate. More precisely, it can be shown (Exercise 22) that $z = \bar{z}$ if and only if z is a real number.

If a complex number z is viewed as a vector in R^2, then the norm or length of the vector is called the modulus (or *absolute value*) of z. More precisely:

Definition

The ***modulus*** of a complex number $z = a + bi$, denoted by $|z|$, is defined by

$$|z| = \sqrt{a^2 + b^2} \tag{1}$$

Paul Adrien Maurice Dirac (1902–1984) was a British theoretical physicist who devised a new form of quantum mechanics and a theory that predicted electron spin and the existence of a fundamental atomic particle called a positron. In 1933 he received the Nobel prize for physics and in 1939, the medal of the Royal Society.

If $b = 0$, then $z = a$ is a real number, and

$$|z| = \sqrt{a^2 + 0^2} = \sqrt{a^2} = |a|$$

so the modulus of a real number is simply its absolute value. Thus, the modulus of z is also called the ***absolute value*** of z.

EXAMPLE 2 Modulus of a Complex Number

Find $|z|$ if $z = 3 - 4i$.

Solution.

From (1) with $a = 3$ and $b = -4$, $|z| = \sqrt{(3)^2 + (-4)^2} = \sqrt{25} = 5$. ♦

The following theorem establishes a basic relationship between $\bar{z}$ and $|z|$.

Theorem 10.2.1

For any complex number z,

$$z\bar{z} = |z|^2$$

Proof. If $z = a + bi$, then

$$z\bar{z} = (a + bi)(a - bi) = a^2 - abi + bai - b^2i^2 = a^2 + b^2 = |z|^2$$ ■

Division of Complex Numbers We now turn to the division of complex numbers. Our objective is to define division as the inverse of multiplication. Thus, if $z_2 \neq 0$, then our definition of $z = z_1/z_2$ should be such that

$$z_1 = z_2 z \tag{2}$$

Our procedure will be to prove that (2) has a unique solution for z if $z_2 \neq 0$, and then define z_1/z_2 to be this value of z. As with real numbers, division by zero is not allowed.

Theorem 10.2.2

If $z_2 \neq 0$, then Equation (2) *has a unique solution, which is*

$$z = \frac{1}{|z_2|^2} z_1 \bar{z}_2 \tag{3}$$

Proof. Let $z = x + iy$, $z_1 = x_1 + iy_1$, and $z_2 = x_2 + iy_2$. Then (2) can be written as

$$x_1 + iy_1 = (x_2 + iy_2)(x + iy)$$

or

$$x_1 + iy_1 = (x_2 x - y_2 y) + i(y_2 x + x_2 y)$$

or, on equating real and imaginary parts,

$$x_2 x - y_2 y = x_1$$
$$y_2 x + x_2 y = y_1$$

or

$$\begin{bmatrix} x_2 & -y_2 \\ y_2 & x_2 \end{bmatrix} \begin{bmatrix} x \\ y \end{bmatrix} = \begin{bmatrix} x_1 \\ y_1 \end{bmatrix} \tag{4}$$

Since $z_2 = x_2 + iy_2 \neq 0$, it follows that x_2 and y_2 are not both zero, so

$$\begin{vmatrix} x_2 & -y_2 \\ y_2 & x_2 \end{vmatrix} = x_2^2 + y_2^2 \neq 0$$

Thus, by Cramer's rule (Theorem 2.4.3), system (4) has the unique solution

$$x = \frac{\begin{vmatrix} x_1 & -y_2 \\ y_1 & x_2 \end{vmatrix}}{\begin{vmatrix} x_2 & -y_2 \\ y_2 & x_2 \end{vmatrix}} = \frac{x_1x_2 + y_1y_2}{x_2^2 + y_2^2} = \frac{x_1x_2 + y_1y_2}{|z_2|^2}$$

$$y = \frac{\begin{vmatrix} x_2 & x_1 \\ y_2 & y_1 \end{vmatrix}}{\begin{vmatrix} x_2 & -y_2 \\ y_2 & x_2 \end{vmatrix}} = \frac{y_1x_2 - x_1y_2}{x_2^2 + y_2^2} = \frac{y_1x_2 - x_1y_2}{|z_2|^2}$$

Thus,

$$\begin{aligned} z = x + iy &= \frac{1}{|z_2|^2}\left[(x_1x_2 + y_1y_2) + i(y_1x_2 - x_1y_2)\right] \\ &= \frac{1}{|z_2|^2}(x_1 + iy_1)(x_2 - iy_2) = \frac{1}{|z_2|^2}z_1\bar{z}_2 \end{aligned}$$ ■

Thus, for $z_2 \neq 0$ we define

$$\frac{z_1}{z_2} = \frac{1}{|z_2|^2}z_1\bar{z}_2 \tag{5}$$

REMARK. To remember this formula, multiply numerator and denominator of z_1/z_2 by $\bar{z}_2$:

$$\frac{z_1}{z_2} = \frac{z_1\bar{z}_2}{z_2\bar{z}_2} = \frac{z_1\bar{z}_2}{|z_2|^2} = \frac{1}{|z_2|^2}z_1\bar{z}_2$$

EXAMPLE 3 Quotient in the Form $a + bi$

Express

$$\frac{3 + 4i}{1 - 2i}$$

in the form $a + bi$.

Solution.

From (5) with $z_1 = 3 + 4i$ and $z_2 = 1 - 2i$,

$$\begin{aligned} \frac{3 + 4i}{1 - 2i} &= \frac{1}{|1 - 2i|^2}(3 + 4i)(\overline{1 - 2i}) = \frac{1}{5}(3 + 4i)(1 + 2i) \\ &= \frac{1}{5}(-5 + 10i) = -1 + 2i \end{aligned}$$

Alternative Solution. As in the remark above, multiply numerator and denominator by the conjugate of the denominator:

$$\frac{3 + 4i}{1 - 2i} = \frac{3 + 4i}{1 - 2i} \cdot \frac{1 + 2i}{1 + 2i} = \frac{-5 + 10i}{5} = -1 + 2i$$ ◆

Systems of linear equations with complex coefficients arise in various applications. Without going into detail we note that all the results about linear systems studied in Chapters 1 and 2 carry over without change to systems with complex coefficients.

EXAMPLE 4 A Linear System with Complex Coefficients

Use Cramer's rule to solve

$$\begin{aligned} ix + 2y &= 1 - 2i \\ 4x - iy &= -1 + 3i \end{aligned}$$

Solution.

$$x = \frac{\begin{vmatrix} 1-2i & 2 \\ -1+3i & -i \end{vmatrix}}{\begin{vmatrix} i & 2 \\ 4 & -i \end{vmatrix}} = \frac{(-i)(1-2i) - 2(-1+3i)}{i(-i) - 2(4)} = \frac{-7i}{-7} = i$$

$$y = \frac{\begin{vmatrix} i & 1-2i \\ 4 & -1+3i \end{vmatrix}}{\begin{vmatrix} i & 2 \\ 4 & -i \end{vmatrix}} = \frac{(i)(-1+3i) - 4(1-2i)}{i(-i) - 2(4)} = \frac{-7+7i}{-7} = 1 - i$$

Thus, the solution is $x = i$, $y = 1 - i$. ◆

We conclude this section by listing some properties of the complex conjugate that will be useful in later sections.

Theorem 10.2.3 Properties of the Conjugate

For any complex numbers z, z_1, and z_2

(*a*) $\overline{z_1 + z_2} = \overline{z}_1 + \overline{z}_2$ (*b*) $\overline{z_1 - z_2} = \overline{z}_1 - \overline{z}_2$

(*c*) $\overline{z_1 z_2} = \overline{z}_1 \overline{z}_2$ (*d*) $\overline{(z_1/z_2)} = \overline{z}_1/\overline{z}_2$

(*e*) $\overline{\overline{z}} = z$

We prove (*a*) and leave the rest as exercises.

Proof (*a*). Let $z_1 = a_1 + b_1 i$ and $z_2 = a_2 + b_2 i$; then

$$\begin{aligned} \overline{z_1 + z_2} &= \overline{(a_1 + a_2) + (b_1 + b_2)i} \\ &= (a_1 + a_2) - (b_1 + b_2)i \\ &= (a_1 - b_1 i) + (a_2 - b_2 i) \\ &= \overline{z}_1 + \overline{z}_2 \end{aligned}$$

■

REMARK. It is possible to extend part (*a*) of Theorem 10.2.3 to n terms and part (*c*) to n factors. More precisely,

$$\begin{aligned} \overline{z_1 + z_2 + \cdots + z_n} &= \overline{z}_1 + \overline{z}_2 + \cdots + \overline{z}_n \\ \overline{z_1 z_2 \cdots z_n} &= \overline{z}_1 \overline{z}_2 \cdots \overline{z}_n \end{aligned}$$

Exercise Set 10.2

1. In each part find $\bar{z}$.

(a) $z = 2 + 7i$ (b) $z = -3 - 5i$ (c) $z = 5i$ (d) $z = -i$ (e) $z = -9$ (f) $z = 0$

2. In each part find $|z|$.

(a) $z = i$ (b) $z = -7i$ (c) $z = -3 - 4i$ (d) $z = 1 + i$ (e) $z = -8$ (f) $z = 0$

3. Verify that $z\bar{z} = |z|^2$ for

(a) $z = 2 - 4i$ (b) $z = -3 + 5i$ (c) $z = \sqrt{2} - \sqrt{2}i$

4. Given that $z_1 = 1 - 5i$ and $z_2 = 3 + 4i$, find

(a) z_1/z_2 (b) $\bar{z}_1/z_2$ (c) $z_1/\bar{z}_2$ (d) $\overline{(z_1/z_2)}$ (e) $z_1/|z_2|$ (f) $|z_1/z_2|$

5. In each part find $1/z$.

(a) $z = i$ (b) $z = 1 - 5i$ (c) $z = \dfrac{-i}{7}$

6. Given that $z_1 = 1 + i$ and $z_2 = 1 - 2i$, find

(a) $z_1 - \left(\dfrac{z_1}{z_2}\right)$ (b) $\dfrac{z_1 - 1}{z_2}$ (c) $z_1^2 - \left(\dfrac{iz_1}{z_2}\right)$ (d) $\dfrac{z_1}{iz_2}$

In Exercises 7–14 perform the calculations and express the result in the form $a + bi$.

7. $\dfrac{i}{1+i}$ **8.** $\dfrac{2}{(1-i)(3+i)}$ **9.** $\dfrac{1}{(3+4i)^2}$ **10.** $\dfrac{2+i}{i(-3+4i)}$

11. $\dfrac{\sqrt{3}+i}{(1-i)(\sqrt{3}-i)}$ **12.** $\dfrac{1}{i(3-2i)(1+i)}$ **13.** $\dfrac{i}{(1-i)(1-2i)(1+2i)}$ **14.** $\dfrac{1-2i}{3+4i} - \dfrac{2+i}{5i}$

15. In each part solve for z.

(a) $iz = 2 - i$ (b) $(4 - 3i)\bar{z} = i$

16. Use Theorem 10.2.3 to prove the following identities:

(a) $\overline{\bar{z} + 5i} = z - 5i$ (b) $\overline{iz} = -i\bar{z}$ (c) $\dfrac{\overline{i + \bar{z}}}{i - z} = -1$

17. In each part sketch the set of points in the complex plane that satisfies the equation.

(a) $|z| = 2$ (b) $|z - (1 + i)| = 1$ (c) $|z - i| = |z + i|$ (d) $\text{Im}(\bar{z} + i) = 3$

18. In each part sketch the set of points in the complex plane that satisfies the given condition(s).

(a) $|z + i| \le 1$ (b) $1 < |z| < 2$ (c) $|2z - 4i| < 1$ (d) $|z| \le |z + i|$

19. Given that $z = x + iy$, find

(a) $\text{Re}(\overline{iz})$ (b) $\text{Im}(\overline{iz})$ (c) $\text{Re}(i\bar{z})$ (d) $\text{Im}(i\bar{z})$

20. (a) Show that if n is a positive integer, then the only possible values for $(1/i)^n$ are $1, -1, i$, and $-i$.

(b) Find $(1/i)^{2509}$. [***Hint.*** See Exercise 23(b) of Section 10.1.]

21. Prove:

(a) $\dfrac{1}{2}(z + \bar{z}) = \text{Re}(z)$ (b) $\dfrac{1}{2i}(z - \bar{z}) = \text{Im}(z)$

22. Prove: $z = \bar{z}$ if and only if z is a real number.

23. Given that $z_1 = x_1 + iy_1$ and $z_2 = x_2 + iy_2 \neq 0$, find

(a) $\text{Re}\left(\dfrac{z_1}{z_2}\right)$ (b) $\text{Im}\left(\dfrac{z_1}{z_2}\right)$

24. Prove: If $(\bar{z})^2 = z^2$, then z is either real or pure imaginary.

25. Prove that $|z| = |\bar{z}|$.

26. Prove:

(a) $\overline{z_1 - z_2} = \bar{z}_1 - \bar{z}_2$ (b) $\overline{z_1 z_2} = \bar{z}_1 \bar{z}_2$ (c) $\overline{(z_1/z_2)} = \bar{z}_1/\bar{z}_2$ (d) $\bar{\bar{z}} = z$

27. (a) Prove that $\overline{z^2} = (\bar{z})^2$.
(b) Prove that if n is a positive integer, then $\overline{z^n} = (\bar{z})^n$.
(c) Is the result in (b) true if n is a negative integer? Explain.

In Exercises 28–31 solve the system of linear equations by Cramer's rule.

28. $$\begin{aligned} ix_1 - ix_2 &= -2 \\ 2x_1 + x_2 &= i \end{aligned}$$

29. $$\begin{aligned} x_1 + x_2 &= 2 \\ x_1 - x_2 &= 2i \end{aligned}$$

30. $$\begin{aligned} x_1 + x_2 + x_3 &= 3 \\ x_1 + x_2 - x_3 &= 2 + 2i \\ x_1 - x_2 + x_3 &= -1 \end{aligned}$$

31. $$\begin{aligned} ix_1 + 3x_2 + (1+i)x_3 &= -i \\ x_1 + ix_2 + 3x_3 &= -2i \\ x_1 + x_2 + x_3 &= 0 \end{aligned}$$

In Exercises 32 and 33 solve the system of linear equations by Gauss–Jordan elimination.

32. $$\begin{bmatrix} -1 & -1-i \\ -1+i & -2 \end{bmatrix} \begin{bmatrix} x_1 \\ x_2 \end{bmatrix} = \begin{bmatrix} 0 \\ 0 \end{bmatrix}$$

33. $$\begin{bmatrix} 2 & -1-i \\ -1+i & 1 \end{bmatrix} \begin{bmatrix} x_1 \\ x_2 \end{bmatrix} = \begin{bmatrix} 0 \\ 0 \end{bmatrix}$$

34. Solve the following system of linear equations by Gauss–Jordan elimination.

$$\begin{aligned} x_1 + ix_2 - ix_3 &= 0 \\ -x_1 + (1-i)x_2 + 2ix_3 &= 0 \\ 2x_1 + (-1+2i)x_2 - 3ix_3 &= 0 \end{aligned}$$

35. In each part use the formula in Theorem 1.4.5 to compute the inverse of the matrix, and check your result by showing that $AA^{-1} = A^{-1}A = I$.

(a) $A = \begin{bmatrix} i & -2 \\ 1 & i \end{bmatrix}$ (b) $A = \begin{bmatrix} 2 & i \\ 1 & 0 \end{bmatrix}$

36. Let $p(x) = a_0 + a_1 x + a_2 x^2 + \cdots + a_n x^n$ be a polynomial for which the coefficients $a_0, a_1, a_2, \ldots, a_n$ are real. Prove that if z is a solution of the equation $p(z) = 0$, then so is $\bar{z}$.

37. Prove: For any complex number z, $|\text{Re}(z)| \le |z|$ and $|\text{Im}(z)| \le |z|$.

38. Prove that

$$\frac{|\text{Re}(z)| + |\text{Im}(z)|}{\sqrt{2}} \le |z|$$

[***Hint.*** Let $z = x + iy$ and use the fact that $(|x| - |y|)^2 \ge 0$.]

39. In each part use the method of Example 4 in Section 1.5 to find A^{-1}, and check your result by showing that $AA^{-1} = A^{-1}A = I$.

(a) $A = \begin{bmatrix} 1 & 1+i & 0 \\ 0 & 1 & i \\ -i & 1-2i & 2 \end{bmatrix}$ (b) $A = \begin{bmatrix} i & 0 & -i \\ 0 & 1 & -1-4i \\ 2-i & i & 3 \end{bmatrix}$

40. Show that $|z - 1| = |\bar{z} - 1|$. Discuss the geometric interpretation of the result.

41. (a) If $z_1 = a_1 + b_1 i$ and $z_2 = a_2 + b_2 i$, find $|z_1 - z_2|$ and interpret the result geometrically.
(b) Use part (a) to show that the complex numbers 12, $6 + 2i$, and $8 + 8i$ are vertices of a right triangle.

42. Use Theorem 10.2.3 to show that if the coefficients a, b, and c in a quadratic polynomial are real, then the solutions of the equation $az^2 + bz + c = 0$ are complex conjugates. What can you conclude if a, b, and c are complex?

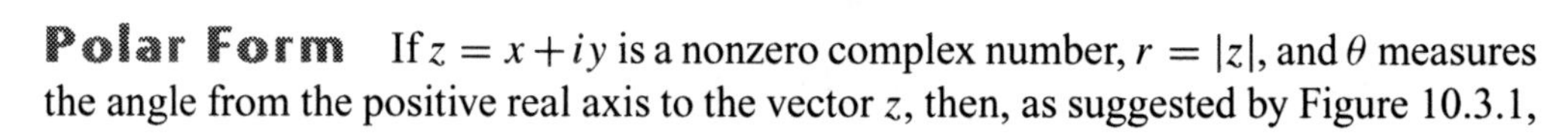

10.3 POLAR FORM OF A COMPLEX NUMBER

In this section we shall discuss a way to represent complex numbers using trigonometric properties. Our work will lead to an important formula for powers of complex numbers and a method for finding nth roots of complex numbers.

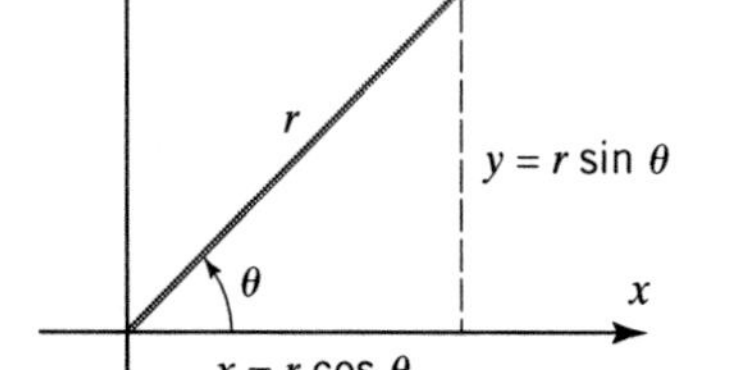

Figure 10.3.1

Polar Form If $z = x + iy$ is a nonzero complex number, $r = |z|$, and θ measures the angle from the positive real axis to the vector z, then, as suggested by Figure 10.3.1,

$$x = r\cos\theta, \qquad y = r\sin\theta \tag{1}$$

so that $z = x + iy$ can be written as $z = r\cos\theta + ir\sin\theta$ or

$$z = r(\cos\theta + i\sin\theta) \tag{2}$$

This is called a ***polar form of z***.

Argument of a Complex Number The angle θ is called an ***argument of z*** and is denoted by

$$\theta = \arg z$$

The argument of z is not uniquely determined because we can add or subtract any multiple of 2π from θ to produce another value of the argument. However, there is only one value of the argument in radians that satisfies

$$-\pi < \theta \leq \pi$$

This is called the ***principal argument of z*** and is denoted by

$$\theta = \operatorname{Arg} z$$

EXAMPLE 1 Polar Forms

Express the following complex numbers in polar form using their principal arguments:

$$\text{(a) } z = 1 + \sqrt{3}i \qquad \text{(b) } z = -1 - i$$

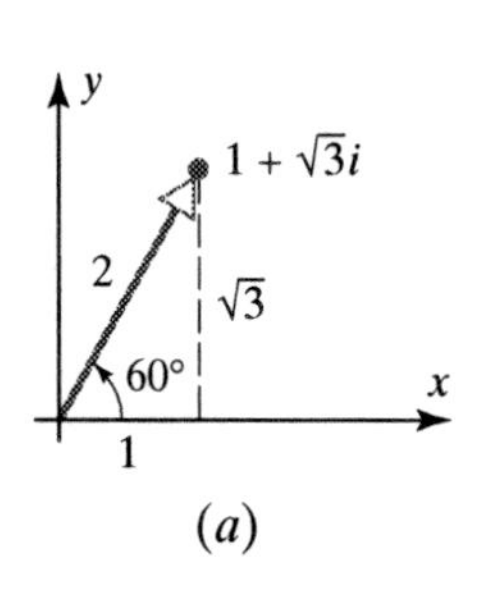

(a)

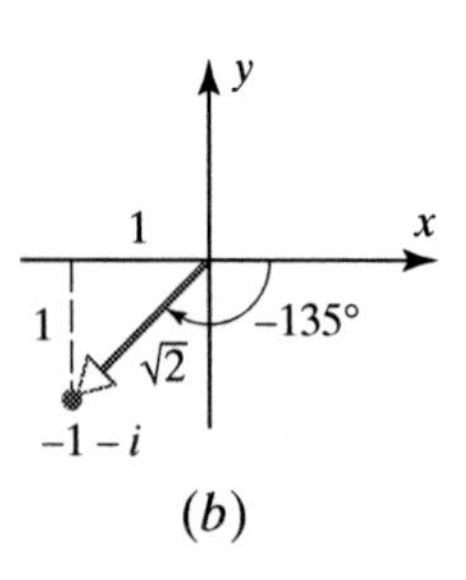

(b)

Figure 10.3.2

Solution (a). The value of r is

$$r = |z| = \sqrt{1^2 + (\sqrt{3})^2} = \sqrt{4} = 2$$

and since $x = 1$ and $y = \sqrt{3}$, it follows from (1) that

$$1 = 2\cos\theta \quad \text{and} \quad \sqrt{3} = 2\sin\theta$$

so $\cos\theta = 1/2$ and $\sin\theta = \sqrt{3}/2$. The only value of θ that satisfies these relations and meets the requirement $-\pi < \theta \leq \pi$ is $\theta = \pi/3$ $(= 60^\circ)$ (see Figure 10.3.2*a*). Thus, a polar form of z is

$$z = 2\left(\cos\frac{\pi}{3} + i\sin\frac{\pi}{3}\right)$$

Solution (b). The value of r is

$$r = |z| = \sqrt{(-1)^2 + (-1)^2} = \sqrt{2}$$

and since $x = -1$, $y = -1$, it follows from (1) that

$$-1 = \sqrt{2}\cos\theta \quad \text{and} \quad -1 = \sqrt{2}\sin\theta$$

so $\cos\theta = -1/\sqrt{2}$ and $\sin\theta = -1/\sqrt{2}$. The only value of θ that satisfies these relations and meets the requirement $-\pi < \theta \le \pi$ is $\theta = -3\pi/4\,(= -135°)$ (Figure 10.3.2*b*). Thus, a polar form of z is

$$z = \sqrt{2}\left(\cos\frac{-3\pi}{4} + i\sin\frac{-3\pi}{4}\right)$$ ♦

Multiplication and Division Interpreted Geometrically

We now show how polar forms can be used to give geometric interpretations of multiplication and division of complex numbers. Let

$$z_1 = r_1(\cos\theta_1 + i\sin\theta_1) \quad \text{and} \quad z_2 = r_2(\cos\theta_2 + i\sin\theta_2)$$

Multiplying, we obtain

$$z_1z_2 = r_1r_2[(\cos\theta_1\cos\theta_2 - \sin\theta_1\sin\theta_2) + i(\sin\theta_1\cos\theta_2 + \cos\theta_1\sin\theta_2)]$$

Recalling the trigonometric identities

$$\cos(\theta_1+\theta_2) = \cos\theta_1\cos\theta_2 - \sin\theta_1\sin\theta_2$$
$$\sin(\theta_1+\theta_2) = \sin\theta_1\cos\theta_2 + \cos\theta_1\sin\theta_2$$

we obtain

$$z_1z_2 = r_1r_2[\cos(\theta_1+\theta_2) + i\sin(\theta_1+\theta_2)] \tag{3}$$

which is a polar form of the complex number with modulus r_1r_2 and argument $\theta_1+\theta_2$. Thus, we have shown that

$$|z_1z_2| = |z_1||z_2| \tag{4}$$

and

$$\arg(z_1z_2) = \arg z_1 + \arg z_2$$

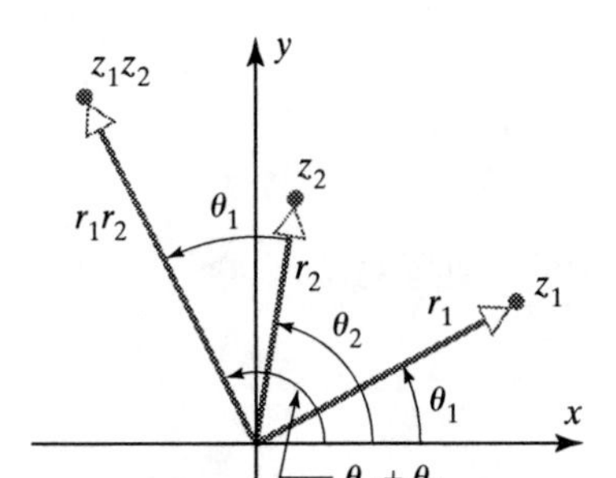

Figure 10.3.3 The product of two complex numbers.

(Why?) In words, *the product of two complex numbers is obtained by multiplying their moduli and adding their arguments* (Figure 10.3.3).

We leave it as an exercise to show that if $z_2 \ne 0$, then

$$\frac{z_1}{z_2} = \frac{r_1}{r_2}[\cos(\theta_1-\theta_2) + i\sin(\theta_1-\theta_2)] \tag{5}$$

from which it follows that

$$\left|\frac{z_1}{z_2}\right| = \frac{|z_1|}{|z_2|} \quad \text{if } z_2 \ne 0$$

and

$$\arg\left(\frac{z_1}{z_2}\right) = \arg z_1 - \arg z_2$$

In words, *the quotient of two complex numbers is obtained by dividing their moduli and subtracting their arguments (in the appropriate order).*

EXAMPLE 2 A Quotient Using Polar Forms

Let

$$z_1 = 1 + \sqrt{3}i \quad \text{and} \quad z_2 = \sqrt{3} + i$$

Polar forms of these complex numbers are

$$z_1 = 2\left(\cos\frac{\pi}{3} + i\sin\frac{\pi}{3}\right) \quad \text{and} \quad z_2 = 2\left(\cos\frac{\pi}{6} + i\sin\frac{\pi}{6}\right)$$

(verify) so that from (3)

$$z_1 z_2 = 4\left[\cos\left(\frac{\pi}{3}+\frac{\pi}{6}\right) + i\sin\left(\frac{\pi}{3}+\frac{\pi}{6}\right)\right]$$
$$= 4\left[\cos\frac{\pi}{2} + i\sin\frac{\pi}{2}\right] = 4[0+i] = 4i$$

and from (5)

$$\frac{z_1}{z_2} = 1\cdot\left[\cos\left(\frac{\pi}{3}-\frac{\pi}{6}\right) + i\sin\left(\frac{\pi}{3}-\frac{\pi}{6}\right)\right]$$
$$= \cos\frac{\pi}{6} + i\sin\frac{\pi}{6} = \frac{\sqrt{3}}{2} + \frac{1}{2}i$$

As a check, we calculate $z_1 z_2$ and z_1/z_2 directly without using polar forms for z_1 and z_2:

$$z_1 z_2 = (1+\sqrt{3}i)(\sqrt{3}+i) = (\sqrt{3}-\sqrt{3}) + (3+1)i = 4i$$

$$\frac{z_1}{z_2} = \frac{1+\sqrt{3}i}{\sqrt{3}+i}\cdot\frac{\sqrt{3}-i}{\sqrt{3}-i} = \frac{(\sqrt{3}+\sqrt{3})+(-i+3i)}{4} = \frac{\sqrt{3}}{2}+\frac{1}{2}i$$

which agrees with our previous results. ♦

The complex number i has a modulus of 1 and an argument of $\pi/2$ ($=90°$), so the product iz has the same modulus as z, but its argument is 90° greater than that of z. In short, *multiplying z by i rotates z counterclockwise by* 90° (Figure 10.3.4).

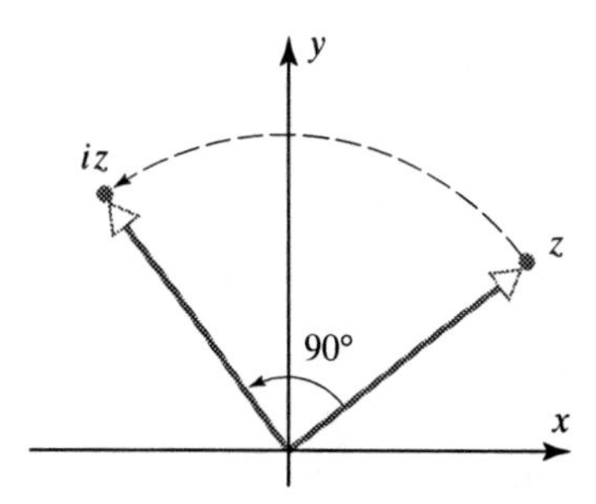

Figure 10.3.4 Multiplying by i rotates z counterclockwise by 90°.

Demoivre's Formula If n is a positive integer and $z = r(\cos\theta + i\sin\theta)$, then from Formula (3),

$$z^n = \underbrace{z\cdot z\cdot z\cdots z}_{n\text{-factors}} = r^n[\cos\underbrace{(\theta+\theta+\cdots+\theta)}_{n\text{-terms}} + i\sin\underbrace{(\theta+\theta+\cdots+\theta)}_{n\text{-terms}}]$$

or

$$z^n = r^n(\cos n\theta + i\sin n\theta) \tag{6}$$

Moreover, (6) also holds for negative integers if $z \neq 0$ (see Exercise 23).

In the special case where $r = 1$, we have $z = \cos\theta + i\sin\theta$, so (6) becomes

$$(\cos\theta + i\sin\theta)^n = \cos n\theta + i\sin n\theta \tag{7}$$

which is called ***DeMoivre's formula***. Although we derived (7) assuming n to be a positive integer, it will be shown in the exercises that this formula is valid for all integers n.

Finding nth Roots We now show how DeMoivre's formula can be used to obtain roots of complex numbers. If n is a positive integer, and z is any complex number, then we define an ***nth root of z*** to be any complex number w that satisfies the equation

$$w^n = z \tag{8}$$

Abraham DeMoivre (1667–1754) was a French mathematician who made important contributions to probability, statistics, and trigonometry. He developed the concept of statistically independent events, wrote a major and influential treatise on probability, and helped transform trigonometry from a branch of geometry into a branch of analysis through his use of complex numbers. In spite of his important work, he barely managed to eke out a living as a tutor and a consultant on gambling and insurance.

We denote an nth root of z by $z^{1/n}$. If $z \neq 0$, then we can derive formulas for the nth roots of z as follows. Let

$$w = \rho(\cos\alpha + i\sin\alpha) \quad \text{and} \quad z = r(\cos\theta + i\sin\theta)$$

If we assume that w satisfies (8), then it follows from (6) that

$$\rho^n(\cos n\alpha + i\sin n\alpha) = r(\cos\theta + i\sin\theta) \tag{9}$$

Comparing the moduli of the two sides, we see that $\rho^n = r$ or

$$\rho = \sqrt[n]{r}$$

where $\sqrt[n]{r}$ denotes the real positive nth root of r. Moreover, in order to have the equalities $\cos n\alpha = \cos\theta$ and $\sin n\alpha = \sin\theta$ in (9), the angles $n\alpha$ and θ must either be equal or differ by a multiple of 2π. That is,

$$n\alpha = \theta + 2k\pi \quad \text{or} \quad \alpha = \frac{\theta}{n} + \frac{2k\pi}{n}, \qquad k = 0, \pm 1, \pm 2, \ldots$$

Thus, the values of $w = \rho(\cos\alpha + i\sin\alpha)$ that satisfy (8) are given by

$$w = \sqrt[n]{r}\left[\cos\left(\frac{\theta}{n} + \frac{2k\pi}{n}\right) + i\sin\left(\frac{\theta}{n} + \frac{2k\pi}{n}\right)\right], \qquad k = 0, \pm 1, \pm 2, \ldots$$

Although there are infinitely many values of k, it can be shown (see Exercise 16) that $k = 0, 1, 2, \ldots, n-1$ produce distinct values of w satisfying (8), but all other choices of k yield duplicates of these. Therefore, there are exactly n different nth roots of $z = r(\cos\theta + i\sin\theta)$, and these are given by

$$z^{1/n} = \sqrt[n]{r}\left[\cos\left(\frac{\theta}{n} + \frac{2k\pi}{n}\right) + i\sin\left(\frac{\theta}{n} + \frac{2k\pi}{n}\right)\right], \qquad k = 0, 1, 2, \ldots, n-1 \tag{10}$$

EXAMPLE 3 Cube Roots of a Complex Number

Find all cube roots of -8.

Solution.

Since -8 lies on the negative real axis, we can use $\theta = \pi$ as an argument. Moreover, $r = |z| = |-8| = 8$, so a polar form of -8 is

$$-8 = 8(\cos\pi + i\sin\pi)$$

From (10) with $n = 3$ it follows that

$$(-8)^{1/3} = \sqrt[3]{8}\left[\cos\left(\frac{\pi}{3} + \frac{2k\pi}{3}\right) + i\sin\left(\frac{\pi}{3} + \frac{2k\pi}{3}\right)\right], \qquad k = 0, 1, 2$$

Thus, the cube roots of -8 are

$$2\left(\cos\frac{\pi}{3} + i\sin\frac{\pi}{3}\right) = 2\left(\frac{1}{2} + \frac{\sqrt{3}}{2}i\right) = 1 + \sqrt{3}i$$

$$2(\cos\pi + i\sin\pi) = 2(-1) = -2$$

$$2\cos\left(\frac{5\pi}{3} + i\sin\frac{5\pi}{3}\right) = 2\left(\frac{1}{2} - \frac{\sqrt{3}}{2}i\right) = 1 - \sqrt{3}i$$

◆

As shown in Figure 10.3.5, the three cube roots of -8 obtained in Example 3 are equally spaced $\pi/3$ radians ($= 120°$) apart around the circle of radius 2 centered at the

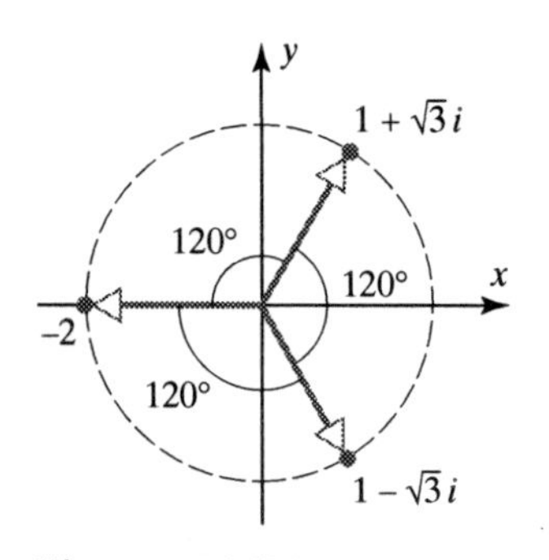

Figure 10.3.5
The cube roots of −8.

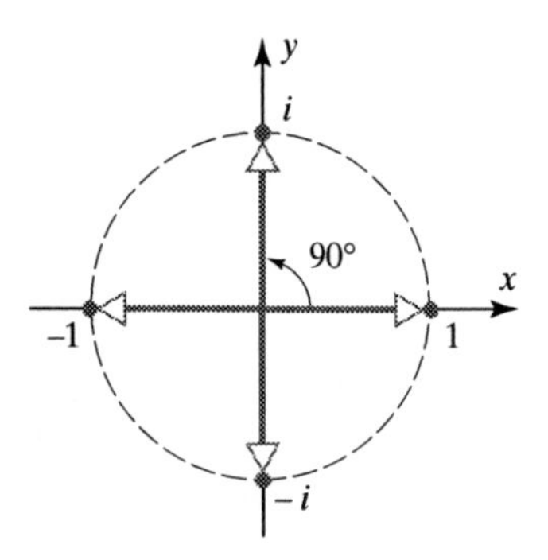

Figure 10.3.6
The fourth roots of 1.

origin. This is not accidental. In general, it follows from Formula (10) that the nth roots of z lie on the circle of radius $\sqrt[n]{r}\ (=\sqrt[n]{|z|})$, and are equally spaced $2\pi/n$ radians apart. (Can you see why?) Thus, once one nth root of z is found, the remaining $n-1$ roots can be generated by rotating this root successively through increments of $2\pi/n$ radians.

EXAMPLE 4 Fourth Roots of a Complex Number

Find all fourth roots of 1.

Solution.

We could apply Formula (10). Instead, we observe that $w=1$ is one fourth root of 1, so that the remaining three roots can be generated by rotating this root through increments of $2\pi/4=\pi/2$ radians ($=90°$). From Figure 10.3.6 we see that the fourth roots of 1 are

$$1,\quad i,\quad -1,\quad -i$$ ◆

Complex Exponents We conclude this section with some comments on notation.

In more detailed studies of complex numbers, complex exponents are defined, and it is proved that

$$\cos\theta + i\sin\theta = e^{i\theta} \tag{11}$$

where e is an irrational real number given approximately by $e\approx 2.71828\ldots$. (For readers who have studied calculus, a proof of this result is given in Exercise 18.)

It follows from (11) that the polar form

$$z = r(\cos\theta + i\sin\theta)$$

can be written more briefly as

$$z = re^{i\theta} \tag{12}$$

EXAMPLE 5 Expressing a Complex Number in Form (12)

In Example 1 it was shown that

$$1+\sqrt{3}i = 2\left(\cos\frac{\pi}{3} + i\sin\frac{\pi}{3}\right)$$

From (12) this can also be written as

$$1+\sqrt{3}i = 2e^{i\pi/3}$$ ◆

It can be proved that complex exponents follow the same laws as real exponents, so that if

$$z_1 = r_1e^{i\theta_1} \quad\text{and}\quad z_2 = r_2e^{i\theta_2}$$

are nonzero complex numbers, then

$$z_1z_2 = r_1r_2e^{i\theta_1+i\theta_2} = r_1r_2e^{i(\theta_1+\theta_2)}$$

$$\frac{z_1}{z_2} = \frac{r_1}{r_2}e^{i\theta_1-i\theta_2} = \frac{r_1}{r_2}e^{i(\theta_1-\theta_2)}$$

But these are just Formulas (3) and (5) in a different notation.

We conclude this section with a useful formula for $\bar{z}$ in polar notation. If

$$z = re^{i\theta} = r(\cos\theta + i\sin\theta)$$

then

$$\bar{z} = r(\cos\theta - i\sin\theta) \tag{13}$$

Recalling the trigonometric identities

$$\sin(-\theta) = -\sin\theta \quad \text{and} \quad \cos(-\theta) = \cos\theta$$

we can rewrite (13) as

$$\bar{z} = r[\cos(-\theta) + i\sin(-\theta)] = re^{i(-\theta)}$$

or equivalently,

$$\bar{z} = re^{-i\theta} \tag{14}$$

In the special case where $r = 1$, the polar form of z is $z = e^{i\theta}$, and (14) yields the formula

$$\overline{e^{i\theta}} = e^{-i\theta} \tag{15}$$

Exercise Set 10.3

1. In each part find the principal argument of z.

(a) $z = 1$ (b) $z = i$ (c) $z = -i$ (d) $z = 1 + i$ (e) $z = -1 + \sqrt{3}i$ (f) $z = 1 - i$

2. In each part find the value of $\theta = \arg(1 - \sqrt{3}i)$ that satisfies the given condition.

(a) $0 < \theta \leq 2\pi$ (b) $-\pi < \theta \leq \pi$ (c) $-\dfrac{\pi}{6} \leq \theta < \dfrac{11\pi}{6}$

3. In each part express the complex number in polar form using its principal argument.

(a) $2i$ (b) -4 (c) $5 + 5i$ (d) $-6 + 6\sqrt{3}i$ (e) $-3 - 3i$ (f) $2\sqrt{3} - 2i$

4. Given that $z_1 = 2(\cos\pi/4 + i\sin\pi/4)$ and $z_2 = 3(\cos\pi/6 + i\sin\pi/6)$, find a polar form of

(a) z_1z_2 (b) $\dfrac{z_1}{z_2}$ (c) $\dfrac{z_2}{z_1}$ (d) $\dfrac{z_1^5}{z_2^2}$

5. Express $z_1 = i$, $z_2 = 1 - \sqrt{3}i$, and $z_3 = \sqrt{3} + i$ in polar form, and use your results to find z_1z_2/z_3. Check your results by performing the calculations without using polar forms.

6. Use Formula (6) to find

(a) $(1 + i)^{12}$ (b) $\left(\dfrac{1}{\sqrt{2}} - \dfrac{1}{\sqrt{2}}i\right)^{-6}$ (c) $(\sqrt{3} + i)^7$ (d) $(1 - i\sqrt{3})^{-10}$

7. In each part find all the roots and sketch them as vectors in the complex plane.

(a) $(-i)^{1/2}$ (b) $(1 + \sqrt{3}i)^{1/2}$ (c) $(-27)^{1/3}$ (d) $(i)^{1/3}$ (e) $(-1)^{1/4}$ (f) $(-8 + 8\sqrt{3}i)^{1/4}$

8. Use the method of Example 4 to find all cube roots of 1.

9. Use the method of Example 4 to find all sixth roots of 1.

10. Find all square roots of $1 + i$ and express your results in polar form.

11. Find all solutions of the equation $z^4 - 16 = 0$.

12. Find four solutions of the equation $z^4 + 8 = 0$ and use your results to factor $z^4 + 8$ into two quadratic factors with real coefficients.

7

Eigenvalues, Eigenvectors

Chapter Contents

INTRODUCTION: If A is an $n \times n$ matrix and $\mathbf{x}$ is a vector in R^n, then $A\mathbf{x}$ is also a vector in R^n, but usually there is no simple geometric relationship between $\mathbf{x}$ and $A\mathbf{x}$. However, in the special case where $\mathbf{x}$ is a nonzero vector and $A\mathbf{x}$ is a scalar multiple of $\mathbf{x}$, a simple geometric relationship occurs. For example, if A is a 2×2 matrix, and if $\mathbf{x}$ is a nonzero vector such that $A\mathbf{x}$ is a scalar multiple of $\mathbf{x}$, say $A\mathbf{x} = \lambda\mathbf{x}$, then each vector on the line through the origin determined by $\mathbf{x}$ gets mapped back onto the same line under multiplication by A.

Nonzero vectors that get mapped into scalar multiples of themselves under a linear operator arise naturally in the study of vibrations, genetics, population dynamics, quantum mechanics, and economics, as well as in geometry. In this chapter we will study such vectors and their applications.

7.1 EIGENVALUES AND EIGENVECTORS

In Section 2.3 we introduced the concepts of eigenvalue and eigenvector. In this section we will study those ideas in more detail to set the stage for applications of them in later sections.

Review We begin with a review of some concepts that were developed in Sections 2.3 and 4.3.

Definition

If A is an $n \times n$ matrix, then a nonzero vector $\mathbf{x}$ in R^n is called an ***eigenvector*** of A if $A\mathbf{x}$ is a scalar multiple of $\mathbf{x}$; that is,

$$A\mathbf{x} = \lambda\mathbf{x}$$

for some scalar λ. The scalar λ is called an ***eigenvalue*** of A, and $\mathbf{x}$ is said to be an eigenvector of A ***corresponding*** to λ.

In R^2 and R^3 multiplication by A maps each eigenvector $\mathbf{x}$ of A (if any) onto the same line through the origin as $\mathbf{x}$. Depending on the sign and the magnitude of the eigenvalue λ corresponding to $\mathbf{x}$, the linear operator $A\mathbf{x} = \lambda\mathbf{x}$ compresses or stretches $\mathbf{x}$ by a factor of λ, with a reversal of direction in the case where λ is negative (Figure 7.1.1).

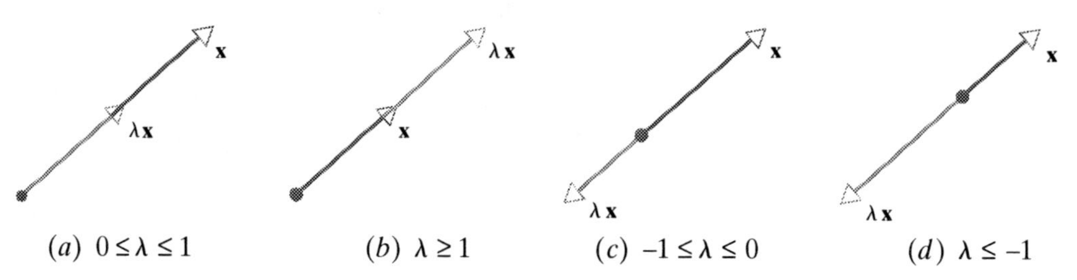

(*a*) $0 \le \lambda \le 1$ (*b*) $\lambda \ge 1$ (*c*) $-1 \le \lambda \le 0$ (*d*) $\lambda \le -1$

Figure 7.1.1

EXAMPLE 1 Eigenvector of a 2 × 2 Matrix

The vector $\mathbf{x} = \begin{bmatrix} 1 \\ 2 \end{bmatrix}$ is an eigenvector of

$$A = \begin{bmatrix} 3 & 0 \\ 8 & -1 \end{bmatrix}$$

corresponding to the eigenvalue $\lambda = 3$, since

$$A\mathbf{x} = \begin{bmatrix} 3 & 0 \\ 8 & -1 \end{bmatrix}\begin{bmatrix} 1 \\ 2 \end{bmatrix} = \begin{bmatrix} 3 \\ 6 \end{bmatrix} = 3\mathbf{x}$$ ◆

To find the eigenvalues of an $n \times n$ matrix A we rewrite $A\mathbf{x} = \lambda\mathbf{x}$ as

$$A\mathbf{x} = \lambda I\mathbf{x}$$

or equivalently,

$$(\lambda I - A)\mathbf{x} = \mathbf{0} \tag{1}$$

For λ to be an eigenvalue, there must be a nonzero solution of this equation. However, by Theorem 6.4.5, Equation (1) has a nonzero solution if and only if

$$\det(\lambda I - A) = 0$$

This is called the ***characteristic equation*** of A; the scalars satisfying this equation are the eigenvalues of A. When expanded, the determinant $\det(\lambda I - A)$ is a polynomial p in λ called the ***characteristic polynomial*** of A.

It can be shown (Exercise 15) that if A is an $n \times n$ matrix, then the characteristic polynomial of A has degree n and the coefficient of λ^n is 1; that is, the characteristic polynomial $p(x)$ of an $n \times n$ matrix has the form

$$p(\lambda) = \det(\lambda I - A) = \lambda^n + c_1\lambda^{n-1} + \cdots + c_n$$

It follows from the Fundamental Theorem of Algebra that the characteristic equation

$$\lambda^n + c_1\lambda^{n-1} + \cdots + c_n = 0$$

has at most n distinct solutions, so an $n \times n$ matrix has at most n distinct eigenvalues.

The reader may wish to review Example 6 of Section 2.3, where we found the eigenvalues of a 2×2 matrix by solving the characteristic equation. The following example involves a 3×3 matrix.

EXAMPLE 2 Eigenvalues of a 3 x 3 Matrix

Find the eigenvalues of

$$A = \begin{bmatrix} 0 & 1 & 0 \\ 0 & 0 & 1 \\ 4 & -17 & 8 \end{bmatrix}$$

Solution.

The characteristic polynomial of A is

$$\det(\lambda I - A) = \det\begin{bmatrix} \lambda & -1 & 0 \\ 0 & \lambda & -1 \\ -4 & 17 & \lambda - 8 \end{bmatrix} = \lambda^3 - 8\lambda^2 + 17\lambda - 4$$

The eigenvalues of A must therefore satisfy the cubic equation

$$\lambda^3 - 8\lambda^2 + 17\lambda - 4 = 0 \tag{2}$$

To solve this equation, we shall begin by searching for integer solutions. This task can be greatly simplified by exploiting the fact that all integer solutions (if there are any) to a polynomial equation with integer coefficients

$$\lambda^n + c_1\lambda^{n-1} + \cdots + c_n = 0$$

must be divisors of the constant term, c_n. Thus, the only possible integer solutions of (2) are the divisors of -4, that is, $\pm 1, \pm 2, \pm 4$. Successively substituting these values in (2) shows that $\lambda = 4$ is an integer solution. As a consequence, $\lambda - 4$ must be a factor of the left side of (2). Dividing $\lambda - 4$ into $\lambda^3 - 8\lambda^2 + 17\lambda - 4$ shows that (2) can be rewritten as

$$(\lambda - 4)(\lambda^2 - 4\lambda + 1) = 0$$

Thus, the remaining solutions of (2) satisfy the quadratic equation

$$\lambda^2 - 4\lambda + 1 = 0$$

which can be solved by the quadratic formula. Thus, the eigenvalues of A are

$$\lambda = 4, \quad \lambda = 2 + \sqrt{3}, \quad \text{and} \quad \lambda = 2 - \sqrt{3}$$

◆

EXAMPLE 3 Eigenvalues of an Upper Triangular Matrix

Find the eigenvalues of the upper triangular matrix

$$A = \begin{bmatrix} a_{11} & a_{12} & a_{13} & a_{14} \\ 0 & a_{22} & a_{23} & a_{24} \\ 0 & 0 & a_{33} & a_{34} \\ 0 & 0 & 0 & a_{44} \end{bmatrix}$$

Solution.

Recalling that the determinant of a triangular matrix is the product of the entries on the main diagonal (Theorem 2.2.2), we obtain

$$\det(\lambda I - A) = \det \begin{bmatrix} \lambda - a_{11} & -a_{12} & -a_{13} & -a_{14} \\ 0 & \lambda - a_{22} & -a_{23} & -a_{24} \\ 0 & 0 & \lambda - a_{33} & -a_{34} \\ 0 & 0 & 0 & \lambda - a_{44} \end{bmatrix}$$

$$= (\lambda - a_{11})(\lambda - a_{22})(\lambda - a_{33})(\lambda - a_{44})$$

Thus, the characteristic equation is

$$(\lambda - a_{11})(\lambda - a_{22})(\lambda - a_{33})(\lambda - a_{44}) = 0$$

and the eigenvalues are

$$\lambda = a_{11}, \quad \lambda = a_{22}, \quad \lambda = a_{33}, \quad \lambda = a_{44}$$

which are precisely the diagonal entries of A. ◆

The following general theorem should be evident from the computations in the preceding example.

Theorem 7.1.1

If A is an $n \times n$ triangular matrix (upper triangular, lower triangular, or diagonal), then the eigenvalues of A are the entries on the main diagonal of A.

EXAMPLE 4 Eigenvalues of a Lower Triangular Matrix

By inspection, the eigenvalues of the lower triangular matrix

$$A = \begin{bmatrix} \frac{1}{2} & 0 & 0 \\ -1 & \frac{2}{3} & 0 \\ 5 & -8 & -\frac{1}{4} \end{bmatrix}$$

are $\lambda = \frac{1}{2}$, $\lambda = \frac{2}{3}$, and $\lambda = -\frac{1}{4}$. ◆

REMARK. In practical problems, the matrix A is often so large that computing the characteristic equation is not practical. As a result, various approximation methods are used to obtain eigenvalues.

Complex Eigenvalues It is possible for the characteristic equation of a matrix with real entries to have complex solutions. For example, the characteristic polynomial of the matrix

$$A = \begin{bmatrix} -2 & -1 \\ 5 & 2 \end{bmatrix}$$

is

$$\det(\lambda I - A) = \det \begin{bmatrix} \lambda + 2 & 1 \\ -5 & \lambda - 2 \end{bmatrix} = \lambda^2 + 1$$

so the characteristic equation is $\lambda^2 + 1 = 0$, the solutions of which are the imaginary numbers $\lambda = i$ and $\lambda = -i$. Thus, we are forced to consider complex eigenvalues, even for real matrices. This, in turn, leads us to consider the possibility of complex vector spaces, that is, vector spaces in which scalars are allowed to have complex values. Such vector spaces will be considered in Chapter 10. For now, we will allow complex eigenvalues, but our discussion of eigenvectors will be limited to matrices with real eigenvalues.

The following theorem summarizes our discussion thus far.

Theorem 7.1.2 Equivalent Statements

If A is an $n \times n$ matrix and λ is a real number, then the following are equivalent.

(a) *λ is an eigenvalue of A.*
(b) *The system of equations $(\lambda I - A)\mathbf{x} = \mathbf{0}$ has nontrivial solutions.*
(c) *There is a nonzero vector $\mathbf{x}$ in R^n such that $A\mathbf{x} = \lambda\mathbf{x}$.*
(d) *λ is a solution of the characteristic equation* $\det(\lambda I - A) = 0$.

Finding Bases for Eigenspaces Now that we know how to find eigenvalues, we turn to the problem of finding eigenvectors. The eigenvectors of A corresponding to an eigenvalue λ are the nonzero vectors $\mathbf{x}$ that satisfy $A\mathbf{x} = \lambda\mathbf{x}$. Equivalently, the eigenvectors corresponding to λ are the nonzero vectors in the solution space of $(\lambda I - A)\mathbf{x} = \mathbf{0}$. We call this solution space the ***eigenspace*** of A corresponding to λ.

EXAMPLE 5 Bases for Eigenspaces

Find bases for the eigenspaces of

$$A = \begin{bmatrix} 0 & 0 & -2 \\ 1 & 2 & 1 \\ 1 & 0 & 3 \end{bmatrix}$$

Solution.

The characteristic equation of matrix A is $\lambda^3 - 5\lambda^2 + 8\lambda - 4 = 0$, or in factored form, $(\lambda - 1)(\lambda - 2)^2 = 0$ (verify); thus, the eigenvalues of A are $\lambda = 1$ and $\lambda = 2$, so there are two eigenspaces of A.

By definition,

$$\mathbf{x} = \begin{bmatrix} x_1 \\ x_2 \\ x_3 \end{bmatrix}$$

is an eigenvector of A corresponding to λ if and only if $\mathbf{x}$ is a nontrivial solution of $(\lambda I - A)\mathbf{x} = \mathbf{0}$, that is, of

$$\begin{bmatrix} \lambda & 0 & 2 \\ -1 & \lambda - 2 & -1 \\ -1 & 0 & \lambda - 3 \end{bmatrix} \begin{bmatrix} x_1 \\ x_2 \\ x_3 \end{bmatrix} = \begin{bmatrix} 0 \\ 0 \\ 0 \end{bmatrix} \tag{3}$$

If $\lambda = 2$, then (3) becomes

$$\begin{bmatrix} 2 & 0 & 2 \\ -1 & 0 & -1 \\ -1 & 0 & -1 \end{bmatrix} \begin{bmatrix} x_1 \\ x_2 \\ x_3 \end{bmatrix} = \begin{bmatrix} 0 \\ 0 \\ 0 \end{bmatrix}$$

Solving this system yields (verify)

$$x_1 = -s, \qquad x_2 = t, \qquad x_3 = s$$

Thus, the eigenvectors of A corresponding to $\lambda = 2$ are the nonzero vectors of the form

$$\mathbf{x} = \begin{bmatrix} -s \\ t \\ s \end{bmatrix} = \begin{bmatrix} -s \\ 0 \\ s \end{bmatrix} + \begin{bmatrix} 0 \\ t \\ 0 \end{bmatrix} = s \begin{bmatrix} -1 \\ 0 \\ 1 \end{bmatrix} + t \begin{bmatrix} 0 \\ 1 \\ 0 \end{bmatrix}$$

Since

$$\begin{bmatrix} -1 \\ 0 \\ 1 \end{bmatrix} \quad \text{and} \quad \begin{bmatrix} 0 \\ 1 \\ 0 \end{bmatrix}$$

are linearly independent, these vectors form a basis for the eigenspace corresponding to $\lambda = 2$.

If $\lambda = 1$, then (3) becomes

$$\begin{bmatrix} 1 & 0 & 2 \\ -1 & -1 & -1 \\ -1 & 0 & -2 \end{bmatrix} \begin{bmatrix} x_1 \\ x_2 \\ x_3 \end{bmatrix} = \begin{bmatrix} 0 \\ 0 \\ 0 \end{bmatrix}$$

Solving this system yields (verify)

$$x_1 = -2s, \qquad x_2 = s, \qquad x_3 = s$$

Thus, the eigenvectors corresponding to $\lambda = 1$ are the nonzero vectors of the form

$$\begin{bmatrix} -2s \\ s \\ s \end{bmatrix} = s \begin{bmatrix} -2 \\ 1 \\ 1 \end{bmatrix} \quad \text{so that} \quad \begin{bmatrix} -2 \\ 1 \\ 1 \end{bmatrix}$$

is a basis for the eigenspace corresponding to $\lambda = 1$. ◆

Powers of a Matrix Once the eigenvalues and eigenvectors of a matrix A are found, it is a simple matter to find the eigenvalues and eigenvectors of any positive integer power of A; for example, if λ is an eigenvalue of A and $\mathbf{x}$ is a corresponding eigenvector, then

$$A^2\mathbf{x} = A(A\mathbf{x}) = A(\lambda\mathbf{x}) = \lambda(A\mathbf{x}) = \lambda(\lambda\mathbf{x}) = \lambda^2\mathbf{x}$$

which shows that λ^2 is an eigenvalue of A^2 and $\mathbf{x}$ is a corresponding eigenvector. In general, we have the following result.

Theorem 7.1.3

If k is a positive integer, λ is an eigenvalue of a matrix A, and $\mathbf{x}$ is a corresponding eigenvector, then λ^k is an eigenvalue of A^k and $\mathbf{x}$ is a corresponding eigenvector.

EXAMPLE 6 Using Theorem 7.1.3

In Example 5 we showed that the eigenvalues of

$$A = \begin{bmatrix} 0 & 0 & -2 \\ 1 & 2 & 1 \\ 1 & 0 & 3 \end{bmatrix}$$

are $\lambda = 2$ and $\lambda = 1$, so from Theorem 7.1.3 both $\lambda = 2^7 = 128$ and $\lambda = 1^7 = 1$ are eigenvalues of A^7. We also showed that

$$\begin{bmatrix} -1 \\ 0 \\ 1 \end{bmatrix} \quad \text{and} \quad \begin{bmatrix} 0 \\ 1 \\ 0 \end{bmatrix}$$

are eigenvectors of A corresponding to the eigenvalue $\lambda = 2$, so from Theorem 7.1.3 they are also eigenvectors of A^7 corresponding to $\lambda = 2^7 = 128$. Similarly, the eigenvector

$$\begin{bmatrix} -2 \\ 1 \\ 1 \end{bmatrix}$$

of A corresponding to the eigenvalue $\lambda = 1$ is also an eigenvector of A^7 corresponding to $\lambda = 1^7 = 1$. ◆

Eigenvalues and Invertibility The next theorem establishes a relationship between the eigenvalues and the invertibility of a matrix.

Theorem 7.1.4

A square matrix A is invertible if and only if $\lambda = 0$ is not an eigenvalue of A.

Proof. Assume that A is an $n \times n$ matrix and observe first that $\lambda = 0$ is a solution of the characteristic equation

$$\lambda^n + c_1\lambda^{n-1} + \cdots + c_n = 0$$

if and only if the constant term c_n is zero. Thus, it suffices to prove that A is invertible if and only if $c_n \neq 0$. But

$$\det(\lambda I - A) = \lambda^n + c_1\lambda^{n-1} + \cdots + c_n$$

or on setting $\lambda = 0$,

$$\det(-A) = c_n \quad \text{or} \quad (-1)^n \det(A) = c_n$$

It follows from the last equation that $\det(A) = 0$ if and only if $c_n = 0$, and this in turn implies that A is invertible if and only if $c_n \neq 0$. ■

EXAMPLE 7 Using Theorem 7.1.4

The matrix A in Example 5 is invertible since it has eigenvalues $\lambda = 1$ and $\lambda = 2$, neither of which is zero. We leave it for the reader to check this conclusion by showing that $\det(A) \neq 0$. ♦

Summary Theorem 7.1.4 enables us to add an additional result to Theorem 6.4.5.

Theorem 7.1.5 Equivalent Statements

If A is an $n \times n$ matrix, and if $T_A\colon R^n \to R^n$ is multiplication by A, then the following are equivalent.

(a) *A is invertible.*
(b) $A\mathbf{x} = \mathbf{0}$ *has only the trivial solution.*
(c) *The reduced row-echelon form of A is I_n.*
(d) *A is expressible as a product of elementary matrices.*
(e) $A\mathbf{x} = \mathbf{b}$ *is consistent for every $n \times 1$ matrix* $\mathbf{b}$.
(f) $A\mathbf{x} = \mathbf{b}$ *has exactly one solution for every $n \times 1$ matrix* $\mathbf{b}$.
(g) $\det(A) \neq 0$.
(h) *The range of T_A is R^n.*
(i) *T_A is one-to-one.*
(j) *The column vectors of A are linearly independent.*
(k) *The row vectors of A are linearly independent.*
(l) *The column vectors of A span R^n.*
(m) *The row vectors of A span R^n.*
(n) *The column vectors of A form a basis for R^n.*
(o) *The row vectors of A form a basis for R^n.*
(p) *A has rank n.*
(q) *A has nullity* 0.
(r) *The orthogonal complement of the nullspace of A is R^n.*
(s) *The orthogonal complement of the row space of A is* $\{\mathbf{0}\}$.
(t) *$A^T A$ is invertible.*
(u) *$\lambda = 0$ is not an eigenvalue of A.*

This theorem relates all of the major topics we have studied thus far.

Exercise Set 7.1

1. Find the characteristic equations of the following matrices:

(a) $\begin{bmatrix} 3 & 0 \\ 8 & -1 \end{bmatrix}$ (b) $\begin{bmatrix} 10 & -9 \\ 4 & -2 \end{bmatrix}$ (c) $\begin{bmatrix} 0 & 3 \\ 4 & 0 \end{bmatrix}$ (d) $\begin{bmatrix} -2 & -7 \\ 1 & 2 \end{bmatrix}$ (e) $\begin{bmatrix} 0 & 0 \\ 0 & 0 \end{bmatrix}$ (f) $\begin{bmatrix} 1 & 0 \\ 0 & 1 \end{bmatrix}$

2. Find the eigenvalues of the matrices in Exercise 1.

3. Find bases for the eigenspaces of the matrices in Exercise 1.

4. Find the characteristic equations of the following matrices:

(a) $\begin{bmatrix} 4 & 0 & 1 \\ -2 & 1 & 0 \\ -2 & 0 & 1 \end{bmatrix}$ (b) $\begin{bmatrix} 3 & 0 & -5 \\ \frac{1}{5} & -1 & 0 \\ 1 & 1 & -2 \end{bmatrix}$ (c) $\begin{bmatrix} -2 & 0 & 1 \\ -6 & -2 & 0 \\ 19 & 5 & -4 \end{bmatrix}$

(d) $\begin{bmatrix} -1 & 0 & 1 \\ -1 & 3 & 0 \\ -4 & 13 & -1 \end{bmatrix}$ (e) $\begin{bmatrix} 5 & 0 & 1 \\ 1 & 1 & 0 \\ -7 & 1 & 0 \end{bmatrix}$ (f) $\begin{bmatrix} 5 & 6 & 2 \\ 0 & -1 & -8 \\ 1 & 0 & -2 \end{bmatrix}$

5. Find the eigenvalues of the matrices in Exercise 4.

6. Find bases for the eigenspaces of the matrices in Exercise 4.

7. Find the characteristic equations of the following matrices:

(a) $\begin{bmatrix} 0 & 0 & 2 & 0 \\ 1 & 0 & 1 & 0 \\ 0 & 1 & -2 & 0 \\ 0 & 0 & 0 & 1 \end{bmatrix}$ (b) $\begin{bmatrix} 10 & -9 & 0 & 0 \\ 4 & -2 & 0 & 0 \\ 0 & 0 & -2 & -7 \\ 0 & 0 & 1 & 2 \end{bmatrix}$

8. Find the eigenvalues of the matrices in Exercise 7.

9. Find bases for the eigenspaces of the matrices in Exercise 7.

10. By inspection, find the eigenvalues of the following matrices:

(a) $\begin{bmatrix} -1 & 6 \\ 0 & 5 \end{bmatrix}$ (b) $\begin{bmatrix} 3 & 0 & 0 \\ -2 & 7 & 0 \\ 4 & 8 & 1 \end{bmatrix}$ (c) $\begin{bmatrix} -\frac{1}{3} & 0 & 0 & 0 \\ 0 & -\frac{1}{3} & 0 & 0 \\ 0 & 0 & 1 & 0 \\ 0 & 0 & 0 & \frac{1}{2} \end{bmatrix}$

11. Find the eigenvalues of A^9 for

$$A = \begin{bmatrix} 1 & 3 & 7 & 11 \\ 0 & \frac{1}{2} & 3 & 8 \\ 0 & 0 & 0 & 4 \\ 0 & 0 & 0 & 2 \end{bmatrix}$$

12. Find the eigenvalues and bases for the eigenspaces of A^{25} for

$$A = \begin{bmatrix} -1 & -2 & -2 \\ 1 & 2 & 1 \\ -1 & -1 & 0 \end{bmatrix}$$

13. Let A be a 2×2 matrix, and call a line through the origin of R^2 ***invariant*** under A if $A\mathbf{x}$ lies on the line when $\mathbf{x}$ does. Find equations for all lines in R^2, if any, that are invariant under the given matrix.

(a) $A = \begin{bmatrix} 4 & -1 \\ 2 & 1 \end{bmatrix}$ (b) $A = \begin{bmatrix} 0 & 1 \\ -1 & 0 \end{bmatrix}$ (c) $A = \begin{bmatrix} 2 & 3 \\ 0 & 2 \end{bmatrix}$

14. Find $\det(A)$ given that A has $p(\lambda)$ as its characteristic polynomial.

(a) $p(\lambda) = \lambda^3 - 2\lambda^2 + \lambda + 5$ (b) $p(\lambda) = \lambda^4 - \lambda^3 + 7$

[***Hint.*** See the proof of Theorem 7.1.4.]

15. Let A be an $n \times n$ matrix.

(a) Prove that the characteristic polynomial of A has degree n.

(b) Prove that the coefficient of λ^n in the characteristic polynomial is 1.

16. Show that the characteristic equation of a 2×2 matrix A can be expressed as $\lambda^2 - \text{tr}(A)\lambda + \det(A) = 0$, where $\text{tr}(A)$ is the trace of A.

17. Use the result in Exercise 16 to show that if

$$A = \begin{bmatrix} a & b \\ c & d \end{bmatrix}$$

then the solutions of the characteristic equation of A are

$$\lambda = \tfrac{1}{2}\left[(a+d) \pm \sqrt{(a-d)^2 + 4bc}\,\right]$$

Use this result to show that A has

(a) two distinct real eigenvalues if $(a-d)^2 + 4bc > 0$ (b) one real eigenvalue if $(a-d)^2 + 4bc = 0$
(c) no real eigenvalues if $(a-d)^2 + 4bc < 0$

18. Let A be the matrix in Exercise 17. Show that if $(a-d)^2 + 4bc > 0$ and $b \neq 0$, then eigenvectors of A corresponding to the eigenvalues

$$\lambda_1 = \tfrac{1}{2}\left[(a+d) + \sqrt{(a-d)^2 + 4bc}\,\right] \quad \text{and} \quad \lambda_2 = \tfrac{1}{2}\left[(a+d) - \sqrt{(a-d)^2 + 4bc}\,\right]$$

are

$$\begin{bmatrix} -b \\ a - \lambda_1 \end{bmatrix} \quad \text{and} \quad \begin{bmatrix} -b \\ a - \lambda_2 \end{bmatrix}$$

respectively.

19. Prove: If a, b, c, and d are integers such that $a + b = c + d$, then

$$A = \begin{bmatrix} a & b \\ c & d \end{bmatrix}$$

has integer eigenvalues, namely, $\lambda_1 = a + b$ and $\lambda_2 = a - c$. [***Hint.*** See Exercise 17.]

20. Prove: If λ is an eigenvalue of an invertible matrix A and $\mathbf{x}$ is a corresponding eigenvector, then $1/\lambda$ is an eigenvalue of A^{-1}, and $\mathbf{x}$ is a corresponding eigenvector.

21. Prove: If λ is an eigenvalue of A, $\mathbf{x}$ is a corresponding eigenvector, and s is a scalar, then $\lambda - s$ is an eigenvalue of $A - sI$, and $\mathbf{x}$ is a corresponding eigenvector.

22. Find the eigenvalues and bases for the eigenspaces of

$$A = \begin{bmatrix} -2 & 2 & 3 \\ -2 & 3 & 2 \\ -4 & 2 & 5 \end{bmatrix}$$

Then use Exercises 20 and 21 to find the eigenvalues and bases for the eigenspaces of

(a) A^{-1} (b) $A - 3I$ (c) $A + 2I$

23. (a) Prove that if A is a square matrix, then A and A^T have the same eigenvalues. [***Hint.*** Look at the characteristic equation $\det(\lambda I - A) = 0$.]

(b) Show that A and A^T need not have the same eigenspaces. [***Hint.*** Use the result in Exercise 18 to find a 2×2 matrix for which A and A^T have different eigenspaces.]

Discussion and Discovery

24. Indicate whether the statement is always true or sometimes false. Justify your answer by giving a logical argument or a counterexample. In each part, A is an $n \times n$ matrix.

(a) If $A\mathbf{x} = \lambda\mathbf{x}$ for some nonzero scalar λ, then $\mathbf{x}$ is an eigenvector of A.

(b) If λ is not an eigenvalue of A, then the linear system $(\lambda I - A)\mathbf{x} = \mathbf{0}$ has only the trivial solution.

(c) If $\lambda = 0$ is an eigenvalue of A, then A^2 is singular.

(d) If the characteristic polynomial of A is $p(\lambda) = \lambda^n + 1$, then A is invertible.

25. Suppose that the characteristic polynomial of some matrix A is $p(\lambda) = (\lambda-1)(\lambda-3)^2(\lambda-4)^3$. In each part answer the question and explain your reasoning.

(a) What is the size of A?

(b) Is A invertible?

(c) How many eigenspaces does A have?

7.2 DIAGONALIZATION

In this section we shall be concerned with the problem of finding a basis for R^n that consists of eigenvectors of a given $n \times n$ matrix A. Such bases can be used to study geometric properties of A and to simplify various numerical computations involving A. These bases are also of physical significance in a wide variety of applications, some of which will be considered later in this text.

The Matrix Diagonalization Problem Our first objective in this section is to show that the following two problems, which on the surface seem quite different, are actually equivalent.

The Eigenvector Problem. Given an $n \times n$ matrix A, does there exist a basis for R^n consisting of eigenvectors of A?

The Diagonalization Problem (Matrix Form). Given an $n \times n$ matrix A, does there exist an invertible matrix P such that $P^{-1}AP$ is a diagonal matrix?

The latter problem suggests the following terminology.

Definition

A square matrix A is called ***diagonalizable*** if there is an invertible matrix P such that $P^{-1}AP$ is a diagonal matrix; the matrix P is said to ***diagonalize*** A.

The following theorem shows that the eigenvector problem and the diagonalization problem are equivalent.

Theorem 7.2.1

If A is an $n \times n$ matrix, then the following are equivalent.

(a) A is diagonalizable.

(b) A has n linearly independent eigenvectors.

Proof (a) $\Rightarrow$ (b). Since A is assumed diagonalizable, there is an invertible matrix

$$P = \begin{bmatrix} p_{11} & p_{12} & \cdots & p_{1n} \\ p_{21} & p_{22} & \cdots & p_{2n} \\ \vdots & \vdots & & \vdots \\ p_{n1} & p_{n2} & \cdots & p_{nn} \end{bmatrix}$$

such that $P^{-1}AP$ is diagonal, say $P^{-1}AP = D$, where

$$D = \begin{bmatrix} \lambda_1 & 0 & \cdots & 0 \\ 0 & \lambda_2 & \cdots & 0 \\ \vdots & \vdots & & \vdots \\ 0 & 0 & \cdots & \lambda_n \end{bmatrix}$$

It follows from the formula $P^{-1}AP = D$ that $AP = PD$; that is,

$$AP = \begin{bmatrix} p_{11} & p_{12} & \cdots & p_{1n} \\ p_{21} & p_{22} & \cdots & p_{2n} \\ \vdots & \vdots & & \vdots \\ p_{n1} & p_{n2} & \cdots & p_{nn} \end{bmatrix} \begin{bmatrix} \lambda_1 & 0 & \cdots & 0 \\ 0 & \lambda_2 & \cdots & 0 \\ \vdots & \vdots & & \vdots \\ 0 & 0 & \cdots & \lambda_n \end{bmatrix} = \begin{bmatrix} \lambda_1 p_{11} & \lambda_2 p_{12} & \cdots & \lambda_n p_{1n} \\ \lambda_1 p_{21} & \lambda_2 p_{22} & \cdots & \lambda_n p_{2n} \\ \vdots & \vdots & & \vdots \\ \lambda_1 p_{n1} & \lambda_2 p_{n2} & \cdots & \lambda_n p_{nn} \end{bmatrix} \tag{1}$$

If we now let $\mathbf{p}_1, \mathbf{p}_2, \ldots, \mathbf{p}_n$ denote the column vectors of P, then from (1) the successive columns of AP are $\lambda_1\mathbf{p}_1, \lambda_2\mathbf{p}_2, \ldots, \lambda_n\mathbf{p}_n$. However, from Formula (6) of Section 1.3, the successive columns of AP are $A\mathbf{p}_1, A\mathbf{p}_2, \ldots, A\mathbf{p}_n$. Thus, we must have

$$A\mathbf{p}_1 = \lambda_1\mathbf{p}_1, \quad A\mathbf{p}_2 = \lambda_2\mathbf{p}_2, \ldots, \quad A\mathbf{p}_n = \lambda_n\mathbf{p}_n \tag{2}$$

Since P is invertible, its column vectors are all nonzero; thus, it follows from (2) that $\lambda_1, \lambda_2, \ldots, \lambda_n$ are eigenvalues of A, and $\mathbf{p}_1, \mathbf{p}_2, \ldots, \mathbf{p}_n$ are corresponding eigenvectors. Since P is invertible, it follows from Theorem 7.1.5 that $\mathbf{p}_1, \mathbf{p}_2, \ldots, \mathbf{p}_n$ are linearly independent. Thus, A has n linearly independent eigenvectors.

(b) $\Rightarrow$ *(a)*. Assume that A has n linearly independent eigenvectors, $\mathbf{p}_1, \mathbf{p}_2, \ldots, \mathbf{p}_n$, with corresponding eigenvalues $\lambda_1, \lambda_2, \ldots, \lambda_n$, and let

$$P = \begin{bmatrix} p_{11} & p_{12} & \cdots & p_{1n} \\ p_{21} & p_{22} & \cdots & p_{2n} \\ \vdots & \vdots & & \vdots \\ p_{n1} & p_{n2} & \cdots & p_{nn} \end{bmatrix}$$

be the matrix whose column vectors are $\mathbf{p}_1, \mathbf{p}_2, \ldots, \mathbf{p}_n$. By Formula (6) of Section 1.3, the column vectors of the product AP are

$$A\mathbf{p}_1, A\mathbf{p}_2, \ldots, A\mathbf{p}_n$$

But

$$A\mathbf{p}_1 = \lambda_1\mathbf{p}_1, \quad A\mathbf{p}_2 = \lambda_2\mathbf{p}_2, \ldots, \quad A\mathbf{p}_n = \lambda_n\mathbf{p}_n$$

so that

$$AP = \begin{bmatrix} \lambda_1 p_{11} & \lambda_2 p_{12} & \cdots & \lambda_n p_{1n} \\ \lambda_1 p_{21} & \lambda_2 p_{22} & \cdots & \lambda_n p_{2n} \\ \vdots & \vdots & & \vdots \\ \lambda_1 p_{n1} & \lambda_2 p_{n2} & \cdots & \lambda_n p_{nn} \end{bmatrix}$$

$$= \begin{bmatrix} p_{11} & p_{12} & \cdots & p_{1n} \\ p_{21} & p_{22} & \cdots & p_{2n} \\ \vdots & \vdots & & \vdots \\ p_{n1} & p_{n2} & \cdots & p_{nn} \end{bmatrix} \begin{bmatrix} \lambda_1 & 0 & \cdots & 0 \\ 0 & \lambda_2 & \cdots & 0 \\ \vdots & \vdots & & \vdots \\ 0 & 0 & \cdots & \lambda_n \end{bmatrix} = PD \tag{3}$$

where D is the diagonal matrix having the eigenvalues $\lambda_1, \lambda_2, \ldots, \lambda_n$ on the main diagonal. Since the column vectors of P are linearly independent, P is invertible; thus, (3) can be rewritten as $P^{-1}AP = D$; that is, A is diagonalizable. ■

Procedure for Diagonalizing a Matrix The preceding theorem guarantees that an $n \times n$ matrix A with n linearly independent eigenvectors is diagonalizable, and the proof provides the following method for diagonalizing A.

Step 1. Find n linearly independent eigenvectors of A, say, $\mathbf{p}_1, \mathbf{p}_2, \ldots, \mathbf{p}_n$.

Step 2. Form the matrix P having $\mathbf{p}_1, \mathbf{p}_2, \ldots, \mathbf{p}_n$ as its column vectors.

Step 3. The matrix $P^{-1}AP$ will then be diagonal with $\lambda_1, \lambda_2, \ldots, \lambda_n$ as its successive diagonal entries, where λ_i is the eigenvalue corresponding to $\mathbf{p}_i$, for $i = 1, 2, \ldots, n$.

In order to carry out Step 1 of this procedure, one first needs a way of determining whether a given $n \times n$ matrix A has n linearly independent eigenvectors, and then one needs a method for finding them. One can address both problems at once by finding bases for the eigenspaces of A. Later in this section we will show that those basis vectors, as a combined set, are linearly independent, so that if there is a total of n such vectors, then A is diagonalizable, and the n basis vectors can be used as the column vectors of the diagonalizing matrix P. If there are fewer than n basis vectors, then A is not diagonalizable.

EXAMPLE 1 Finding a Matrix P That Diagonalizes a Matrix A

Find a matrix P that diagonalizes

$$A = \begin{bmatrix} 0 & 0 & -2 \\ 1 & 2 & 1 \\ 1 & 0 & 3 \end{bmatrix}$$

Solution.

From Example 5 of the preceding section we found the characteristic equation of A to be

$$(\lambda - 1)(\lambda - 2)^2 = 0$$

and we found the following bases for the eigenspaces:

$$\lambda = 2: \quad \mathbf{p}_1 = \begin{bmatrix} -1 \\ 0 \\ 1 \end{bmatrix}, \quad \mathbf{p}_2 = \begin{bmatrix} 0 \\ 1 \\ 0 \end{bmatrix} \qquad \lambda = 1: \quad \mathbf{p}_3 = \begin{bmatrix} -2 \\ 1 \\ 1 \end{bmatrix}$$

There are three basis vectors in total, so the matrix A is diagonalizable and

$$P = \begin{bmatrix} -1 & 0 & -2 \\ 0 & 1 & 1 \\ 1 & 0 & 1 \end{bmatrix}$$

diagonalizes A. As a check, the reader should verify that

$$P^{-1}AP = \begin{bmatrix} 1 & 0 & 2 \\ 1 & 1 & 1 \\ -1 & 0 & -1 \end{bmatrix} \begin{bmatrix} 0 & 0 & -2 \\ 1 & 2 & 1 \\ 1 & 0 & 3 \end{bmatrix} \begin{bmatrix} -1 & 0 & -2 \\ 0 & 1 & 1 \\ 1 & 0 & 1 \end{bmatrix} = \begin{bmatrix} 2 & 0 & 0 \\ 0 & 2 & 0 \\ 0 & 0 & 1 \end{bmatrix} \quad \blacklozenge$$

There is no preferred order for the columns of P. Since the ith diagonal entry of $P^{-1}AP$ is an eigenvalue for the ith column vector of P, changing the order of the columns of P just changes the order of the eigenvalues on the diagonal of $P^{-1}AP$. Thus, had we written

$$P = \begin{bmatrix} -1 & -2 & 0 \\ 0 & 1 & 1 \\ 1 & 1 & 0 \end{bmatrix}$$

in Example 1, we would have obtained

$$P^{-1}AP = \begin{bmatrix} 2 & 0 & 0 \\ 0 & 1 & 0 \\ 0 & 0 & 2 \end{bmatrix}$$

EXAMPLE 2 A Matrix That Is Not Diagonalizable

Find a matrix P that diagonalizes

$$A = \begin{bmatrix} 1 & 0 & 0 \\ 1 & 2 & 0 \\ -3 & 5 & 2 \end{bmatrix}$$

Solution.

The characteristic polynomial of A is

$$\det(\lambda I - A) = \begin{vmatrix} \lambda - 1 & 0 & 0 \\ -1 & \lambda - 2 & 0 \\ 3 & -5 & \lambda - 2 \end{vmatrix} = (\lambda - 1)(\lambda - 2)^2$$

so the characteristic equation is

$$(\lambda - 1)(\lambda - 2)^2 = 0$$

Thus, the eigenvalues of A are $\lambda = 1$ and $\lambda = 2$. We leave it for the reader to show that bases for the eigenspaces are

$$\lambda = 1: \quad \mathbf{p}_1 = \begin{bmatrix} \frac{1}{8} \\ -\frac{1}{8} \\ 1 \end{bmatrix} \qquad \lambda = 2: \quad \mathbf{p}_2 = \begin{bmatrix} 0 \\ 0 \\ 1 \end{bmatrix}$$

Since A is a 3×3 matrix and there are only two basis vectors in total, A is not diagonalizable.

Alternative Solution.

If one is interested only in determining whether a matrix is diagonalizable and is not concerned with actually finding a diagonalizing matrix P, then it is not necessary to compute bases for the eigenspaces; it suffices to find the dimensions of the eigenspaces. For this example, the eigenspace corresponding to $\lambda = 1$ is the solution space of the system

$$\begin{bmatrix} 0 & 0 & 0 \\ -1 & -1 & 0 \\ 3 & -5 & -1 \end{bmatrix} \begin{bmatrix} x_1 \\ x_2 \\ x_3 \end{bmatrix} = \begin{bmatrix} 0 \\ 0 \\ 0 \end{bmatrix}$$

The coefficient matrix has rank 2 (verify). Thus, the nullity of this matrix is 1 by Theorem 5.6.3, and hence the solution space is one-dimensional.

The eigenspace corresponding to $\lambda = 2$ is the solution space of the system

$$\begin{bmatrix} 1 & 0 & 0 \\ -1 & 0 & 0 \\ 3 & -5 & 0 \end{bmatrix} \begin{bmatrix} x_1 \\ x_2 \\ x_3 \end{bmatrix} = \begin{bmatrix} 0 \\ 0 \\ 0 \end{bmatrix}$$

This coefficient matrix also has rank 2 and nullity 1 (verify), so the eigenspace corresponding to $\lambda = 2$ is also one-dimensional. Since the eigenspaces produce a total of two basis vectors, the matrix A is not diagonalizable. ♦

There is an assumption in Example 1 that the column vectors of P, which are made up of basis vectors from the various eigenspaces of A, are linearly independent. The following theorem addresses this issue.

Theorem 7.2.2

If $\mathbf{v}_1, \mathbf{v}_2, \ldots, \mathbf{v}_k$ are eigenvectors of A corresponding to distinct eigenvalues $\lambda_1, \lambda_2, \ldots, \lambda_k$, then $\{\mathbf{v}_1, \mathbf{v}_2, \ldots, \mathbf{v}_k\}$ is a linearly independent set.

Proof. Let $\mathbf{v}_1, \mathbf{v}_2, \ldots, \mathbf{v}_k$ be eigenvectors of A corresponding to distinct eigenvalues $\lambda_1, \lambda_2, \ldots, \lambda_k$. We shall assume that $\mathbf{v}_1, \mathbf{v}_2, \ldots, \mathbf{v}_k$ are linearly dependent and obtain a contradiction. We can then conclude that $\mathbf{v}_1, \mathbf{v}_2, \ldots, \mathbf{v}_k$ are linearly independent.

Since an eigenvector is nonzero by definition, $\{\mathbf{v}_1\}$ is linearly independent. Let r be the largest integer such that $\{\mathbf{v}_1, \mathbf{v}_2, \ldots, \mathbf{v}_r\}$ is linearly independent. Since we are assuming that $\{\mathbf{v}_1, \mathbf{v}_2, \ldots, \mathbf{v}_k\}$ is linearly dependent, r satisfies $1 \leq r < k$. Moreover, by definition of r, $\{\mathbf{v}_1, \mathbf{v}_2, \ldots, \mathbf{v}_{r+1}\}$ is linearly dependent. Thus, there are scalars $c_1, c_2, \ldots, c_{r+1}$, not all zero, such that

$$c_1\mathbf{v}_1 + c_2\mathbf{v}_2 + \cdots + c_{r+1}\mathbf{v}_{r+1} = \mathbf{0} \tag{4}$$

Multiplying both sides of (4) by A and using

$$A\mathbf{v}_1 = \lambda_1\mathbf{v}_1, \quad A\mathbf{v}_2 = \lambda_2\mathbf{v}_2, \ldots, \quad A\mathbf{v}_{r+1} = \lambda_{r+1}\mathbf{v}_{r+1}$$

we obtain

$$c_1\lambda_1\mathbf{v}_1 + c_2\lambda_2\mathbf{v}_2 + \cdots + c_{r+1}\lambda_{r+1}\mathbf{v}_{r+1} = \mathbf{0} \tag{5}$$

Multiplying both sides of (4) by λ_{r+1} and subtracting the resulting equation from (5) yields

$$c_1(\lambda_1 - \lambda_{r+1})\mathbf{v}_1 + c_2(\lambda_2 - \lambda_{r+1})\mathbf{v}_2 + \cdots + c_r(\lambda_r - \lambda_{r+1})\mathbf{v}_r = \mathbf{0}$$

Since $\{\mathbf{v}_1, \mathbf{v}_2, \ldots, \mathbf{v}_r\}$ is a linearly independent set, this equation implies that

$$c_1(\lambda_1 - \lambda_{r+1}) = c_2(\lambda_2 - \lambda_{r+1}) = \cdots = c_r(\lambda_r - \lambda_{r+1}) = 0$$

and since $\lambda_1, \lambda_2, \ldots, \lambda_{r+1}$ are distinct, it follows that

$$c_1 = c_2 = \cdots = c_r = 0 \tag{6}$$

Substituting these values in (4) yields

$$c_{r+1}\mathbf{v}_{r+1} = \mathbf{0}$$

Since the eigenvector $\mathbf{v}_{r+1}$ is nonzero, it follows that

$$c_{r+1} = 0 \tag{7}$$

Equations 6 and 7 contradict the fact that $c_1, c_2, \ldots, c_{r+1}$ are not all zero; this completes the proof. ■

REMARK. Theorem 7.2.2 is a special case of a more general result: Suppose that $\lambda_1, \lambda_2, \ldots, \lambda_k$ are distinct eigenvalues and that we choose a linearly independent set in each of the corresponding eigenspaces. If we then merge all these vectors into a single set, the result will still be a linearly independent set. For example, if we choose three linearly independent vectors from one eigenspace and two linearly independent vectors from another eigenspace, then the five vectors together form a linearly independent set. We omit the proof.

As a consequence of Theorem 7.2.2, we obtain the following important result.

Theorem 7.2.3

If an $n \times n$ matrix A has n distinct eigenvalues, then A is diagonalizable.

Proof. If $\mathbf{v}_1, \mathbf{v}_2, \ldots, \mathbf{v}_n$ are eigenvectors corresponding to the distinct eigenvalues $\lambda_1, \lambda_2, \ldots, \lambda_n$, then by Theorem 7.2.2, $\mathbf{v}_1, \mathbf{v}_2, \ldots, \mathbf{v}_n$ are linearly independent. Thus, A is diagonalizable by Theorem 7.2.1. ■

EXAMPLE 3 Using Theorem 7.2.3

We saw in Example 2 of the preceding section that

$$A = \begin{bmatrix} 0 & 1 & 0 \\ 0 & 0 & 1 \\ 4 & -17 & 8 \end{bmatrix}$$

has three distinct eigenvalues, $\lambda = 4$, $\lambda = 2 + \sqrt{3}$, $\lambda = 2 - \sqrt{3}$. Therefore, A is diagonalizable. Further,

$$P^{-1}AP = \begin{bmatrix} 4 & 0 & 0 \\ 0 & 2+\sqrt{3} & 0 \\ 0 & 0 & 2-\sqrt{3} \end{bmatrix}$$

for some invertible matrix P. If desired, the matrix P can be found using the method shown in Example 1 of this section. ♦

EXAMPLE 4 A Diagonalizable Matrix

From Theorem 7.1.1 the eigenvalues of a triangular matrix are the entries on its main diagonal. Thus, a triangular matrix with distinct entries on the main diagonal is diagonalizable. For example,

$$A = \begin{bmatrix} -1 & 2 & 4 & 0 \\ 0 & 3 & 1 & 7 \\ 0 & 0 & 5 & 8 \\ 0 & 0 & 0 & -2 \end{bmatrix}$$

is a diagonalizable matrix. ♦

Geometric and Algebraic Multiplicity Theorem 7.2.3 does not completely settle the diagonalization problem, since it is possible for an $n \times n$ matrix A to be diagonalizable without having n distinct eigenvalues. We saw this in Example 1, where the given 3×3 matrix had only two distinct eigenvalues, yet was diagonalizable. What really matters for diagonalizability are the dimensions of the eigenspaces—those dimensions must add up to n in order for an $n \times n$ matrix to be diagonalizable. Examples 1 and 2 illustrate this; the matrices in those examples have the same characteristic equation and the same eigenvalues, but the matrix in Example 1 is diagonalizable because the dimensions of the eigenspaces add to 3 and the matrix in Example 2 is not diagonalizable because the dimensions only add to 2.

A full excursion into the study of diagonalizability is left for more advanced courses, but we shall touch on one theorem that is important to a fuller understanding of diagonalizability. It can be proved that if λ_0 is an eigenvalue of A, then the dimension of the eigenspace corresponding to λ_0 cannot exceed the number of times that $\lambda - \lambda_0$ appears as a factor in the characteristic polynomial of A. For example, in Examples 1 and 2 the characteristic polynomial is

$$(\lambda - 1)(\lambda - 2)^2$$

Thus, the eigenspace corresponding to $\lambda = 1$ is at most (hence exactly) one-dimensional and the eigenspace corresponding to $\lambda = 2$ is at most two-dimensional. In Example 1, the eigenspace corresponding to $\lambda = 2$ actually had dimension 2, resulting in diagonalizability, but in Example 2 that eigenspace had only dimension 1, resulting in the failure of diagonalizability.

There is some terminology that relates to these ideas. If λ_0 is an eigenvalue of an $n \times n$ matrix A, then the dimension of the eigenspace corresponding to λ_0 is called the ***geometric multiplicity*** of λ_0, and the number of times that $\lambda - \lambda_0$ appears as a factor in the characteristic polynomial of A is called the ***algebraic multiplicity*** of A. The following theorem, which we state without proof, summarizes the preceding discussion.

Theorem 7.2.4 Geometric and Algebraic Multiplicity

If A is a square matrix, then:

(*a*) *For every eigenvalue of A the geometric multiplicity is less than or equal to the algebraic multiplicity.*

(*b*) *A is diagonalizable if and only if the geometric multiplicity is equal to the algebraic multiplicity for every eigenvalue.*

Computing Powers of a Matrix There are numerous problems in applied mathematics that require the computation of high powers of a square matrix. We shall conclude this section by showing how diagonalization can be used to simplify such computations for diagonalizable matrices.

If A is an $n \times n$ matrix and P is an invertible matrix, then

$$(P^{-1}AP)^2 = P^{-1}APP^{-1}AP = P^{-1}AIAP = P^{-1}A^2P$$

More generally, for any positive integer k

$$(P^{-1}AP)^k = P^{-1}A^kP \tag{8}$$

It follows from this equation that if A is diagonalizable, and $P^{-1}AP = D$ is a diagonal matrix, then

$$P^{-1}A^kP = (P^{-1}AP)^k = D^k \tag{9}$$

Solving this equation for A^k yields

$$A^k = PD^kP^{-1} \tag{10}$$

This last equation expresses the kth power of A in terms of the kth power of the diagonal matrix D. But D^k is easy to compute; for example, if

$$D = \begin{bmatrix} d_1 & 0 & \cdots & 0 \\ 0 & d_2 & \cdots & 0 \\ \vdots & \vdots & & \vdots \\ 0 & 0 & \cdots & d_n \end{bmatrix}, \quad \text{then} \quad D^k = \begin{bmatrix} d_1^k & 0 & \cdots & 0 \\ 0 & d_2^k & \cdots & 0 \\ \vdots & \vdots & & \vdots \\ 0 & 0 & \cdots & d_n^k \end{bmatrix}$$

EXAMPLE 5 Power of a Matrix

Use (10) to find A^{13}, where

$$A = \begin{bmatrix} 0 & 0 & -2 \\ 1 & 2 & 1 \\ 1 & 0 & 3 \end{bmatrix}$$

Solution.

We showed in Example 1 that the matrix A is diagonalized by

$$P = \begin{bmatrix} -1 & 0 & -2 \\ 0 & 1 & 1 \\ 1 & 0 & 1 \end{bmatrix}$$

and that

$$D = P^{-1}AP = \begin{bmatrix} 2 & 0 & 0 \\ 0 & 2 & 0 \\ 0 & 0 & 1 \end{bmatrix}$$

Thus, from (10)

$$\begin{aligned} A^{13} = PD^{13}P^{-1} &= \begin{bmatrix} -1 & 0 & -2 \\ 0 & 1 & 1 \\ 1 & 0 & 1 \end{bmatrix} \begin{bmatrix} 2^{13} & 0 & 0 \\ 0 & 2^{13} & 0 \\ 0 & 0 & 1^{13} \end{bmatrix} \begin{bmatrix} 1 & 0 & 2 \\ 1 & 1 & 1 \\ -1 & 0 & -1 \end{bmatrix} \\ &= \begin{bmatrix} -8190 & 0 & -16382 \\ 8191 & 8192 & 8191 \\ 8191 & 0 & 16383 \end{bmatrix} \end{aligned} \tag{11}$$

♦

REMARK. With the method in the preceding example, most of the work is in diagonalizing A. Once that work is done, it can be used to compute any power of A. Thus, to compute A^{1000} we need only change the exponents from 13 to 1000 in (11).

Exercise Set 7.2

1. Let A be a 6×6 matrix with characteristic equation $\lambda^2(\lambda - 1)(\lambda - 2)^3 = 0$. What are the possible dimensions for eigenspaces of A?

2. Let

$$A = \begin{bmatrix} 4 & 0 & 1 \\ 2 & 3 & 2 \\ 1 & 0 & 4 \end{bmatrix}$$

(a) Find the eigenvalues of A.
(b) For each eigenvalue λ find the rank of the matrix $\lambda I - A$.
(c) Is A diagonalizable? Justify your conclusion.

In Exercises 3–7 use the method of Exercise 2 to determine whether the matrix is diagonalizable.

3. $\begin{bmatrix} 2 & 0 \\ 1 & 2 \end{bmatrix}$ **4.** $\begin{bmatrix} 2 & -3 \\ 1 & -1 \end{bmatrix}$ **5.** $\begin{bmatrix} 3 & 0 & 0 \\ 0 & 2 & 0 \\ 0 & 1 & 2 \end{bmatrix}$ **6.** $\begin{bmatrix} -1 & 0 & 1 \\ -1 & 3 & 0 \\ -4 & 13 & -1 \end{bmatrix}$ **7.** $\begin{bmatrix} 2 & -1 & 0 & 1 \\ 0 & 2 & 1 & -1 \\ 0 & 0 & 3 & 2 \\ 0 & 0 & 0 & 3 \end{bmatrix}$

In Exercises 8–11 find a matrix P that diagonalizes A, and determine $P^{-1}AP$.

8. $A = \begin{bmatrix} -14 & 12 \\ -20 & 17 \end{bmatrix}$ **9.** $A = \begin{bmatrix} 1 & 0 \\ 6 & -1 \end{bmatrix}$ **10.** $A = \begin{bmatrix} 1 & 0 & 0 \\ 0 & 1 & 1 \\ 0 & 1 & 1 \end{bmatrix}$ **11.** $A = \begin{bmatrix} 2 & 0 & -2 \\ 0 & 3 & 0 \\ 0 & 0 & 3 \end{bmatrix}$

In Exercises 12–17 determine whether A is diagonalizable. If so, find a matrix P that diagonalizes A, and determine $P^{-1}AP$.

12. $A = \begin{bmatrix} 19 & -9 & -6 \\ 25 & -11 & -9 \\ 17 & -9 & -4 \end{bmatrix}$ **13.** $A = \begin{bmatrix} -1 & 4 & -2 \\ -3 & 4 & 0 \\ -3 & 1 & 3 \end{bmatrix}$ **14.** $A = \begin{bmatrix} 5 & 0 & 0 \\ 1 & 5 & 0 \\ 0 & 1 & 5 \end{bmatrix}$

15. $A = \begin{bmatrix} 0 & 0 & 0 \\ 0 & 0 & 0 \\ 3 & 0 & 1 \end{bmatrix}$ **16.** $A = \begin{bmatrix} -2 & 0 & 0 & 0 \\ 0 & -2 & 0 & 0 \\ 0 & 0 & 3 & 0 \\ 0 & 0 & 1 & 3 \end{bmatrix}$ **17.** $A = \begin{bmatrix} -2 & 0 & 0 & 0 \\ 0 & -2 & 5 & -5 \\ 0 & 0 & 3 & 0 \\ 0 & 0 & 0 & 3 \end{bmatrix}$

18. Use the method of Example 5 to compute A^{10}, where

$$A = \begin{bmatrix} 1 & 0 \\ -1 & 2 \end{bmatrix}$$

19. Use the method of Example 5 to compute A^{11}, where

$$A = \begin{bmatrix} -1 & 7 & -1 \\ 0 & 1 & 0 \\ 0 & 15 & -2 \end{bmatrix}$$

20. In each part compute the stated power of

$$A = \begin{bmatrix} 1 & -2 & 8 \\ 0 & -1 & 0 \\ 0 & 0 & -1 \end{bmatrix}$$

(a) A^{1000} (b) A^{-1000} (c) A^{2301} (d) A^{-2301}

21. Find A^n if n is a positive integer and

$$A = \begin{bmatrix} 3 & -1 & 0 \\ -1 & 2 & -1 \\ 0 & -1 & 3 \end{bmatrix}$$

Linear Transformations from R^n to R^m In the special case where the equations in (1) are linear, the transformation $T: R^n \to R^m$ defined by those equations is called a ***linear transformation*** (or a ***linear operator*** if $m = n$). Thus, a linear transformation $T: R^n \to R^m$ is defined by equations of the form

$$\begin{aligned} w_1 &= a_{11}x_1 + a_{12}x_2 + \cdots + a_{1n}x_n \\ w_2 &= a_{21}x_1 + a_{22}x_2 + \cdots + a_{2n}x_n \\ \vdots \quad & \quad \vdots \qquad\quad \vdots \qquad\qquad\quad \vdots \\ w_m &= a_{m1}x_1 + a_{m2}x_2 + \cdots + a_{mn}x_n \end{aligned} \tag{2}$$

or in matrix notation

$$\begin{bmatrix} w_1 \\ w_2 \\ \vdots \\ w_m \end{bmatrix} = \begin{bmatrix} a_{11} & a_{12} & \cdots & a_{1n} \\ a_{21} & a_{22} & \cdots & a_{2n} \\ \vdots & \vdots & & \vdots \\ a_{m1} & a_{m2} & \cdots & a_{mn} \end{bmatrix} \begin{bmatrix} x_1 \\ x_2 \\ \vdots \\ x_n \end{bmatrix} \tag{3}$$

or more briefly by

$$\mathbf{w} = A\mathbf{x} \tag{4}$$

The matrix $A = [a_{ij}]$ is called the ***standard matrix*** for the linear transformation T, and T is called ***multiplication by*** A.

EXAMPLE 2 A Linear Transformation from R^4 to R^3

The linear transformation $T: R^4 \to R^3$ defined by the equations

$$\begin{aligned} w_1 &= 2x_1 - 3x_2 + x_3 - 5x_4 \\ w_2 &= 4x_1 + x_2 - 2x_3 + x_4 \\ w_3 &= 5x_1 - x_2 + 4x_3 \end{aligned} \tag{5}$$

can be expressed in matrix form as

$$\begin{bmatrix} w_1 \\ w_2 \\ w_3 \end{bmatrix} = \begin{bmatrix} 2 & -3 & 1 & -5 \\ 4 & 1 & -2 & 1 \\ 5 & -1 & 4 & 0 \end{bmatrix} \begin{bmatrix} x_1 \\ x_2 \\ x_3 \\ x_4 \end{bmatrix} \tag{6}$$

so the standard matrix for T is

$$A = \begin{bmatrix} 2 & -3 & 1 & -5 \\ 4 & 1 & -2 & 1 \\ 5 & -1 & 4 & 0 \end{bmatrix}$$

The image of a point (x_1, x_2, x_3, x_4) can be computed directly from the defining equations (5) or from (6) by matrix multiplication. For example, if $(x_1, x_2, x_3, x_4) = (1, -3, 0, 2)$, then substituting in (5) yields

$$w_1 = 1, \qquad w_2 = 3, \qquad w_3 = 8$$

(verify) or alternatively from (6)

$$\begin{bmatrix} w_1 \\ w_2 \\ w_3 \end{bmatrix} = \begin{bmatrix} 2 & -3 & 1 & -5 \\ 4 & 1 & -2 & 1 \\ 5 & -1 & 4 & 0 \end{bmatrix} \begin{bmatrix} 1 \\ -3 \\ 0 \\ 2 \end{bmatrix} = \begin{bmatrix} 1 \\ 3 \\ 8 \end{bmatrix}$$

◆

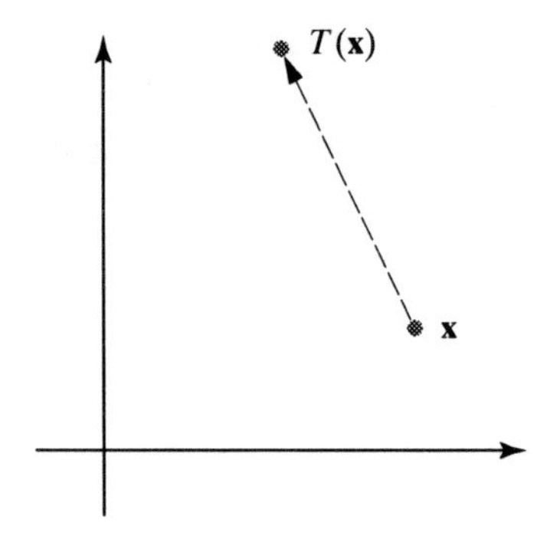

(*a*) *T* maps points to points

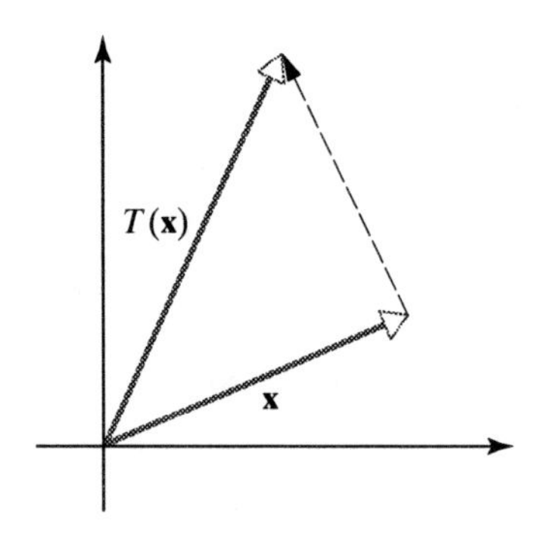

(*b*) *T* maps vectors to vectors

Figure 4.2.1

Some Notational Matters If $T: R^n \to R^m$ is multiplication by A, and if it is important to emphasize that A is the standard matrix for T, we shall denote the linear transformation $T: R^n \to R^m$ by $T_A: R^n \to R^m$. Thus,

$$T_A(\mathbf{x}) = A\mathbf{x} \tag{7}$$

It is understood in this equation that the vector $\mathbf{x}$ in R^n is expressed as a column matrix.

Sometimes it is awkward to introduce a new letter to denote the standard matrix for a linear transformation $T: R^n \to R^m$. In such cases we will denote the standard matrix for T by the symbol $[T]$. With this notation equation (7) would take the form

$$T(\mathbf{x}) = [T]\mathbf{x} \tag{8}$$

Occasionally, the two notations for a standard matrix will be mixed, in which case we have the relationship

$$[T_A] = A \tag{9}$$

REMARK. Amidst all of this notation it is important to keep in mind that we have established a correspondence between $m \times n$ matrices and linear transformations from R^n to R^m: To each matrix A there corresponds a linear transformation T_A (multiplication by A), and to each linear transformation $T: R^n \to R^m$, there corresponds an $m \times n$ matrix $[T]$ (the standard matrix for T).

Geometry of Linear Transformations Depending on whether n-tuples are regarded as points or vectors, the geometric effect of an operator $T: R^n \to R^n$ is to transform each point (or vector) in R^n into some new point (or vector) (Figure 4.2.1).

EXAMPLE 3 Zero Transformation from R^n to R^m

If 0 is the $m \times n$ zero matrix and $\mathbf{0}$ is the zero vector in R^n, then for every vector $\mathbf{x}$ in R^n

$$T_0(\mathbf{x}) = 0\mathbf{x} = \mathbf{0}$$

so multiplication by zero maps every vector in R^n into the zero vector in R^m. We call T_0 the ***zero transformation*** from R^n to R^m. Sometimes the zero transformation is denoted by 0. Although this is the same notation used for the zero matrix, the appropriate interpretation will usually be clear from the context. ◆

EXAMPLE 4 Identity Operator on R^n

If I is the $n \times n$ identity matrix, then for every vector $\mathbf{x}$ in R^n

$$T_I(\mathbf{x}) = I\mathbf{x} = \mathbf{x}$$

so multiplication by I maps every vector in R^n into itself. We call T_I the ***identity operator*** on R^n. Sometimes the identity operator is denoted by I. Although this is the same notation used for the identity matrix, the appropriate interpretation will usually be clear from the context. ◆

Among the most important linear operators on R^2 and R^3 are those that produce reflections, projections, and rotations. We shall now discuss such operators.

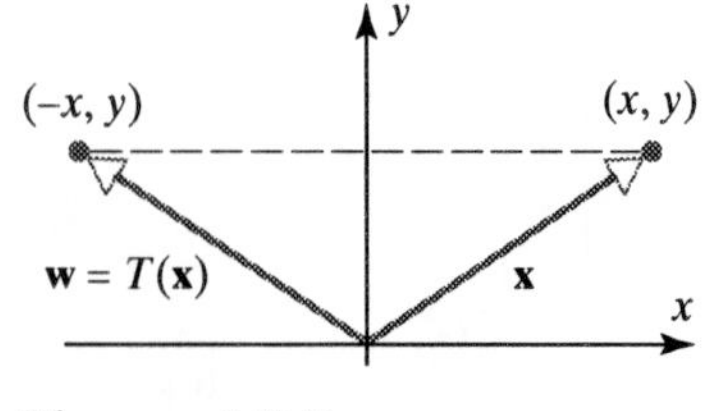

Figure 4.2.2

Reflection Operators Consider the operator $T: R^2 \to R^2$ that maps each vector into its symmetric image about the y-axis (Figure 4.2.2).

If we let $\mathbf{w} = T(\mathbf{x})$, then the equations relating the components of $\mathbf{x}$ and $\mathbf{w}$ are

$$\begin{aligned} w_1 &= -x = -x + 0y \\ w_2 &= y = 0x + y \end{aligned} \tag{10}$$

or in matrix form,

$$\begin{bmatrix} w_1 \\ w_2 \end{bmatrix} = \begin{bmatrix} -1 & 0 \\ 0 & 1 \end{bmatrix} \begin{bmatrix} x \\ y \end{bmatrix} \tag{11}$$

Since the equations in (10) are linear, T is a linear operator and from (11) the standard matrix for T is

$$[T] = \begin{bmatrix} -1 & 0 \\ 0 & 1 \end{bmatrix}$$

In general, operators on R^2 and R^3 that map each vector into its symmetric image about some line or plane are called ***reflection operators***. Such operators are linear. Tables 2 and 3 list some of the common reflection operators.

Projection Operators Consider the operator $T: R^2 \to R^2$ that maps each vector into its orthogonal projection on the x-axis (Figure 4.2.3). The equations relating the components of $\mathbf{x}$ and $\mathbf{w} = T(\mathbf{x})$ are

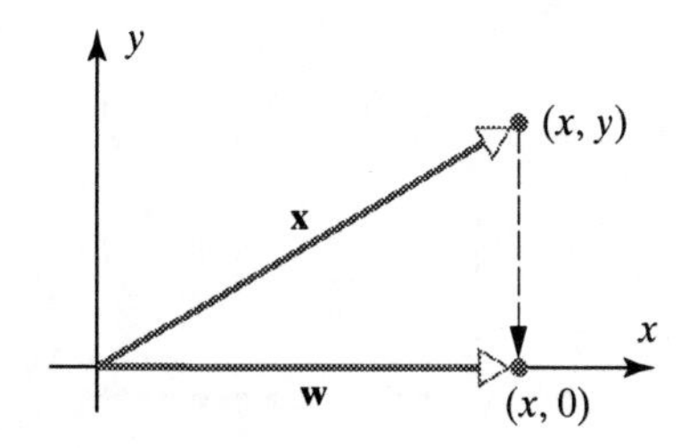

Figure 4.2.3

$$\begin{aligned} w_1 &= x = x + 0y \\ w_2 &= 0 = 0x + 0y \end{aligned} \tag{12}$$

or in matrix form,

$$\begin{bmatrix} w_1 \\ w_2 \end{bmatrix} = \begin{bmatrix} 1 & 0 \\ 0 & 0 \end{bmatrix} \begin{bmatrix} x \\ y \end{bmatrix} \tag{13}$$

The equations in (12) are linear, so T is a linear operator and from (13) the standard

TABLE 2

Operator	Illustration	Equations	Standard Matrix
Reflection about the y-axis	$(-x, y)$, (x, y), $\mathbf{w} = T(\mathbf{x})$, $\mathbf{x}$	$w_1 = -x$ $w_2 = y$	$\begin{bmatrix} -1 & 0 \\ 0 & 1 \end{bmatrix}$
Reflection about the x-axis	(x, y), $\mathbf{x}$, $\mathbf{w} = T(\mathbf{x})$, $(x, -y)$	$w_1 = x$ $w_2 = -y$	$\begin{bmatrix} 1 & 0 \\ 0 & -1 \end{bmatrix}$
Reflection about the line $y = x$	$y = x$, (y, x), $\mathbf{w} = T(\mathbf{x})$, $\mathbf{x}$, (x, y)	$w_1 = y$ $w_2 = x$	$\begin{bmatrix} 0 & 1 \\ 1 & 0 \end{bmatrix}$

TABLE 3

Operator	Illustration	Equations	Standard Matrix
Reflection about the xy-plane	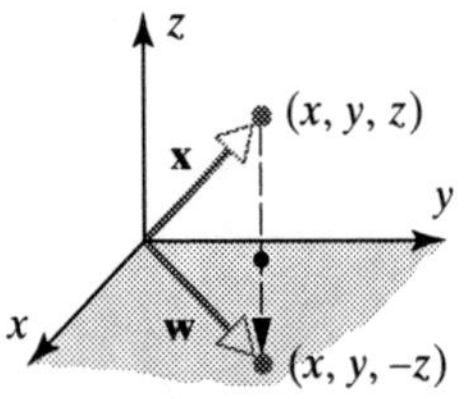	$w_1 = x$ $w_2 = y$ $w_3 = -z$	$\begin{bmatrix} 1 & 0 & 0 \\ 0 & 1 & 0 \\ 0 & 0 & -1 \end{bmatrix}$
Reflection about the xz-plane	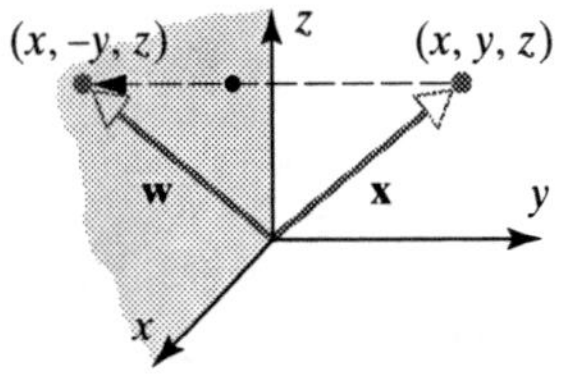	$w_1 = x$ $w_2 = -y$ $w_3 = z$	$\begin{bmatrix} 1 & 0 & 0 \\ 0 & -1 & 0 \\ 0 & 0 & 1 \end{bmatrix}$
Reflection about the yz-plane	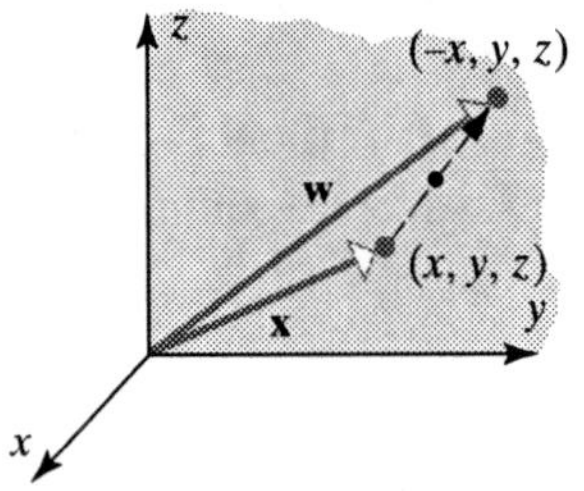	$w_1 = -x$ $w_2 = y$ $w_3 = z$	$\begin{bmatrix} -1 & 0 & 0 \\ 0 & 1 & 0 \\ 0 & 0 & 1 \end{bmatrix}$

matrix for T is

$$[T] = \begin{bmatrix} 1 & 0 \\ 0 & 0 \end{bmatrix}$$

In general, a ***projection operator*** (or more precisely an ***orthogonal projection operator***) on R^2 or R^3 is any operator that maps each vector into its orthogonal projection on a line or plane through the origin. It can be shown that such operators are linear. Some of the basic projection operators on R^2 and R^3 are listed in Tables 4 and 5.

TABLE 4

Operator	Illustration	Equations	Standard Matrix
Orthogonal projection on the x-axis	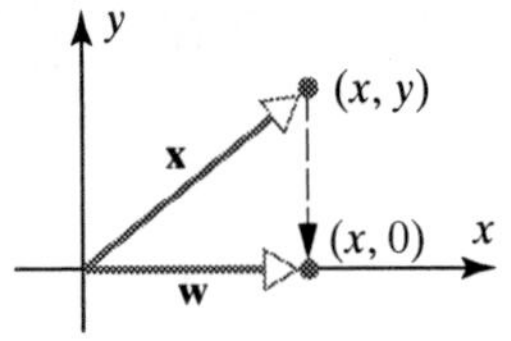	$w_1 = x$ $w_2 = 0$	$\begin{bmatrix} 1 & 0 \\ 0 & 0 \end{bmatrix}$
Orthogonal projection on the y-axis	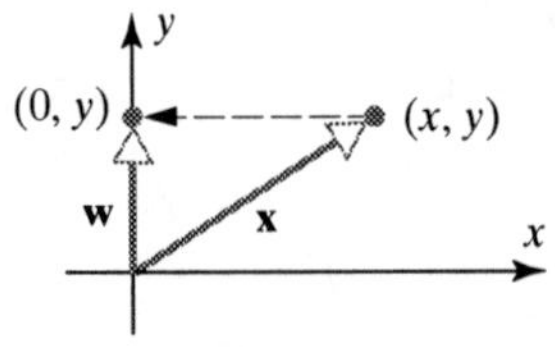	$w_1 = 0$ $w_2 = y$	$\begin{bmatrix} 0 & 0 \\ 0 & 1 \end{bmatrix}$

TABLE 5

Operator	Illustration	Equations	Standard Matrix
Orthogonal projection on the xy-plane		$w_1 = x$ $w_2 = y$ $w_3 = 0$	$\begin{bmatrix} 1 & 0 & 0 \\ 0 & 1 & 0 \\ 0 & 0 & 0 \end{bmatrix}$
Orthogonal projection on the xz-plane		$w_1 = x$ $w_2 = 0$ $w_3 = z$	$\begin{bmatrix} 1 & 0 & 0 \\ 0 & 0 & 0 \\ 0 & 0 & 1 \end{bmatrix}$
Orthogonal projection on the yz-plane		$w_1 = 0$ $w_2 = y$ $w_3 = z$	$\begin{bmatrix} 0 & 0 & 0 \\ 0 & 1 & 0 \\ 0 & 0 & 1 \end{bmatrix}$

Rotation Operators An operator that rotates each vector in R^2 through a fixed angle θ is called a ***rotation operator*** on R^2. Table 6 gives formulas for the rotation operators on R^2. To show how these results are derived, consider the rotation operator that rotates each vector counterclockwise through a fixed positive angle θ. To find equations relating $\mathbf{x}$ and $\mathbf{w} = T(\mathbf{x})$, let ϕ be the angle from the positive x-axis to $\mathbf{x}$, and let r be the common length of $\mathbf{x}$ and $\mathbf{w}$ (Figure 4.2.4).

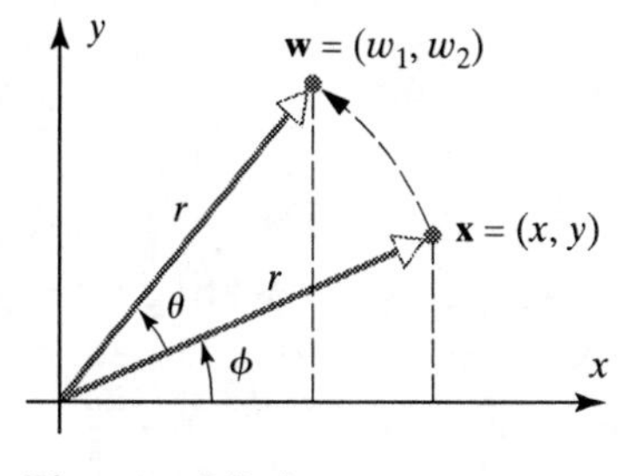

Figure 4.2.4

Then from basic trigonometry

$$x = r\cos\phi, \qquad y = r\sin\phi \tag{14}$$

and

$$w_1 = r\cos(\theta + \phi), \qquad w_2 = r\sin(\theta + \phi) \tag{15}$$

Using trigonometric identities on (15) yields

$$\begin{aligned} w_1 &= r\cos\theta\cos\phi - r\sin\theta\sin\phi \\ w_2 &= r\sin\theta\cos\phi + r\cos\theta\sin\phi \end{aligned}$$

and substituting (14) yields

TABLE 6

Operator	Illustration	Equations	Standard Matrix
Rotation through an angle θ		$w_1 = x\cos\theta - y\sin\theta$ $w_2 = x\sin\theta + y\cos\theta$	$\begin{bmatrix} \cos\theta & -\sin\theta \\ \sin\theta & \cos\theta \end{bmatrix}$

$$\begin{aligned} w_1 &= x\cos\theta - y\sin\theta \\ w_2 &= x\sin\theta + y\cos\theta \end{aligned} \tag{16}$$

The equations in (16) are linear, so T is a linear operator; moreover, it follows from these equations that the standard matrix for T is

$$[T] = \begin{bmatrix} \cos\theta & -\sin\theta \\ \sin\theta & \cos\theta \end{bmatrix}$$

EXAMPLE 5 Rotation

If each vector in R^2 is rotated through an angle of $\pi/6\,(=30°)$, then the image $\mathbf{w}$ of a vector

$$\mathbf{x} = \begin{bmatrix} x \\ y \end{bmatrix}$$

is

$$\mathbf{w} = \begin{bmatrix} \cos\pi/6 & -\sin\pi/6 \\ \sin\pi/6 & \cos\pi/6 \end{bmatrix}\begin{bmatrix} x \\ y \end{bmatrix} = \begin{bmatrix} \sqrt{3}/2 & -1/2 \\ 1/2 & \sqrt{3}/2 \end{bmatrix}\begin{bmatrix} x \\ y \end{bmatrix} = \begin{bmatrix} \frac{\sqrt{3}}{2}x - \frac{1}{2}y \\ \frac{1}{2}x + \frac{\sqrt{3}}{2}y \end{bmatrix}$$

For example, the image of the vector

$$\mathbf{x} = \begin{bmatrix} 1 \\ 1 \end{bmatrix} \quad \text{is} \quad \mathbf{w} = \begin{bmatrix} \dfrac{\sqrt{3}-1}{2} \\ \dfrac{1+\sqrt{3}}{2} \end{bmatrix}$$ ◆

A rotation of vectors in R^3 is usually described in relation to a ray emanating from the origin, called the ***axis of rotation***. As a vector revolves around the axis of rotation it sweeps out some portion of a cone (Figure 4.2.5*a*). The ***angle of rotation***, which is measured in the base of the cone, is described as "clockwise" or "counterclockwise" in relation to a viewpoint that is along the axis of rotation *looking toward the origin*. For example, in Figure 4.2.5*a* the vector $\mathbf{w}$ results from rotating the vector $\mathbf{x}$ counterclockwise around the axis l through an angle θ. As in R^2, angles are *positive* if they are generated by counterclockwise rotations and *negative* if they are generated by clockwise rotations.

The most common way of describing a general axis of rotation is to specify a nonzero vector $\mathbf{u}$ that runs along the axis of rotation and has its initial point at the origin. The counterclockwise direction for a rotation about the axis can then be determined by a "right-hand rule" (Figure 4.2.5*b*): If the thumb of the right hand points in the direction of $\mathbf{u}$, then the cupped fingers point in a counterclockwise direction.

A ***rotation operator*** on R^3 is a linear operator that rotates each vector in R^3 about some rotation axis through a fixed angle θ. In Table 7 we have described the rotation operators on R^3 whose axes of rotation are the positive coordinate axes. For each of these rotations one of the components is unchanged by the rotation, and the relationships between the other components can be derived by the same procedure used to derive (16). For example, in the rotation about the z-axis, the z-components of $\mathbf{x}$ and $\mathbf{w} = T(\mathbf{x})$ are the same, and the x- and y-components are related as in (16). This yields the rotation equations shown in the last row of Table 7.

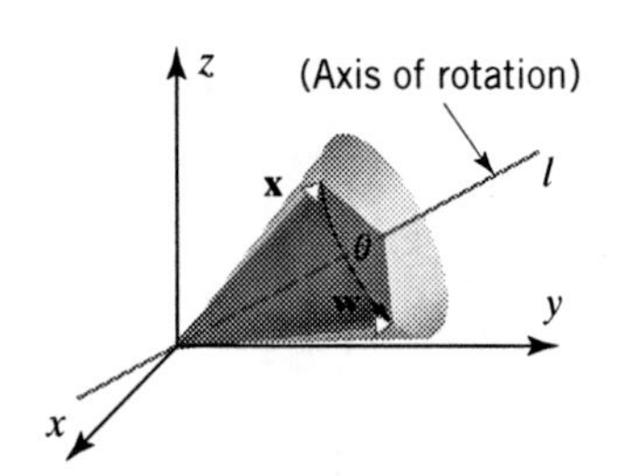

(*a*) Angle of rotation

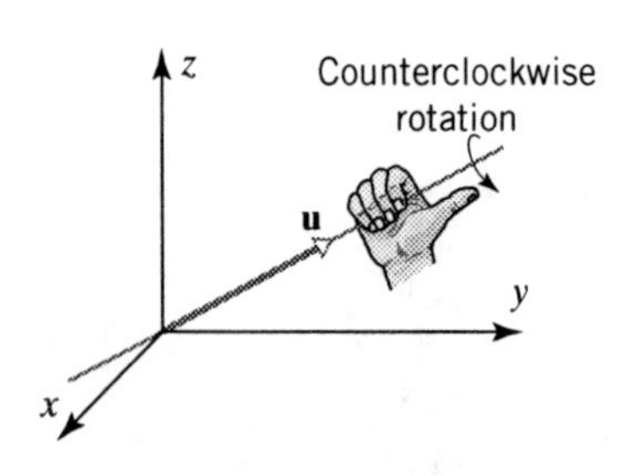

(*b*) Right-hand rule

Figure 4.2.5

TABLE 7

Operator	Illustration	Equations	Standard Matrix
Counterclockwise rotation about the positive x-axis through an angle θ		$w_1 = x$ $w_2 = y\cos\theta - z\sin\theta$ $w_3 = y\sin\theta + z\cos\theta$	$\begin{bmatrix} 1 & 0 & 0 \\ 0 & \cos\theta & -\sin\theta \\ 0 & \sin\theta & \cos\theta \end{bmatrix}$
Counterclockwise rotation about the positive y-axis through an angle θ		$w_1 = x\cos\theta + z\sin\theta$ $w_2 = y$ $w_3 = -x\sin\theta + z\cos\theta$	$\begin{bmatrix} \cos\theta & 0 & \sin\theta \\ 0 & 1 & 0 \\ -\sin\theta & 0 & \cos\theta \end{bmatrix}$
Counterclockwise rotation about the positive z-axis through an angle θ		$w_1 = x\cos\theta - y\sin\theta$ $w_2 = x\sin\theta + y\cos\theta$ $w_3 = z$	$\begin{bmatrix} \cos\theta & -\sin\theta & 0 \\ \sin\theta & \cos\theta & 0 \\ 0 & 0 & 1 \end{bmatrix}$

For completeness, we note that the standard matrix for a counterclockwise rotation through an angle θ about an axis in R^3, which is determined by an arbitrary *unit vector* $\mathbf{u} = (a, b, c)$ that has its initial point at the origin, is

$$\begin{bmatrix} a^2(1-\cos\theta)+\cos\theta & ab(1-\cos\theta)-c\sin\theta & ac(1-\cos\theta)+b\sin\theta \\ ab(1-\cos\theta)+c\sin\theta & b^2(1-\cos\theta)+\cos\theta & bc(1-\cos\theta)-a\sin\theta \\ ac(1-\cos\theta)-b\sin\theta & bc(1-\cos\theta)+a\sin\theta & c^2(1-\cos\theta)+\cos\theta \end{bmatrix} \tag{17}$$

The derivation can be found in the book *Principles of Interactive Computer Graphics*, by W. M. Newman and R. F. Sproull, New York, McGraw-Hill, 1979. The reader may find it instructive to derive the results in Table 7 as special cases of this more general result.

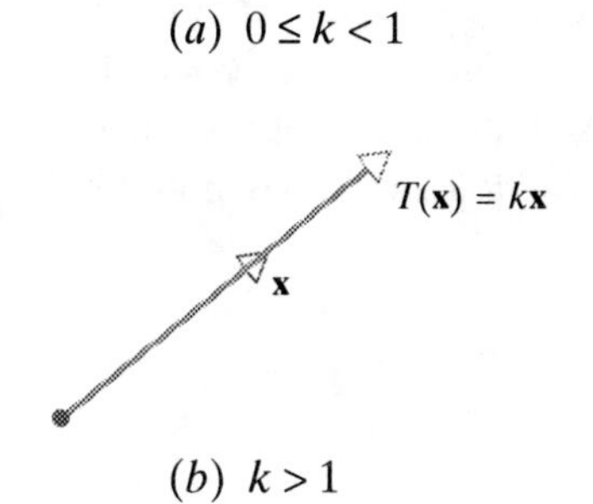

Figure 4.2.6

Dilation and Contraction Operators If k is a nonnegative scalar, then the operator $T(\mathbf{x}) = k\mathbf{x}$ on R^2 or R^3 is called a ***contraction with factor k*** if $0 \le k \le 1$ and a ***dilation with factor k*** if $k \ge 1$. The geometric effect of a contraction is to compress each vector by a factor of k (Figure 4.2.6*a*), and the effect of a dilation is to stretch each vector by a factor of k (Figure 4.2.6*b*). A contraction compresses R^2 or R^3 uniformly toward the origin from all directions, and a dilation stretches R^2 or R^3 uniformly away from the origin in all directions.

TABLE 8

Operator	Illustration	Equations	Standard Matrix
Contraction with factor k on R^2 $(0 \le k \le 1)$	y, x, (x, y), (kx, ky)	$w_1 = kx$ $w_2 = ky$	$\begin{bmatrix} k & 0 \\ 0 & k \end{bmatrix}$
Dilation with factor k on R^2 $(k \ge 1)$	y, x, (kx, ky), (x, y)	$w_1 = kx$ $w_2 = ky$	

TABLE 9

Operator	Illustration	Equations	Standard Matrix
Contraction with factor k on R^3 $(0 \le k \le 1)$	z, y, x, (x, y, z), (kx, ky, kz)	$w_1 = kx$ $w_2 = ky$ $w_3 = kz$	$\begin{bmatrix} k & 0 & 0 \\ 0 & k & 0 \\ 0 & 0 & k \end{bmatrix}$
Dilation with factor k on R^3 $(k \ge 1)$	z, y, x, (kx, ky, kz), (x, y, z)	$w_1 = kx$ $w_2 = ky$ $w_3 = kz$	

The most extreme contraction occurs when $k = 0$, in which case $T(\mathbf{x}) = k\mathbf{x}$ reduces to the zero operator $T(\mathbf{x}) = \mathbf{0}$, which compresses every vector into a single point (the origin). If $k = 1$, then $T(\mathbf{x}) = k\mathbf{x}$ reduces to the identity operator $T(\mathbf{x}) = \mathbf{x}$, which leaves each vector unchanged; this can be regarded as either a contraction or a dilation. Tables 8 and 9 list the dilation and contraction operators on R^2 and R^3.

Compositions of Linear Transformations If $T_A: R^n \to R^k$ and $T_B: R^k \to R^m$ are linear transformations, then for each $\mathbf{x}$ in R^n one can first compute $T_A(\mathbf{x})$, which is a vector in R^k, and then one can compute $T_B(T_A(\mathbf{x}))$, which is a vector in R^m. Thus, the application of T_A followed by T_B produces a transformation from R^n to R^m. This transformation is called the ***composition of T_B with T_A*** and is denoted by $T_B \circ T_A$ (read "T_B circle T_A"). Thus,

$$(T_B \circ T_A)(\mathbf{x}) = T_B(T_A(\mathbf{x})) \tag{18}$$

The composition $T_B \circ T_A$ is linear since

$$(T_B \circ T_A)(\mathbf{x}) = T_B(T_A(\mathbf{x})) = B(A\mathbf{x}) = (BA)\mathbf{x} \tag{19}$$

so $T_B \circ T_A$ is multiplication by BA, which is a linear transformation. Formula (19) also tells us that the standard matrix for $T_B \circ T_A$ is BA. This is expressed by the formula

$$T_B \circ T_A = T_{BA} \tag{20}$$

REMARK. Formula (20) captures an important idea: *Multiplying matrices is equivalent to composing the corresponding linear transformations in the right-to-left order of the factors.*

There is an alternative form of Formula (20): If $T_1\colon R^n \to R^k$ and $T_2\colon R^k \to R^m$ are linear transformations, then because the standard matrix for the composition $T_2 \circ T_1$ is the product of the standard matrices of T_2 and T_1, we have

$$[T_2 \circ T_1] = [T_2][T_1] \tag{21}$$

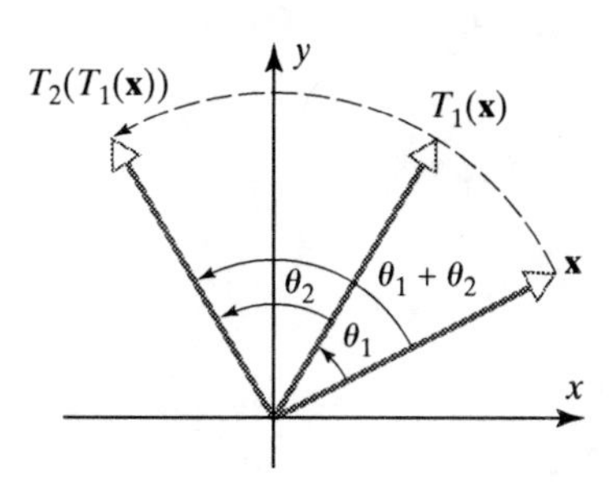

Figure 4.2.7

EXAMPLE 6 Composition of Two Rotations

Let $T_1\colon R^2 \to R^2$ and $T_2\colon R^2 \to R^2$ be the linear operators that rotate vectors through the angles θ_1 and θ_2, respectively. Thus, the operation

$$(T_2 \circ T_1)(\mathbf{x}) = T_2(T_1(\mathbf{x}))$$

first rotates $\mathbf{x}$ through the angle θ_1, then rotates $T_1(\mathbf{x})$ through the angle θ_2. It follows that the net effect of $T_2 \circ T_1$ is to rotate each vector in R^2 through the angle $\theta_1 + \theta_2$ (Figure 4.2.7).

Thus, the standard matrices for these linear operators are

$$[T_1] = \begin{bmatrix} \cos\theta_1 & -\sin\theta_1 \\ \sin\theta_1 & \cos\theta_1 \end{bmatrix}, \qquad [T_2] = \begin{bmatrix} \cos\theta_2 & -\sin\theta_2 \\ \sin\theta_2 & \cos\theta_2 \end{bmatrix},$$

$$[T_2 \circ T_1] = \begin{bmatrix} \cos(\theta_1+\theta_2) & -\sin(\theta_1+\theta_2) \\ \sin(\theta_1+\theta_2) & \cos(\theta_1+\theta_2) \end{bmatrix}$$

These matrices should satisfy (21). With the help of some basic trigonometric identities, we can show that this is so as follows:

$$\begin{aligned}
[T_2][T_1] &= \begin{bmatrix} \cos\theta_2 & -\sin\theta_2 \\ \sin\theta_2 & \cos\theta_2 \end{bmatrix}\begin{bmatrix} \cos\theta_1 & -\sin\theta_1 \\ \sin\theta_1 & \cos\theta_1 \end{bmatrix} \\
&= \begin{bmatrix} \cos\theta_2\cos\theta_1 - \sin\theta_2\sin\theta_1 & -(\cos\theta_2\sin\theta_1 + \sin\theta_2\cos\theta_1) \\ \sin\theta_2\cos\theta_1 + \cos\theta_2\sin\theta_1 & -\sin\theta_2\sin\theta_1 + \cos\theta_2\cos\theta_1 \end{bmatrix} \\
&= \begin{bmatrix} \cos(\theta_1+\theta_2) & -\sin(\theta_1+\theta_2) \\ \sin(\theta_1+\theta_2) & \cos(\theta_1+\theta_2) \end{bmatrix} \\
&= [T_2 \circ T_1]
\end{aligned}$$

◆

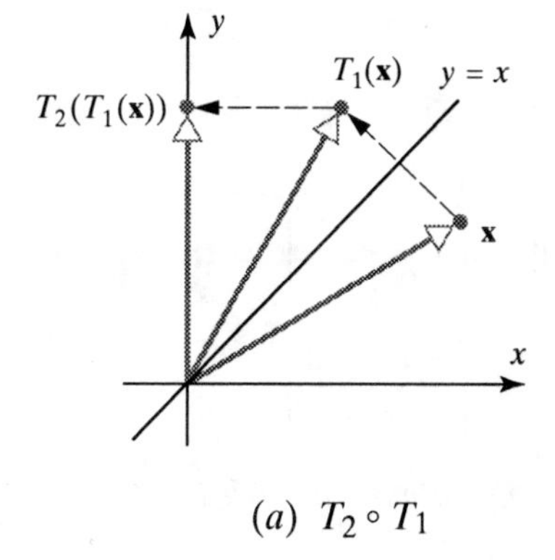

(a) $T_2 \circ T_1$

(b) $T_1 \circ T_2$

Figure 4.2.8

REMARK. In general, the order in which linear transformations are composed matters. This is to be expected, since the composition of two linear transformations corresponds to the multiplication of their standard matrices, and we know that the order in which matrices are multiplied makes a difference.

EXAMPLE 7 Composition Is Not Commutative

Let $T_1\colon R^2 \to R^2$ be the reflection operator about the line $y = x$, and let $T_2\colon R^2 \to R^2$ be the orthogonal projection on the y-axis. Figure 4.2.8 illustrates graphically that $T_1 \circ T_2$ and $T_2 \circ T_1$ have different effects on a vector $\mathbf{x}$. This same conclusion can be reached by showing that the standard matrices for T_1 and T_2 do not commute:

$$[T_1 \circ T_2] = [T_1][T_2] = \begin{bmatrix} 0 & 1 \\ 1 & 0 \end{bmatrix}\begin{bmatrix} 0 & 0 \\ 0 & 1 \end{bmatrix} = \begin{bmatrix} 0 & 1 \\ 0 & 0 \end{bmatrix}$$

$$[T_2 \circ T_1] = [T_2][T_1] = \begin{bmatrix} 0 & 0 \\ 0 & 1 \end{bmatrix}\begin{bmatrix} 0 & 1 \\ 1 & 0 \end{bmatrix} = \begin{bmatrix} 0 & 0 \\ 1 & 0 \end{bmatrix}$$

so $[T_2 \circ T_1] \neq [T_1 \circ T_2]$. ♦

EXAMPLE 8 Composition of Two Reflections

Let $T_1\colon R^2 \to R^2$ be the reflection about the y-axis, and let $T_2\colon R^2 \to R^2$ be the reflection about the x-axis. In this case $T_1 \circ T_2$ and $T_2 \circ T_1$ are the same; both map each vector $\mathbf{x} = (x, y)$ into its negative $-\mathbf{x} = (-x, -y)$ (Figure 4.2.9):

$$(T_1 \circ T_2)(x, y) = T_1(x, -y) = (-x, -y)$$
$$(T_2 \circ T_1)(x, y) = T_2(-x, y) = (-x, -y)$$

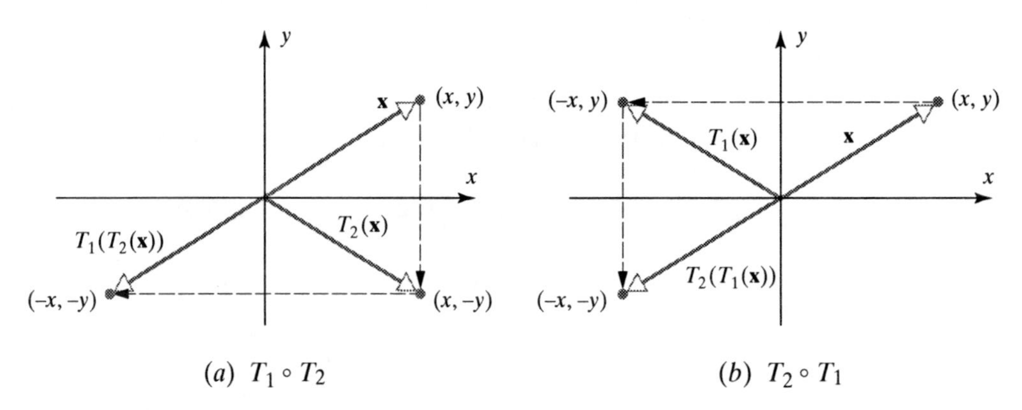

Figure 4.2.9

The equality of $T_1 \circ T_2$ and $T_2 \circ T_1$ can also be deduced by showing that the standard matrices for T_1 and T_2 commute:

$$[T_1 \circ T_2] = [T_1][T_2] = \begin{bmatrix} -1 & 0 \\ 0 & 1 \end{bmatrix}\begin{bmatrix} 1 & 0 \\ 0 & -1 \end{bmatrix} = \begin{bmatrix} -1 & 0 \\ 0 & -1 \end{bmatrix}$$

$$[T_2 \circ T_1] = [T_2][T_1] = \begin{bmatrix} 1 & 0 \\ 0 & -1 \end{bmatrix}\begin{bmatrix} -1 & 0 \\ 0 & 1 \end{bmatrix} = \begin{bmatrix} -1 & 0 \\ 0 & -1 \end{bmatrix}$$

The operator $T(\mathbf{x}) = -\mathbf{x}$ on R^2 or R^3 is called the ***reflection about the origin***. As the computations above show, the standard matrix for this operator on R^2 is

$$[T] = \begin{bmatrix} -1 & 0 \\ 0 & -1 \end{bmatrix}$$ ♦

Compositions of Three or More Linear Transformations

Compositions can be defined for three or more linear transformations. For example, consider the linear transformations

$$T_1\colon R^n \to R^k, \quad T_2\colon R^k \to R^l, \quad T_3\colon R^l \to R^m$$

We define the composition $(T_3 \circ T_2 \circ T_1)\colon R^n \to R^m$ by

$$(T_3 \circ T_2 \circ T_1)(\mathbf{x}) = T_3(T_2(T_1(\mathbf{x})))$$

It can be shown that this composition is a linear transformation and that the standard matrix for $T_3 \circ T_2 \circ T_1$ is related to the standard matrices for T_1, T_2, and T_3 by

$$[T_3 \circ T_2 \circ T_1] = [T_3][T_2][T_1] \tag{22}$$

which is a generalization of (21). If the standard matrices for T_1, T_2, and T_3 are denoted by A, B, and C, respectively, then we also have the following generalization of (20):

$$T_C \circ T_B \circ T_A = T_{CBA} \tag{23}$$

EXAMPLE 9 Composition of Three Transformations

Find the standard matrix for the linear operator $T: R^3 \to R^3$ that first rotates a vector counterclockwise about the z-axis through an angle θ, then reflects the resulting vector about the yz-plane, and then projects that vector orthogonally onto the xy-plane.

Solution.

The linear transformation T can be expressed as the composition

$$T = T_3 \circ T_2 \circ T_1$$

where T_1 is the rotation about the z-axis, T_2 is the reflection about the yz-plane, and T_3 is the orthogonal projection on the xy-plane. From Tables 3, 5, and 7 the standard matrices for these linear transformations are

$$[T_1] = \begin{bmatrix} \cos\theta & -\sin\theta & 0 \\ \sin\theta & \cos\theta & 0 \\ 0 & 0 & 1 \end{bmatrix}, \quad [T_2] = \begin{bmatrix} -1 & 0 & 0 \\ 0 & 1 & 0 \\ 0 & 0 & 1 \end{bmatrix}, \quad [T_3] = \begin{bmatrix} 1 & 0 & 0 \\ 0 & 1 & 0 \\ 0 & 0 & 0 \end{bmatrix}$$

Thus, from (22) the standard matrix for T is $[T] = [T_3][T_2][T_1]$; that is,

$$[T] = \begin{bmatrix} 1 & 0 & 0 \\ 0 & 1 & 0 \\ 0 & 0 & 0 \end{bmatrix} \begin{bmatrix} -1 & 0 & 0 \\ 0 & 1 & 0 \\ 0 & 0 & 1 \end{bmatrix} \begin{bmatrix} \cos\theta & -\sin\theta & 0 \\ \sin\theta & \cos\theta & 0 \\ 0 & 0 & 1 \end{bmatrix}$$

$$= \begin{bmatrix} -\cos\theta & \sin\theta & 0 \\ \sin\theta & \cos\theta & 0 \\ 0 & 0 & 0 \end{bmatrix}$$ ◆

Exercise Set 4.2

1. Find the domain and codomain of the transformation defined by the equations, and determine whether the transformation is linear.

(a) $\begin{aligned} w_1 &= 3x_1 - 2x_2 + 4x_3 \\ w_2 &= 5x_1 - 8x_2 + x_3 \end{aligned}$

(b) $\begin{aligned} w_1 &= 2x_1x_2 - x_2 \\ w_2 &= x_1 + 3x_1x_2 \\ w_3 &= x_1 + x_2 \end{aligned}$

(c) $\begin{aligned} w_1 &= 5x_1 - x_2 + x_3 \\ w_2 &= -x_1 + x_2 + 7x_3 \\ w_3 &= 2x_1 - 4x_2 - x_3 \end{aligned}$

(d) $\begin{aligned} w_1 &= x_1^2 - 3x_2 + x_3 - 2x_4 \\ w_2 &= 3x_1 - 4x_2 - x_3^2 + x_4 \end{aligned}$

2. Find the standard matrix for the linear transformation defined by the equations.

(a) $\begin{aligned} w_1 &= 2x_1 - 3x_2 + x_4 \\ w_2 &= 3x_1 + 5x_2 - x_4 \end{aligned}$

(b) $\begin{aligned} w_1 &= 7x_1 + 2x_2 - 8x_3 \\ w_2 &= - x_2 + 5x_3 \\ w_3 &= 4x_1 + 7x_2 - x_3 \end{aligned}$

(c) $\begin{aligned} w_1 &= -x_1 + x_2 \\ w_2 &= 3x_1 - 2x_2 \\ w_3 &= 5x_1 - 7x_2 \end{aligned}$

(d) $\begin{aligned} w_1 &= x_1 \\ w_2 &= x_1 + x_2 \\ w_3 &= x_1 + x_2 + x_3 \\ w_4 &= x_1 + x_2 + x_3 + x_4 \end{aligned}$

3. Find the standard matrix for the linear transformation $T: R^3 \to R^3$ given by

$$\begin{aligned} w_1 &= 3x_1 + 5x_2 - x_3 \\ w_2 &= 4x_1 - x_2 + x_3 \\ w_3 &= 3x_1 + 2x_2 - x_3 \end{aligned}$$

and then calculate $T(-1, 2, 4)$ by directly substituting in the equations and also by matrix multiplication.

4. Find the standard matrix for the linear operator T defined by the formula.

(a) $T(x_1, x_2) = (2x_1 - x_2, x_1 + x_2)$ (b) $T(x_1, x_2) = (x_1, x_2)$
(c) $T(x_1, x_2, x_3) = (x_1 + 2x_2 + x_3, x_1 + 5x_2, x_3)$ (d) $T(x_1, x_2, x_3) = (4x_1, 7x_2, -8x_3)$

5. Find the standard matrix for the linear transformation T defined by the formula.

(a) $T(x_1, x_2) = (x_2, -x_1, x_1 + 3x_2, x_1 - x_2)$
(b) $T(x_1, x_2, x_3, x_4) = (7x_1 + 2x_2 - x_3 + x_4, x_2 + x_3, -x_1)$
(c) $T(x_1, x_2, x_3) = (0, 0, 0, 0, 0)$
(d) $T(x_1, x_2, x_3, x_4) = (x_4, x_1, x_3, x_2, x_1 - x_3)$

6. In each part the standard matrix $[T]$ of a linear transformation T is given. Use it to find $T(x)$. [Express the answers in matrix form.]

(a) $[T] = \begin{bmatrix} 1 & 2 \\ 3 & 4 \end{bmatrix}$; $\mathbf{x} = \begin{bmatrix} 3 \\ -2 \end{bmatrix}$ (b) $[T] = \begin{bmatrix} -1 & 2 & 0 \\ 3 & 1 & 5 \end{bmatrix}$; $\mathbf{x} = \begin{bmatrix} -1 \\ 1 \\ 3 \end{bmatrix}$

(c) $[T] = \begin{bmatrix} -2 & 1 & 4 \\ 3 & 5 & 7 \\ 6 & 0 & -1 \end{bmatrix}$; $\mathbf{x} = \begin{bmatrix} x_1 \\ x_2 \\ x_3 \end{bmatrix}$ (d) $[T] = \begin{bmatrix} -1 & 1 \\ 2 & 4 \\ 7 & 8 \end{bmatrix}$; $\mathbf{x} = \begin{bmatrix} x_1 \\ x_2 \end{bmatrix}$

7. In each part use the standard matrix for T to find $T(\mathbf{x})$; then check the result by calculating $T(\mathbf{x})$ directly.

(a) $T(x_1, x_2) = (-x_1 + x_2, x_2)$; $\mathbf{x} = (-1, 4)$
(b) $T(x_1, x_2, x_3) = (2x_1 - x_2 + x_3, x_2 + x_3, 0)$; $\mathbf{x} = (2, 1, -3)$

8. Use matrix multiplication to find the reflection of $(-1, 2)$ about

(a) the x-axis (b) the y-axis (c) the line $y = x$

9. Use matrix multiplication to find the reflection of $(2, -5, 3)$ about

(a) the xy-plane (b) the xz-plane (c) the yz-plane

10. Use matrix multiplication to find the orthogonal projection of $(2, -5)$ on

(a) the x-axis (b) the y-axis

11. Use matrix multiplication to find the orthogonal projection of $(-2, 1, 3)$ on

(a) the xy-plane (b) the xz-plane (c) the yz-plane

12. Use matrix multiplication to find the image of the vector $(3, -4)$ when it is rotated through an angle of

(a) $\theta = 30°$ (b) $\theta = -60°$ (c) $\theta = 45°$ (d) $\theta = 90°$

13. Use matrix multiplication to find the image of the vector $(-2, 1, 2)$ if it is rotated

(a) $30°$ about the x-axis (b) $45°$ about the y-axis
(c) $90°$ about the z-axis

14. Find the standard matrix for the linear operator that rotates a vector in R^3 through an angle of $-60°$ about

(a) the x-axis (b) the y-axis (c) the z-axis

15. Use matrix multiplication to find the image of the vector $(-2, 1, 2)$ if it is rotated

(a) $-30°$ about the x-axis (b) $-45°$ about the y-axis
(c) $-90°$ about the z-axis

16. Find the standard matrix for the stated composition of linear operators on R^2.

(a) A rotation of 90°, followed by a reflection about the line $y = x$.
(b) An orthogonal projection on the y-axis, followed by a contraction with factor $k = \frac{1}{2}$.
(c) A reflection about the x-axis, followed by a dilation with factor $k = 3$.

17. Find the standard matrix for the stated composition of linear operators on R^2.

(a) A rotation of 60°, followed by an orthogonal projection on the x-axis, followed by a reflection about the line $y = x$.
(b) A dilation with factor $k = 2$, followed by a rotation of 45°, followed by a reflection about the y-axis.
(c) A rotation of 15°, followed by a rotation of 105°, followed by a rotation of 60°.

18. Find the standard matrix for the stated composition of linear operators on R^3.

(a) A reflection about the yz-plane, followed by an orthogonal projection on the xz-plane.
(b) A rotation of 45° about the y-axis, followed by a dilation with factor $k = \sqrt{2}$.
(c) An orthogonal projection on the xy-plane, followed by a reflection about the yz-plane.

19. Find the standard matrix for the stated composition of linear operators on R^3.

(a) A rotation of 30° about the x-axis, followed by a rotation of 30° about the z-axis, followed by a contraction with factor $k = \frac{1}{4}$.
(b) A reflection about the xy-plane, followed by a reflection about the xz-plane, followed by an orthogonal projection on the yz-plane.
(c) A rotation of 270° about the x-axis, followed by a rotation of 90° about the y-axis, followed by a rotation of 180° about the z-axis.

20. Determine whether $T_1 \circ T_2 = T_2 \circ T_1$.

(a) $T_1: R^2 \to R^2$ is the orthogonal projection on the x-axis and $T_2: R^2 \to R^2$ is the orthogonal projection on the y-axis.
(b) $T_1: R^2 \to R^2$ is the rotation through an angle θ_1 and $T_2: R^2 \to R^2$ is the rotation through an angle θ_2.
(c) $T_1: R^2 \to R^2$ is the orthogonal projection on the x-axis and $T_2: R^2 \to R^2$ is the rotation through an angle θ.

21. Determine whether $T_1 \circ T_2 = T_2 \circ T_1$.

(a) $T_1: R^3 \to R^3$ is a dilation by a factor k and $T_2: R^3 \to R^3$ is the rotation about the z-axis through an angle θ.
(b) $T_1: R^3 \to R^3$ is the rotation about the x-axis through an angle θ_1 and $T_2: R^3 \to R^3$ is the rotation about the z-axis through an angle θ_2.

22. In R^3 the ***orthogonal projections*** on the x-axis, y-axis, and z-axis are defined by

$$T_1(x, y, z) = (x, 0, 0), \qquad T_2(x, y, z) = (0, y, 0), \qquad T_3(x, y, z) = (0, 0, z)$$

respectively.

(a) Show that the orthogonal projections on the coordinate axes are linear operators and find their standard matrices.
(b) Show that if $T: R^3 \to R^3$ is an orthogonal projection on one of the coordinate axes, then for every vector $\mathbf{x}$ in R^3, the vectors $T(\mathbf{x})$ and $\mathbf{x} - T(\mathbf{x})$ are orthogonal vectors.
(c) Make a sketch showing $\mathbf{x}$ and $\mathbf{x} - T(\mathbf{x})$ in the case where T is the orthogonal projection on the x-axis.

23. Derive the standard matrices for the rotations about the x-axis, y-axis, and z-axis in R^3 from Formula (17).

24. Use Formula (17) to find the standard matrix for a rotation 90° about the axis determined by the vector $\mathbf{v} = (1, 1, 1)$. [***Note.*** Formula (17) requires that the vector defining the axis of rotation have length 1.]

9.2 GEOMETRY OF LINEAR OPERATORS ON R^2

In Section 4.2 we studied some of the geometric properties of linear operators on R^2 and R^3. In this section we shall study linear operators on R^2 in a little more depth. Some of the ideas that will be developed here have important applications to the field of computer graphics.

Vectors or Points If $T: R^2 \to R^2$ is the matrix operator whose standard matrix is

$$A = \begin{bmatrix} a & b \\ c & d \end{bmatrix}$$

then

$$T\left(\begin{bmatrix} x \\ y \end{bmatrix}\right) = \begin{bmatrix} a & b \\ c & d \end{bmatrix}\begin{bmatrix} x \\ y \end{bmatrix} = \begin{bmatrix} ax + by \\ cx + dy \end{bmatrix} \tag{1}$$

There are two equally good geometric interpretations of this formula. We may view the entries in the matrices

$$\begin{bmatrix} x \\ y \end{bmatrix} \quad \text{and} \quad \begin{bmatrix} ax + by \\ cx + dy \end{bmatrix}$$

either as components of vectors or coordinates of points. With the first interpretation, T maps arrows to arrows, and with the second, points to points (Figure 9.2.1). The choice is a matter of taste.

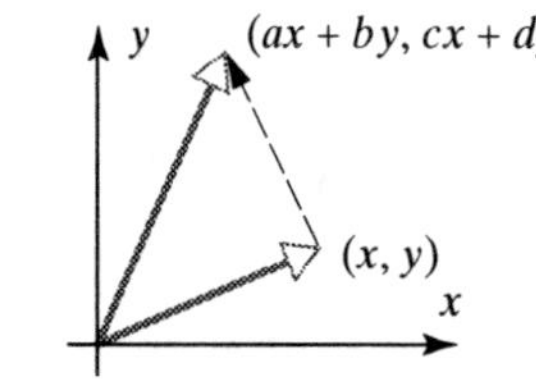

(*a*) T maps vectors to vectors.

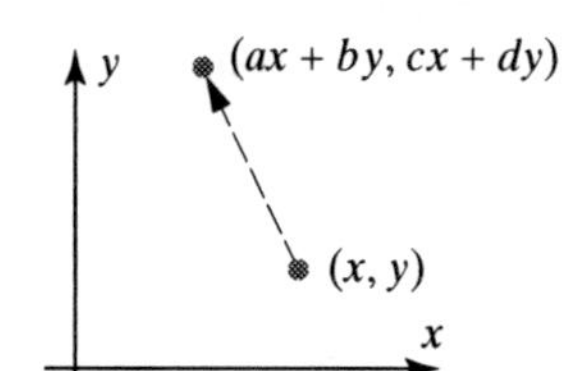

(*b*) T maps points to points.

Figure 9.2.1

In this section we shall view linear operators on R^2 as mapping points to points. One useful device for visualizing the behavior of a linear operator is to observe its effect on the points of simple figures in the plane. For example, Table 1 shows the effect of some basic linear operators on a unit square that has been partially colored.

In Section 4.2 we discussed reflections, projections, rotations, contractions, and dilations of R^2. We shall now consider some other basic linear operators on R^2.

Expansions and Compressions If the x-coordinate of each point in the plane is multiplied by a positive constant k, then the effect is to expand or compress each plane figure in the x-direction. If $0 < k < 1$, the result is a compression, and if $k > 1$, an expansion (Figure 9.2.2). We call such an operator an ***expansion*** (or ***compression***) ***in the x-direction with factor k***. Similarly, if the y-coordinate of each point is multiplied by a positive constant k, we obtain an ***expansion*** (or ***compression***) ***in the y-direction with factor k***. It can be shown that expansions and compressions along the coordinate axes are linear transformations.

TABLE 1

Operator	Standard Matrix	Effect on the Unit Square
Reflection about the y-axis	$\begin{bmatrix} -1 & 0 \\ 0 & 1 \end{bmatrix}$	y, $(1, 1)$, x → y, $(-1, 1)$, x
Reflection about the x-axis	$\begin{bmatrix} 1 & 0 \\ 0 & -1 \end{bmatrix}$	y, $(1, 1)$, x → y, x, $(1, -1)$
Reflection about the line $y = x$	$\begin{bmatrix} 0 & 1 \\ 1 & 0 \end{bmatrix}$	y, $(1, 1)$, x → y, $(1, 1)$, x
Counterclockwise rotation through an angle θ	$\begin{bmatrix} \cos\theta & -\sin\theta \\ \sin\theta & \cos\theta \end{bmatrix}$	y, $(1, 1)$, x → $(\cos\theta - \sin\theta, \sin\theta + \cos\theta)$, y, θ, x

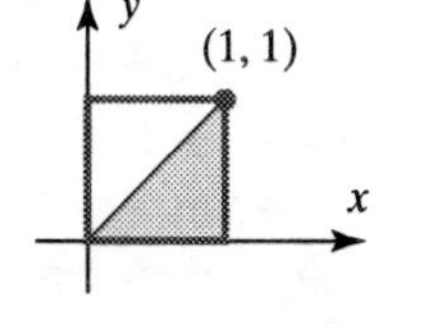

(a) (Unit square)

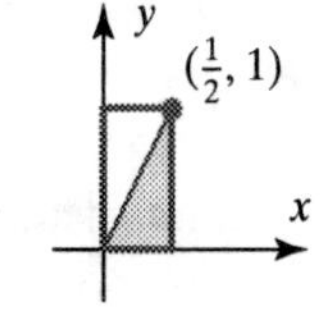

(b) (Compression) $k = \frac{1}{2}$

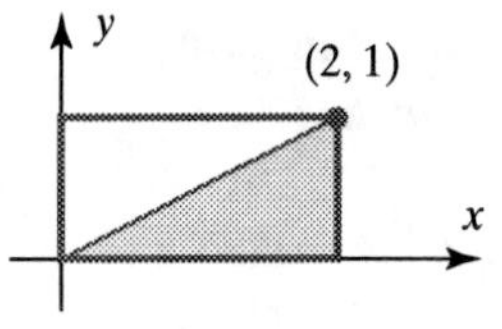

(c) (Expansion) $k = 2$

Figure 9.2.2

If $T: R^2 \to R^2$ is an expansion or compression in the x-direction with factor k, then

$$T(\mathbf{e}_1) = T\left(\begin{bmatrix} 1 \\ 0 \end{bmatrix}\right) = \begin{bmatrix} k \\ 0 \end{bmatrix}, \qquad T(\mathbf{e}_2) = T\left(\begin{bmatrix} 0 \\ 1 \end{bmatrix}\right) = \begin{bmatrix} 0 \\ 1 \end{bmatrix}$$

so the standard matrix for T is

$$\begin{bmatrix} k & 0 \\ 0 & 1 \end{bmatrix}$$

Similarly, the standard matrix for an expansion or compression in the y-direction is

$$\begin{bmatrix} 1 & 0 \\ 0 & k \end{bmatrix}$$

EXAMPLE 1 Operating with Diagonal Matrices

Suppose that the xy-plane is first expanded or compressed by a factor of k_1 in the x-direction and then is expanded or compressed by a factor of k_2 in the y-direction. Find a single matrix operator that performs both operations.

Solution.

The standard matrices for the two operations are

$$\underset{x\text{-expansion (compression)}}{\begin{bmatrix} k_1 & 0 \\ 0 & 1 \end{bmatrix}} \qquad \underset{y\text{-expansion (compression)}}{\begin{bmatrix} 1 & 0 \\ 0 & k_2 \end{bmatrix}}$$

Thus, the standard matrix for the composition of the x-operation followed by the y-operation is

$$A = \begin{bmatrix} 1 & 0 \\ 0 & k_2 \end{bmatrix}\begin{bmatrix} k_1 & 0 \\ 0 & 1 \end{bmatrix} = \begin{bmatrix} k_1 & 0 \\ 0 & k_2 \end{bmatrix} \tag{2}$$

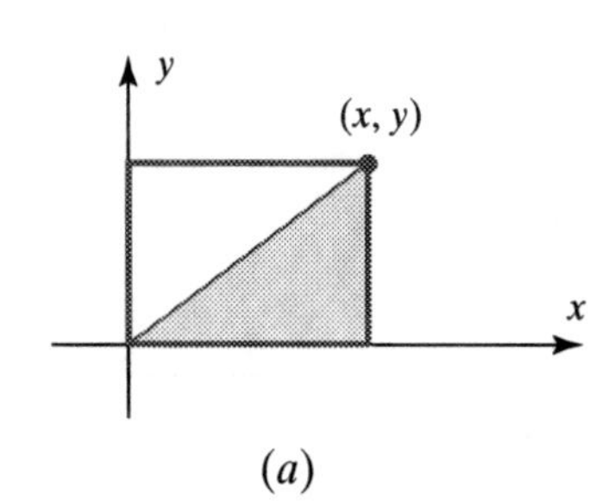

(a)

This shows that multiplication by a diagonal 2×2 matrix expands or compresses the plane in the x-direction and also in the y-direction. In the special case where k_1 and k_2 are the same, say $k_1 = k_2 = k$, note that (2) simplifies to

$$A = \begin{bmatrix} k & 0 \\ 0 & k \end{bmatrix}$$

which is a dilation or a contraction (Table 8 of Section 4.2). ♦

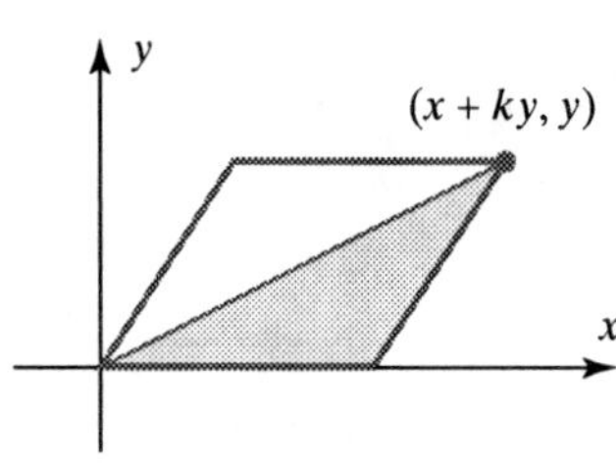

(b) Shear in x-direction with factor $k > 0$

(c) Shear in x-direction with factor $k < 0$

Figure 9.2.3

Shears A ***shear in the x-direction with factor k*** is a transformation that moves each point (x, y) parallel to the x-axis by an amount ky to the new position $(x + ky, y)$. Under such a transformation, points on the x-axis are unmoved since $y = 0$. However, as we progress away from the x-axis, the magnitude of y increases, so that points farther from the x-axis move a greater distance than those closer (Figure 9.2.3).

A ***shear in the y-direction with factor k*** is a transformation that moves each point (x, y) parallel to the y-axis by an amount kx to the new position $(x, y + kx)$. Under such a transformation, points on the y-axis remain fixed, and points farther from the y-axis move a greater distance than do those closer.

It can be shown that shears are linear transformations. If $T: R^2 \to R^2$ is a shear with factor k in the x-direction, then

$$T(\mathbf{e}_1) = T\left(\begin{bmatrix} 1 \\ 0 \end{bmatrix}\right) = \begin{bmatrix} 1 \\ 0 \end{bmatrix}, \qquad T(\mathbf{e}_2) = T\left(\begin{bmatrix} 0 \\ 1 \end{bmatrix}\right) = \begin{bmatrix} k \\ 1 \end{bmatrix}$$

so the standard matrix for T is

$$\begin{bmatrix} 1 & k \\ 0 & 1 \end{bmatrix}$$

Similarly, the standard matrix for a shear in the y-direction with factor k is

$$\begin{bmatrix} 1 & 0 \\ k & 1 \end{bmatrix}$$

REMARK. Multiplication by the 2×2 identity matrix is the identity operator on R^2. This operator can be viewed as a rotation through $0°$, or as a shear along either axis with $k = 0$, or as a compression or expansion along either axis with factor $k = 1$.

EXAMPLE 2 Finding Matrix Transformations

(a) Find a matrix transformation from R^2 to R^2 that first shears by a factor of 2 in the x-direction and then reflects about $y = x$.
(b) Find a matrix transformation from R^2 to R^2 that first reflects about $y = x$ and then shears by a factor of 2 in the x-direction.

Solution (a). The standard matrix for the shear is

$$A_1 = \begin{bmatrix} 1 & 2 \\ 0 & 1 \end{bmatrix}$$

and for the reflection is

$$A_2 = \begin{bmatrix} 0 & 1 \\ 1 & 0 \end{bmatrix}$$

Thus, the standard matrix for the shear followed by the reflection is

$$A_2A_1 = \begin{bmatrix} 0 & 1 \\ 1 & 0 \end{bmatrix}\begin{bmatrix} 1 & 2 \\ 0 & 1 \end{bmatrix} = \begin{bmatrix} 0 & 1 \\ 1 & 2 \end{bmatrix}$$

Solution (b). The reflection followed by the shear is represented by

$$A_1A_2 = \begin{bmatrix} 1 & 2 \\ 0 & 1 \end{bmatrix}\begin{bmatrix} 0 & 1 \\ 1 & 0 \end{bmatrix} = \begin{bmatrix} 2 & 1 \\ 1 & 0 \end{bmatrix}$$ ♦

In the last example, note that $A_1A_2 \neq A_2A_1$, so that the effect of shearing and then reflecting is different from reflecting and then shearing. This is illustrated geometrically in Figure 9.2.4, where we show the effects of the transformations on a unit square.

EXAMPLE 3 Transformations Using Elementary Matrices

Show that if $T: R^2 \rightarrow R^2$ is multiplication by an *elementary matrix*, then the transformation is one of the following:

(a) a shear along a coordinate axis
(b) a reflection about $y = x$
(c) a compression along a coordinate axis
(d) an expansion along a coordinate axis
(e) a reflection about a coordinate axis
(f) a compression or expansion along a coordinate axis followed by a reflection about a coordinate axis.

Solution.

Because a 2×2 elementary matrix results from performing a single elementary row operation on the 2×2 identity matrix, it must have one of the following forms (verify):

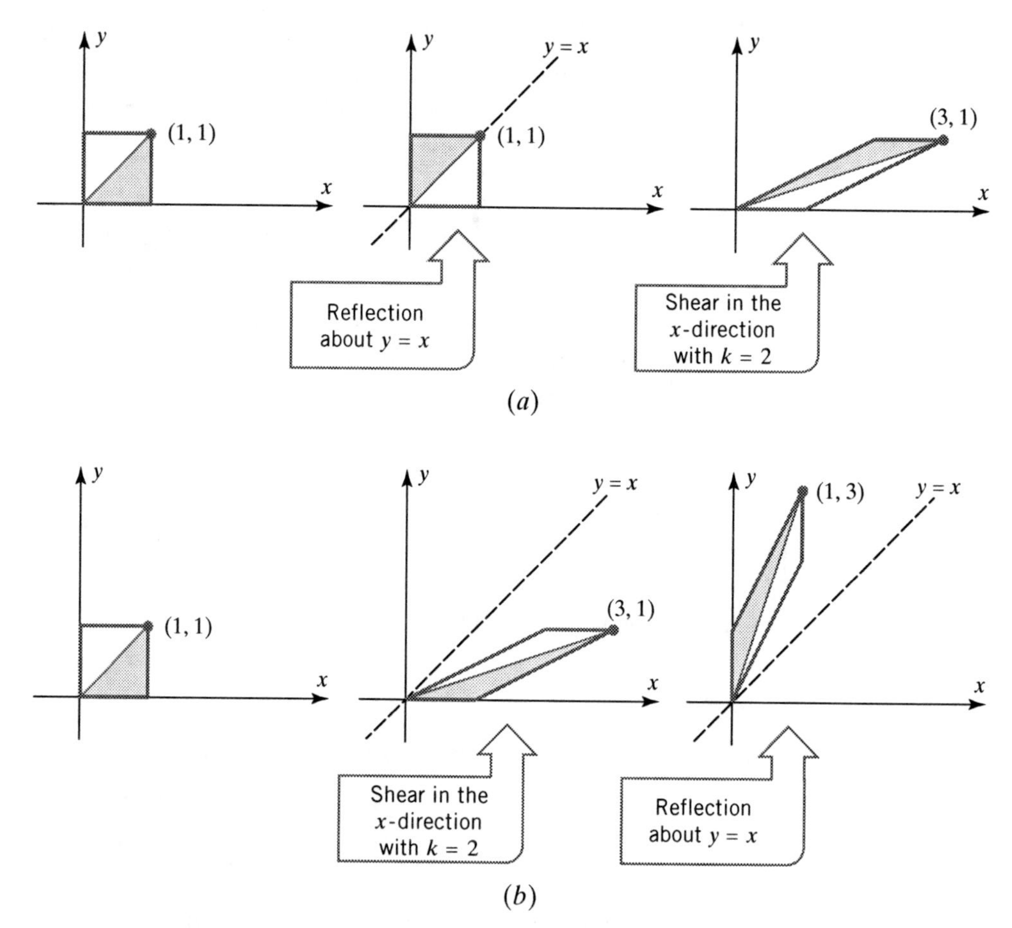

Figure 9.2.4

$$\begin{bmatrix} 1 & 0 \\ k & 1 \end{bmatrix}, \quad \begin{bmatrix} 1 & k \\ 0 & 1 \end{bmatrix}, \quad \begin{bmatrix} 0 & 1 \\ 1 & 0 \end{bmatrix}, \quad \begin{bmatrix} k & 0 \\ 0 & 1 \end{bmatrix}, \quad \begin{bmatrix} 1 & 0 \\ 0 & k \end{bmatrix}$$

The first two matrices represent shears along coordinate axes, and the third, a reflection about $y = x$. If $k > 0$, the last two matrices represent compressions or expansions along coordinate axes depending on whether $0 \leq k \leq 1$ or $k \geq 1$. If $k < 0$, and if we express k in the form $k = -k_1$, where $k_1 > 0$, then the last two matrices can be written as

$$\begin{bmatrix} k & 0 \\ 0 & 1 \end{bmatrix} = \begin{bmatrix} -k_1 & 0 \\ 0 & 1 \end{bmatrix} = \begin{bmatrix} -1 & 0 \\ 0 & 1 \end{bmatrix} \begin{bmatrix} k_1 & 0 \\ 0 & 1 \end{bmatrix} \tag{3}$$

$$\begin{bmatrix} 1 & 0 \\ 0 & k \end{bmatrix} = \begin{bmatrix} 1 & 0 \\ 0 & -k_1 \end{bmatrix} = \begin{bmatrix} 1 & 0 \\ 0 & -1 \end{bmatrix} \begin{bmatrix} 1 & 0 \\ 0 & k_1 \end{bmatrix} \tag{4}$$

Since $k_1 > 0$, the product in (3) represents a compression or expansion along the x-axis followed by a reflection about the y-axis, and (4) represents a compression or expansion along the y-axis followed by a reflection about the x-axis. In the case where $k = -1$, transformations (3) and (4) are simply reflections about the y-axis and x-axis, respectively. ♦

Reflections, rotations, expansions, compressions, and shears are all one-to-one linear operators. This is evident geometrically, since all of those operators map distinct points into distinct points. This can also be checked algebraically by verifying that the standard matrices for those operators are invertible.

EXAMPLE 4 A Transformation and Its Inverse

It is intuitively clear that if we compress the xy-plane by a factor of $\frac{1}{2}$ in the y-direction, then we must expand the xy-plane by a factor of 2 in the y-direction to move each point back to its original position. This is indeed the case, since

$$A = \begin{bmatrix} 1 & 0 \\ 0 & \frac{1}{2} \end{bmatrix}$$

represents a compression of factor $\frac{1}{2}$ in the y-direction, and

$$A^{-1} = \begin{bmatrix} 1 & 0 \\ 0 & 2 \end{bmatrix}$$

is an expansion of factor 2 in the y-direction. ♦

Geometric Properties of Linear Operators on R^2 We conclude this section with two theorems that provide some insight into the geometric properties of linear operators on R^2.

Theorem 9.2.1

If $T: R^2 \to R^2$ is multiplication by an invertible matrix A, then the geometric effect of T is the same as an appropriate succession of shears, compressions, expansions, and reflections.

Proof. Since A is invertible, it can be reduced to the identity by a finite sequence of elementary row operations. An elementary row operation can be performed by multiplying on the left by an elementary matrix, and so there exist elementary matrices $E_1, E_2, \ldots, E_k$ such that

$$E_k \cdots E_2 E_1 A = I$$

Solving for A yields

$$A = E_1^{-1} E_2^{-1} \cdots E_k^{-1} I$$

or equivalently,

$$A = E_1^{-1} E_2^{-1} \cdots E_k^{-1} \tag{5}$$

This equation expresses A as a product of elementary matrices (since the inverse of an elementary matrix is also elementary by Theorem 1.5.2). The result now follows from Example 3. ■

EXAMPLE 5 Geometric Effect of Multiplication by a Matrix

Assuming that k_1 and k_2 are positive, express the diagonal matrix

$$A = \begin{bmatrix} k_1 & 0 \\ 0 & k_2 \end{bmatrix}$$

as a product of elementary matrices and describe the geometric effect of multiplication by A in terms of expansions and compressions.

Solution.

From Example 1 we have

$$A = \begin{bmatrix} k_1 & 0 \\ 0 & k_2 \end{bmatrix} = \begin{bmatrix} 1 & 0 \\ 0 & k_2 \end{bmatrix} \begin{bmatrix} k_1 & 0 \\ 0 & 1 \end{bmatrix}$$

which shows that multiplication by A has the geometric effect of expanding or compressing by a factor of k_1 in the x-direction and then expanding or compressing by a factor of k_2 in the y-direction. ♦

EXAMPLE 6 Analyzing the Geometric Effect of a Matrix Operator

Express

$$A = \begin{bmatrix} 1 & 2 \\ 3 & 4 \end{bmatrix}$$

as a product of elementary matrices, and then describe the geometric effect of multiplication by A in terms of shears, compressions, expansions, and reflections.

Solution.

A can be reduced to I as follows:

$$\begin{bmatrix} 1 & 2 \\ 3 & 4 \end{bmatrix} \rightarrow \begin{bmatrix} 1 & 2 \\ 0 & -2 \end{bmatrix} \rightarrow \begin{bmatrix} 1 & 2 \\ 0 & 1 \end{bmatrix} \rightarrow \begin{bmatrix} 1 & 0 \\ 0 & 1 \end{bmatrix}$$

Add −3 times the first row to the second. | Multiply the second row by $-\frac{1}{2}$. | Add −2 times the second row to the first.

The three successive row operations can be performed by multiplying on the left successively by

$$E_1 = \begin{bmatrix} 1 & 0 \\ -3 & 1 \end{bmatrix}, \quad E_2 = \begin{bmatrix} 1 & 0 \\ 0 & -\frac{1}{2} \end{bmatrix}, \quad E_3 = \begin{bmatrix} 1 & -2 \\ 0 & 1 \end{bmatrix}$$

Inverting these matrices and using (5) yields

$$A = E_1^{-1} E_2^{-1} E_3^{-1} = \begin{bmatrix} 1 & 0 \\ 3 & 1 \end{bmatrix} \begin{bmatrix} 1 & 0 \\ 0 & -2 \end{bmatrix} \begin{bmatrix} 1 & 2 \\ 0 & 1 \end{bmatrix}$$

Reading from right to left and noting that

$$\begin{bmatrix} 1 & 0 \\ 0 & -2 \end{bmatrix} = \begin{bmatrix} 1 & 0 \\ 0 & -1 \end{bmatrix} \begin{bmatrix} 1 & 0 \\ 0 & 2 \end{bmatrix}$$

it follows that the effect of multiplying by A is equivalent to

(1) shearing by a factor of 2 in the x-direction,
(2) then expanding by a factor of 2 in the y-direction,
(3) then reflecting about the x-axis,
(4) then shearing by a factor of 3 in the y-direction. ♦

The proofs for parts of the following theorem are discussed in the exercises.

Theorem 9.2.2 Images of Lines

If $T: R^2 \to R^2$ is multiplication by an invertible matrix, then:

(*a*) *The image of a straight line is a straight line.*
(*b*) *The image of a straight line through the origin is a straight line through the origin.*
(*c*) *The images of parallel straight lines are parallel straight lines.*
(*d*) *The image of the line segment joining points P and Q is the line segment joining the images of P and Q.*
(*e*) *The images of three points lie on a line if and only if the points themselves lie on some line.*

REMARK. It follows from parts (*c*), (*d*), and (*e*) that multiplication by an invertible 2×2 matrix A maps triangles into triangles and parallelograms into parallelograms.

EXAMPLE 7 Image of a Square

The square with vertices $P_1(0, 0)$, $P_2(1, 0)$, $P_3(1, 1)$, and $P_4(0, 1)$ is called the **unit square**. Sketch the image of the unit square under multiplication by

$$A = \begin{bmatrix} -1 & 2 \\ 2 & -1 \end{bmatrix}$$

Solution.

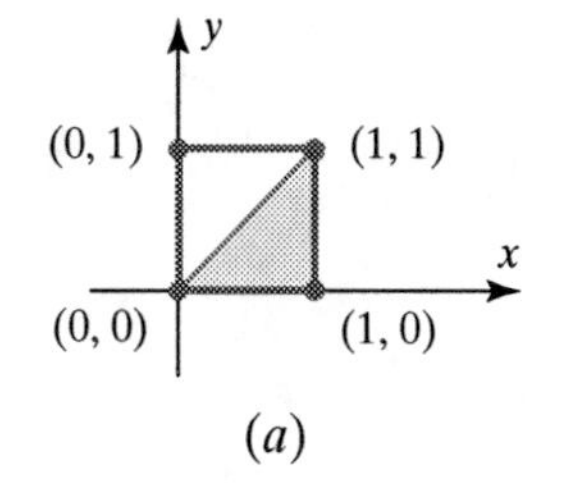

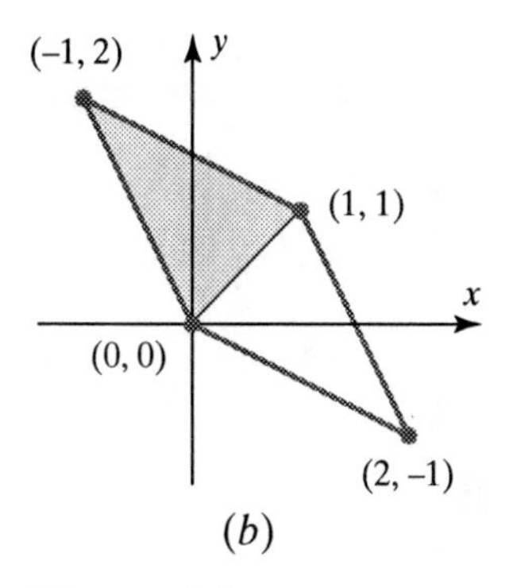

Figure 9.2.5

Since

$$\begin{bmatrix} -1 & 2 \\ 2 & -1 \end{bmatrix}\begin{bmatrix} 0 \\ 0 \end{bmatrix} = \begin{bmatrix} 0 \\ 0 \end{bmatrix} \qquad \begin{bmatrix} -1 & 2 \\ 2 & -1 \end{bmatrix}\begin{bmatrix} 1 \\ 0 \end{bmatrix} = \begin{bmatrix} -1 \\ 2 \end{bmatrix}$$

$$\begin{bmatrix} -1 & 2 \\ 2 & -1 \end{bmatrix}\begin{bmatrix} 0 \\ 1 \end{bmatrix} = \begin{bmatrix} 2 \\ -1 \end{bmatrix} \qquad \begin{bmatrix} -1 & 2 \\ 2 & -1 \end{bmatrix}\begin{bmatrix} 1 \\ 1 \end{bmatrix} = \begin{bmatrix} 1 \\ 1 \end{bmatrix}$$

the image of the square is a parallelogram with vertices $(0, 0)$, $(-1, 2)$, $(2, -1)$, and $(1, 1)$ (Figure 9.2.5). ♦

EXAMPLE 8 Image of a Line

According to Theorem 9.2.2, the invertible matrix

$$A = \begin{bmatrix} 3 & 1 \\ 2 & 1 \end{bmatrix}$$

maps the line $y = 2x + 1$ into another line. Find its equation.

Solution.

Let (x, y) be a point on the line $y = 2x + 1$ and let (x', y') be its image under multiplication by A. Then

$$\begin{bmatrix} x' \\ y' \end{bmatrix} = \begin{bmatrix} 3 & 1 \\ 2 & 1 \end{bmatrix}\begin{bmatrix} x \\ y \end{bmatrix} \quad \text{and} \quad \begin{bmatrix} x \\ y \end{bmatrix} = \begin{bmatrix} 3 & 1 \\ 2 & 1 \end{bmatrix}^{-1}\begin{bmatrix} x' \\ y' \end{bmatrix} = \begin{bmatrix} 1 & -1 \\ -2 & 3 \end{bmatrix}\begin{bmatrix} x' \\ y' \end{bmatrix}$$

so

$$x = x' - y'$$
$$y = -2x' + 3y'$$

Substituting in $y = 2x + 1$ yields

$$-2x' + 3y' = 2(x' - y') + 1 \quad \text{or equivalently,} \quad y' = \tfrac{4}{5}x' + \tfrac{1}{5}$$

Thus, (x', y') satisfies

$$y = \tfrac{4}{5}x + \tfrac{1}{5}$$

which is the equation we want. ◆

Exercise Set 9.2

1. Find the standard matrix for the plane linear transformation $T: R^2 \to R^2$ that maps a point (x, y) into (see accompanying figure)

(a) its reflection about the line $y = -x$ (b) its reflection through the origin
(c) its orthogonal projection on the x-axis (d) its orthogonal projection on the y-axis

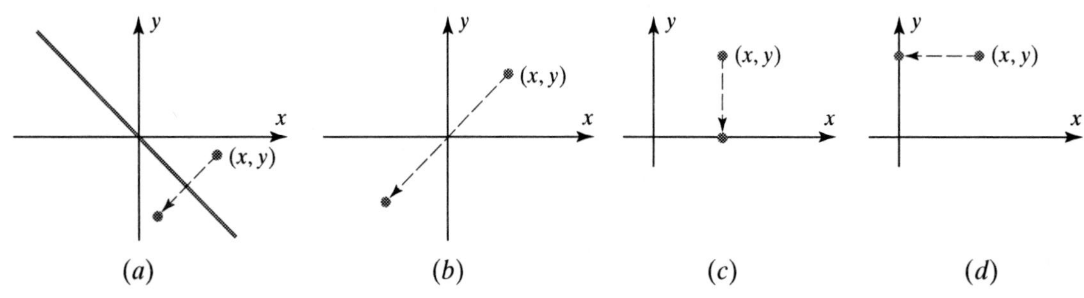

Figure Ex-1

2. For each part of Exercise 1, use the matrix you have obtained to compute $T(2, 1)$. Check your answers geometrically by plotting the points $(2, 1)$ and $T(2, 1)$.

3. Find the standard matrix for the linear operator $T: R^3 \to R^3$ that maps a point (x, y, z) into

(a) its reflection through the xy-plane (b) its reflection through the xz-plane
(c) its reflection through the yz-plane

4. For each part of Exercise 3, use the matrix you have obtained to compute $T(1, 1, 1)$. Check your answers geometrically by sketching the vectors $(1, 1, 1)$ and $T(1, 1, 1)$.

5. Find the standard matrix for the linear operator $T: R^3 \to R^3$ that

(a) rotates each vector 90° counterclockwise about the z-axis (looking along the positive z-axis toward the origin)
(b) rotates each vector 90° counterclockwise about the x-axis (looking along the positive x-axis toward the origin)
(c) rotates each vector 90° counterclockwise about the y-axis (looking along the positive y-axis toward the origin)

6. Sketch the image of the rectangle with vertices $(0, 0)$, $(1, 0)$, $(1, 2)$, and $(0, 2)$ under

(a) a reflection about the x-axis
(b) a reflection about the y-axis
(c) a compression of factor $k = \frac{1}{4}$ in the y-direction
(d) an expansion of factor $k = 2$ in the x-direction
(e) a shear of factor $k = 3$ in the x-direction
(f) a shear of factor $k = 2$ in the y-direction

7. Sketch the image of the square with vertices (0, 0), (1, 0), (0, 1), and (1, 1) under multiplication by

$$A = \begin{bmatrix} -3 & 0 \\ 0 & 1 \end{bmatrix}$$

8. Find the matrix that rotates a point (x, y) about the origin through

(a) 45° (b) 90° (c) 180° (d) 270° (e) −30°

9. Find the matrix that shears by

(a) a factor of $k = 4$ in the y-direction (b) a factor of $k = -2$ in the x-direction

10. Find the matrix that compresses or expands by

(a) a factor of $\frac{1}{3}$ in the y-direction (b) a factor of 6 in the x-direction

11. In each part, describe the geometric effect of multiplication by the given matrix.

(a) $\begin{bmatrix} 3 & 0 \\ 0 & 1 \end{bmatrix}$ (b) $\begin{bmatrix} 1 & 0 \\ 0 & -5 \end{bmatrix}$ (c) $\begin{bmatrix} 1 & 4 \\ 0 & 1 \end{bmatrix}$

12. Express the matrix as a product of elementary matrices, and then describe the effect of multiplication by the given matrix in terms of compressions, expansions, reflections, and shears.

(a) $\begin{bmatrix} 2 & 0 \\ 0 & 3 \end{bmatrix}$ (b) $\begin{bmatrix} 1 & 4 \\ 2 & 9 \end{bmatrix}$ (c) $\begin{bmatrix} 0 & -2 \\ 4 & 0 \end{bmatrix}$ (d) $\begin{bmatrix} 1 & -3 \\ 4 & 6 \end{bmatrix}$

13. In each part, find a single matrix that performs the indicated succession of operations:

(a) compresses by a factor of $\frac{1}{2}$ in the x-direction, then expands by a factor of 5 in the y-direction
(b) expands by a factor of 5 in the y-direction, then shears by a factor of 2 in the y-direction
(c) reflects about $y = x$, then rotates through an angle of 180° about the origin

14. In each part, find a single matrix that performs the indicated succession of operations:

(a) reflects about the y-axis, then expands by a factor of 5 in the x-direction, and then reflects about $y = x$
(b) rotates through 30° about the origin, then shears by a factor of −2 in the y-direction, and then expands by a factor of 3 in the y-direction

15. By matrix inversion, show the following:

(a) The inverse transformation for a reflection about $y = x$ is a reflection about $y = x$.
(b) The inverse transformation for a compression along an axis is an expansion along that axis.
(c) The inverse transformation for a reflection about a coordinate axis is a reflection about that axis.
(d) The inverse transformation for a shear along a coordinate axis is a shear along that axis.

16. Find the equation of the image of the line $y = -4x + 3$ under multiplication by

$$A = \begin{bmatrix} 4 & -3 \\ 3 & -2 \end{bmatrix}$$

17. In parts (a) through (e) find the equation of the image of the line $y = 2x$ under

(a) a shear of factor 3 in the x-direction (b) a compression of factor $\frac{1}{2}$ in the y-direction
(c) a reflection about $y = x$ (d) a reflection about the y-axis
(e) a rotation of 60° about the origin

18. Find the matrix for a shear in the x-direction that transforms the triangle with vertices (0, 0), (2, 1), and (3, 0) into a right triangle with the right angle at the origin.

Answers to Exercises

EXERCISE SET 1.1 [page 6]

1. (a), (c), (f) **2.** (a), (b), (c)

3. **(a)** $x = \frac{3}{7} + \frac{5}{7}t$, $y = t$ **(b)** $x_1 = \frac{5}{3}s - \frac{4}{3}t + \frac{7}{3}$, $x_2 = s$, $x_3 = t$ $x_1 = \frac{1}{4}r - \frac{5}{8}s + \frac{3}{4}t - \frac{1}{8}$, $x_2 = r$, $x_3 = s$, $x_4 = t$ $v = \frac{8}{3}q - \frac{2}{3}r + \frac{1}{3}s - \frac{4}{3}t$, $w = q$, $x = r$, $y = s$, $z = t$

4. **(a)** $\begin{bmatrix} 3 & -2 & -1 \\ 4 & 5 & 3 \\ 7 & 3 & 2 \end{bmatrix}$ **(b)** $\begin{bmatrix} 2 & 0 & 2 & 1 \\ 3 & -1 & 4 & 7 \\ 6 & 1 & -1 & 0 \end{bmatrix}$ **(c)** $\begin{bmatrix} 1 & 2 & 0 & -1 & 1 & 1 \\ 0 & 3 & 1 & 0 & -1 & 2 \\ 0 & 0 & 1 & 7 & 0 & 1 \end{bmatrix}$ **(d)** $\begin{bmatrix} 1 & 0 & 0 & 1 \\ 0 & 1 & 0 & 2 \\ 0 & 0 & 1 & 3 \end{bmatrix}$

5. **(a)** $\begin{aligned} 2x_1 \qquad &= 0 \\ 3x_1 - 4x_2 &= 0 \\ x_2 &= 1 \end{aligned}$ **(b)** $\begin{aligned} 3x_1 \qquad - 2x_3 &= 5 \\ 7x_1 + x_2 + 4x_3 &= -3 \\ -2x_2 + x_3 &= 7 \end{aligned}$ **(c)** $\begin{aligned} 7x_1 + 2x_2 + x_3 - 3x_4 &= 5 \\ x_1 + 2x_2 + 4x_3 \qquad &= 1 \end{aligned}$ **(d)** $\begin{aligned} x_1 &= 7 \\ x_2 &= -2 \\ x_3 &= 3 \\ x_4 &= 4 \end{aligned}$

6. **(a)** $x - 2y = 5$ **(b)** Let $x = t$; then $t - 2y = 5$. Solving for y yields $y = \frac{1}{2}t - \frac{5}{2}$.

10. $k = 6$: infinitely many solutions
$k \neq 6$: no solutions
No value of k yields one solution.

11. **(a)** The lines have no common point of intersection.
(b) The lines intersect in exactly one point.
(c) The three lines coincide.

EXERCISE SET 1.2 [page 19]

1. (a), (b), (c), (d), (h), (i), (j) **2.** (a), (b), (d)

3. **(a)** Both **(b)** Neither **(c)** Both
(d) Row-echelon **(e)** Neither **(f)** Both

4. **(a)** $x_1 = -3, x_2 = 0, x_3 = 7$
(b) $x_1 = 7t + 8, x_2 = -3t + 2, x_3 = -t - 5, x_4 = t$
(c) $x_1 = 6s - 3t - 2, x_2 = s, x_3 = -4t + 7, x_4 = -5t + 8, x_5 = t$
(d) Inconsistent

5. **(a)** $x_1 = -37, x_2 = -8, x_3 = 5$
(b) $x_1 = 13t - 10, x_2 = 13t - 5, x_3 = -t + 2, x_4 = t$
(c) $x_1 = -7s + 2t - 11, x_2 = s, x_3 = -3t - 4, x_4 = -3t + 9, x_5 = t$
(d) Inconsistent

6. **(a)** $x_1 = 3, x_2 = 1, x_3 = 2$
(b) $x_1 = -\frac{1}{7} - \frac{3}{7}t, x_2 = \frac{1}{7} - \frac{4}{7}, x_3 = t$
(c) $x = t - 1, y = 2s, z = s, w = t$
(d) Inconsistent

8. **(a)** Inconsistent
(b) $x_1 = -4, x_2 = 2, x_3 = 7$
(c) $x_1 = 3 + 2t, x_2 = t$
(d) $x = \frac{8}{5} - \frac{3}{5}t - \frac{3}{5}s, y = \frac{1}{10} + \frac{2}{5}t - \frac{1}{10}s, z = t, w = s$

10. **(a)** $x_1 = 2 - 12t, x_2 = 5 - 27t, x_3 = t$
(b) Inconsistent
(c) $u = -2s - 3t - 6, v = s, w = -t - 2, x = t + 3, y = t$

12. (a), (c), (d)

13. **(a)** $x_1 = 0, x_2 = 0, x_3 = 0$
(b) $x_1 = -s, x_2 = -t - s, x_3 = 4s, x_4 = t$
(c) $w = t, x = -t, y = t, z = 0$

14. **(a)** Only the trivial solution
(b) $u = 7s - 5t, v = -6s + 4t, w = 2s, x = 2t$
(c) Only the trivial solution

15. $I_1 = -1, I_2 = 0, I_3 = 1, I_4 = 2$

16. **(a)** $x = \frac{2}{3}a - \frac{1}{9}b, y = -\frac{1}{3}a + \frac{2}{9}b$ **(b)** $x_1 = a - \frac{1}{3}c, x_2 = a - \frac{1}{2}b, x_3 = -a + \frac{1}{2}b + \frac{1}{3}c$

17. $a = -4$, none; $a \neq \pm 4$, exactly one; $a = 4$, infinitely many

19. $\begin{bmatrix} 1 & 3 \\ 0 & 1 \end{bmatrix}$ and $\begin{bmatrix} 1 & 0 \\ 0 & 1 \end{bmatrix}$ are possible answers.

20. $\alpha = \pi/2, \beta = \pi, \gamma = 0$

22. $\lambda = 4, \lambda = 2$

23. If $\lambda = 1$, then $x_1 = x_2 = -\frac{1}{2}s, x_3 = s$
If $\lambda = 2$, then $x_1 = -\frac{1}{2}s, x_2 = 0, x_3 = s$

24. $x = -13/7, y = 91/54, z = -91/8$

25. $a = 1,\ b = -6,\ c = 2,\ d = 10$

26. $a = 1, b = -2, c = -4, d = -29$

30. **(a)** Three distinct lines **(b)** Three identical lines

31. **(a)** False **(b)** True **(c)** False **(d)** False

32. **(a)** False **(b)** False **(c)** False **(d)** False

EXERCISE SET 1.3 (page 33)

1. **(a)** Undefined **(b)** 4×2 **(c)** Undefined **(d)** Undefined
(e) 5×5 **(f)** 5×2 **(g)** Undefined **(h)** 5×2

2. $a = 5,\ b = -3,\ c = 4,\ d = 1$

3. **(a)** $\begin{bmatrix} 7 & 6 & 5 \\ -2 & 1 & 3 \\ 7 & 3 & 7 \end{bmatrix}$ **(b)** $\begin{bmatrix} -5 & 4 & -1 \\ 0 & -1 & -1 \\ -1 & 1 & 1 \end{bmatrix}$ **(c)** $\begin{bmatrix} 15 & 0 \\ -5 & 10 \\ 5 & 5 \end{bmatrix}$ **(d)** $\begin{bmatrix} -7 & -28 & -14 \\ -21 & -7 & -35 \end{bmatrix}$ **(e)** Undefined

(f) $\begin{bmatrix} 22 & -6 & 8 \\ -2 & 4 & 6 \\ 10 & 0 & 4 \end{bmatrix}$ **(g)** $\begin{bmatrix} -39 & -21 & -24 \\ 9 & -6 & -15 \\ -33 & -12 & -30 \end{bmatrix}$ **(h)** $\begin{bmatrix} 0 & 0 \\ 0 & 0 \\ 0 & 0 \end{bmatrix}$ **(i)** 5 **(j)** −25 **(k)** 168 **(l)** Undefined

4. **(a)** $\begin{bmatrix} 7 & 2 & 4 \\ 3 & 5 & 7 \end{bmatrix}$ **(b)** $\begin{bmatrix} -5 & 0 & -1 \\ 4 & -1 & 1 \\ -1 & -1 & 1 \end{bmatrix}$ **(c)** $\begin{bmatrix} -5 & 0 & -1 \\ 4 & -1 & 1 \\ -1 & -1 & 1 \end{bmatrix}$ **(d)** Undefined

(e) $\begin{bmatrix} -\frac{1}{4} & \frac{3}{2} \\ \frac{9}{4} & 0 \\ \frac{3}{4} & \frac{9}{4} \end{bmatrix}$ **(f)** $\begin{bmatrix} 0 & -1 \\ 1 & 0 \end{bmatrix}$ **(g)** $\begin{bmatrix} 9 & 1 & -1 \\ -13 & 2 & -4 \\ 0 & 1 & -6 \end{bmatrix}$ **(h)** $\begin{bmatrix} 9 & -13 & 0 \\ 1 & 2 & 1 \\ -1 & -4 & -6 \end{bmatrix}$

5. **(a)** $\begin{bmatrix} 12 & -3 \\ -4 & 5 \\ 4 & 1 \end{bmatrix}$ **(b)** Undefined **(c)** $\begin{bmatrix} 42 & 108 & 75 \\ 12 & -3 & 21 \\ 36 & 78 & 63 \end{bmatrix}$ **(d)** $\begin{bmatrix} 3 & 45 & 9 \\ 11 & -11 & 17 \\ 7 & 17 & 13 \end{bmatrix}$ **(e)** $\begin{bmatrix} 3 & 45 & 9 \\ 11 & -11 & 17 \\ 7 & 17 & 13 \end{bmatrix}$

(f) $\begin{bmatrix} 21 & 17 \\ 17 & 35 \end{bmatrix}$ **(g)** $\begin{bmatrix} 0 & -2 & 11 \\ 12 & 1 & 8 \end{bmatrix}$ **(h)** $\begin{bmatrix} 12 & 6 & 9 \\ 48 & -20 & 14 \\ 24 & 8 & 16 \end{bmatrix}$ **(i)** 61 **(j)** 35 **(k)** (28)

6. **(a)** $\begin{bmatrix} -6 & -3 \\ 36 & 0 \\ 4 & 7 \end{bmatrix}$ **(b)** Undefined **(c)** $\begin{bmatrix} 2 & -10 & 11 \\ 13 & 2 & 5 \\ 4 & -3 & 13 \end{bmatrix}$ **(d)** $\begin{bmatrix} 10 & -6 \\ -14 & 2 \\ -1 & -8 \end{bmatrix}$ **(e)** $\begin{bmatrix} 40 & 72 \\ 26 & 42 \end{bmatrix}$ **(f)** $\begin{bmatrix} 0 & 0 & 0 \\ 0 & 0 & 0 \\ 0 & 0 & 0 \end{bmatrix}$

7. **(a)** $[67 \quad 41 \quad 41]$ **(b)** $[63 \quad 67 \quad 57]$ **(c)** $\begin{bmatrix} 41 \\ 21 \\ 67 \end{bmatrix}$ **(d)** $\begin{bmatrix} 6 \\ 6 \\ 63 \end{bmatrix}$ **(e)** $[24 \quad 56 \quad 97]$ **(f)** $\begin{bmatrix} 76 \\ 98 \\ 97 \end{bmatrix}$

8. **(a)**

$$\begin{bmatrix} 67 \\ 64 \\ 63 \end{bmatrix} = 6\begin{bmatrix} 3 \\ 6 \\ 0 \end{bmatrix} + 0\begin{bmatrix} -2 \\ 5 \\ 4 \end{bmatrix} + 7\begin{bmatrix} 7 \\ 4 \\ 9 \end{bmatrix}$$

$$\begin{bmatrix} 41 \\ 21 \\ 67 \end{bmatrix} = -2\begin{bmatrix} 3 \\ 6 \\ 0 \end{bmatrix} + 1\begin{bmatrix} -2 \\ 5 \\ 4 \end{bmatrix} + 7\begin{bmatrix} 7 \\ 4 \\ 9 \end{bmatrix}$$

$$\begin{bmatrix} 41 \\ 59 \\ 57 \end{bmatrix} = 4\begin{bmatrix} 3 \\ 6 \\ 0 \end{bmatrix} + 3\begin{bmatrix} -2 \\ 5 \\ 4 \end{bmatrix} + 5\begin{bmatrix} 7 \\ 4 \\ 9 \end{bmatrix}$$

(b)

$$\begin{bmatrix} 6 \\ 6 \\ 63 \end{bmatrix} = 3\begin{bmatrix} 6 \\ 0 \\ 7 \end{bmatrix} + 6\begin{bmatrix} -2 \\ 1 \\ 7 \end{bmatrix} + 0\begin{bmatrix} 4 \\ 3 \\ 5 \end{bmatrix}$$

$$\begin{bmatrix} -6 \\ 17 \\ 41 \end{bmatrix} = -2\begin{bmatrix} 6 \\ 0 \\ 7 \end{bmatrix} + 5\begin{bmatrix} -2 \\ 1 \\ 7 \end{bmatrix} + 4\begin{bmatrix} 4 \\ 3 \\ 5 \end{bmatrix}$$

$$\begin{bmatrix} 70 \\ 31 \\ 122 \end{bmatrix} = 7\begin{bmatrix} 6 \\ 0 \\ 7 \end{bmatrix} + 4\begin{bmatrix} -2 \\ 1 \\ 7 \end{bmatrix} + 9\begin{bmatrix} 4 \\ 3 \\ 5 \end{bmatrix}$$

10. **(a)**

$$[67 \quad 41 \quad 41] = 3[6 \quad -2 \quad 4] - 2[0 \quad 1 \quad 3] + 7[7 \quad 7 \quad 5]$$
$$[64 \quad 21 \quad 59] = 6[6 \quad -2 \quad 4] + 5[0 \quad 1 \quad 3] + 4[7 \quad 7 \quad 5]$$
$$[63 \quad 67 \quad 57] = 0[6 \quad -2 \quad 4] + 4[0 \quad 1 \quad 3] + 9[7 \quad 7 \quad 5]$$

(b)

$$[6 \quad -6 \quad 70] = 6[3 \quad -2 \quad 7] - 2[6 \quad 5 \quad 4] + 4[0 \quad 4 \quad 9]$$
$$[6 \quad 17 \quad 31] = 0[3 \quad -2 \quad 7] + 1[6 \quad 5 \quad 4] + 3[0 \quad 4 \quad 9]$$
$$[63 \quad 41 \quad 122] = 7[3 \quad -2 \quad 7] + 7[6 \quad 5 \quad 4] + 5[0 \quad 4 \quad 9]$$

13. **(a)** $A = \begin{bmatrix} 2 & -3 & 5 \\ 9 & -1 & 1 \\ 1 & 5 & 4 \end{bmatrix}$, $\mathbf{x} = \begin{bmatrix} x_1 \\ x_2 \\ x_3 \end{bmatrix}$, $\mathbf{b} = \begin{bmatrix} 7 \\ -1 \\ 0 \end{bmatrix}$ **(b)** $A = \begin{bmatrix} 4 & 0 & -3 & 1 \\ 5 & 1 & 0 & -8 \\ 2 & -5 & 9 & -1 \\ 0 & 3 & -1 & 7 \end{bmatrix}$, $\mathbf{x} = \begin{bmatrix} x_1 \\ x_2 \\ x_3 \\ x_4 \end{bmatrix}$, $\mathbf{b} = \begin{bmatrix} 1 \\ 3 \\ 0 \\ 2 \end{bmatrix}$

14. **(a)**

$$\begin{aligned} 3x_1 - x_2 + 2x_3 &= 2 \\ 4x_1 + 3x_2 + 7x_3 &= -1 \\ -2x_1 + x_2 + 5x_3 &= 4 \end{aligned}$$

(b)

$$\begin{aligned} 3w - 2x \qquad + z &= 0 \\ 5w \qquad + 2y - 2z &= 0 \\ 3w + x + 4y + 7z &= 0 \\ -2w + 5x + y + 6z &= 0 \end{aligned}$$

15. $\begin{bmatrix} -1 & 23 & -10 \\ 37 & -13 & 8 \\ 29 & 23 & 41 \end{bmatrix}$ **16.** **(a)** $\begin{bmatrix} -3 & -15 & -11 \\ 21 & -15 & 44 \end{bmatrix}$ **(b)** $\begin{bmatrix} 4 & -7 & -19 & -43 \\ 2 & 2 & 18 & 17 \\ 0 & 5 & 25 & 35 \\ 2 & 3 & 23 & 24 \end{bmatrix}$ **(c)** $\begin{bmatrix} 3 & 3 \\ -1 & 4 \\ 1 & 5 \\ 4 & -4 \\ 0 & 14 \end{bmatrix}$

17. **(a)** A_{11} is a 2×3 matrix and B_{11} is a 2×2 matrix. $A_{11}B_{11}$ does not exist. **(b)** $\begin{bmatrix} -1 & 23 & -10 \\ 37 & -13 & 8 \\ 29 & 23 & 41 \end{bmatrix}$

21. **(a)** $\begin{bmatrix} a_{11} & 0 & 0 & 0 & 0 & 0 \\ 0 & a_{22} & 0 & 0 & 0 & 0 \\ 0 & 0 & a_{33} & 0 & 0 & 0 \\ 0 & 0 & 0 & a_{44} & 0 & 0 \\ 0 & 0 & 0 & 0 & a_{55} & 0 \\ 0 & 0 & 0 & 0 & 0 & a_{66} \end{bmatrix}$ **(b)** $\begin{bmatrix} a_{11} & a_{12} & a_{13} & a_{14} & a_{15} & a_{16} \\ 0 & a_{22} & a_{23} & a_{24} & a_{25} & a_{26} \\ 0 & 0 & a_{33} & a_{34} & a_{35} & a_{36} \\ 0 & 0 & 0 & a_{44} & a_{45} & a_{46} \\ 0 & 0 & 0 & 0 & a_{55} & a_{56} \\ 0 & 0 & 0 & 0 & 0 & a_{66} \end{bmatrix}$

(c) $\begin{bmatrix} a_{11} & 0 & 0 & 0 & 0 & 0 \\ a_{21} & a_{22} & 0 & 0 & 0 & 0 \\ a_{31} & a_{32} & a_{33} & 0 & 0 & 0 \\ a_{41} & a_{42} & a_{43} & a_{44} & 0 & 0 \\ a_{51} & a_{52} & a_{53} & a_{54} & a_{55} & 0 \\ a_{61} & a_{62} & a_{63} & a_{64} & a_{65} & a_{66} \end{bmatrix}$ **(d)** $\begin{bmatrix} a_{11} & a_{12} & 0 & 0 & 0 & 0 \\ a_{21} & a_{22} & a_{23} & 0 & 0 & 0 \\ 0 & a_{32} & a_{33} & a_{34} & 0 & 0 \\ 0 & 0 & a_{43} & a_{44} & a_{45} & 0 \\ 0 & 0 & 0 & a_{54} & a_{55} & a_{56} \\ 0 & 0 & 0 & 0 & a_{65} & a_{66} \end{bmatrix}$

22. **(a)** $\begin{bmatrix} 2 & 3 & 4 & 5 \\ 3 & 4 & 5 & 6 \\ 4 & 5 & 6 & 7 \\ 5 & 6 & 7 & 8 \end{bmatrix}$ **(b)** $\begin{bmatrix} 1 & 1 & 1 & 1 \\ 1 & 2 & 4 & 8 \\ 1 & 3 & 9 & 27 \\ 1 & 4 & 16 & 64 \end{bmatrix}$ **(c)** $\begin{bmatrix} -1 & -1 & 1 & 1 \\ -1 & -1 & -1 & 1 \\ 1 & -1 & -1 & -1 \\ 1 & 1 & -1 & -1 \end{bmatrix}$

25. One; namely, $A = \begin{bmatrix} 1 & 1 & 0 \\ 1 & -1 & 0 \\ 0 & 0 & 0 \end{bmatrix}$ **26.** None

27. **(a)** $\pm\begin{bmatrix} 1 & 1 \\ 1 & 1 \end{bmatrix}$ **(b)** Four; namely, $\begin{bmatrix} \pm\sqrt{5} & 0 \\ 0 & \pm 3 \end{bmatrix}$ **(c)** No; for example, $\begin{bmatrix} -1 & 0 \\ 0 & 1 \end{bmatrix}$

28. **(a)** Yes; for example, $\begin{bmatrix} 0 & 1 \\ 0 & 0 \end{bmatrix}$ **(b)** Yes; for example, $\begin{bmatrix} 1 & 0 \\ 0 & 0 \end{bmatrix}$

29. **(a)** True **(b)** True **(c)** False **(d)** True

30. **(a)** True **(b)** False; for example, $A = \begin{bmatrix} 1 & -1 \\ 1 & -1 \end{bmatrix}$ **(c)** True **(d)** True

EXERCISE SET 1.4 [page 47]

4. $A^{-1} = \begin{bmatrix} 2 & -1 \\ -5 & 3 \end{bmatrix}$, $B^{-1} = \begin{bmatrix} \frac{1}{5} & \frac{3}{20} \\ -\frac{1}{5} & \frac{1}{10} \end{bmatrix}$, $C^{-1} = \begin{bmatrix} -\frac{1}{2} & -2 \\ 1 & 3 \end{bmatrix}$, $D^{-1} = \begin{bmatrix} \frac{1}{2} & 0 \\ 0 & \frac{1}{3} \end{bmatrix}$

7. **(a)** $A = \begin{bmatrix} \frac{5}{13} & \frac{1}{13} \\ -\frac{3}{13} & \frac{2}{13} \end{bmatrix}$ **(b)** $A = \begin{bmatrix} \frac{2}{7} & 1 \\ \frac{1}{7} & \frac{3}{7} \end{bmatrix}$ **(c)** $A = \begin{bmatrix} -\frac{2}{5} & 1 \\ -\frac{1}{5} & \frac{3}{5} \end{bmatrix}$ **(d)** $A = \begin{bmatrix} -\frac{9}{13} & \frac{1}{13} \\ \frac{2}{13} & -\frac{6}{13} \end{bmatrix}$

8. $A^3 = \begin{bmatrix} 8 & 0 \\ 28 & 1 \end{bmatrix}$, $A^{-3} = \begin{bmatrix} \frac{1}{8} & 0 \\ -\frac{7}{2} & 1 \end{bmatrix}$, $A^2 - 2A + I = \begin{bmatrix} 1 & 0 \\ 4 & 0 \end{bmatrix}$

9. **(a)** $p(A) = \begin{bmatrix} 1 & 1 \\ 2 & -1 \end{bmatrix}$ **(b)** $p(A) = \begin{bmatrix} 20 & 7 \\ 14 & 6 \end{bmatrix}$ **(c)** $p(A) = \begin{bmatrix} 39 & 13 \\ 26 & 13 \end{bmatrix}$

11. $\begin{bmatrix} \cos\theta & -\sin\theta \\ \sin\theta & \cos\theta \end{bmatrix}$ **12.** $\begin{bmatrix} \frac{1}{2}(e^x + e^{-x}) & \frac{1}{2}(e^{-x} - e^x) \\ \frac{1}{2}(e^{-x} - e^x) & \frac{1}{2}(e^x + e^{-x}) \end{bmatrix}$

13. $A^{-1} = \begin{bmatrix} \frac{1}{a_{11}} & 0 & \cdots & 0 \\ 0 & \frac{1}{a_{22}} & \cdots & 0 \\ \vdots & \vdots & & \vdots \\ 0 & 0 & \cdots & \frac{1}{a_{nn}} \end{bmatrix}$ 16. No 18. $C = -A^{-1}BA^{-1}$

19. **(a)** $\begin{bmatrix} \frac{1}{2} & -\frac{1}{2} & 0 & 0 \\ \frac{1}{2} & \frac{1}{2} & 0 & 0 \\ 0 & 0 & \frac{1}{2} & -\frac{1}{2} \\ -1 & 0 & \frac{1}{2} & \frac{1}{2} \end{bmatrix}$ **(b)** $\begin{bmatrix} 1 & -1 & 0 & 0 \\ 0 & 1 & 0 & 0 \\ 0 & 0 & 1 & -1 \\ 0 & 0 & 0 & 1 \end{bmatrix}$

20. **(a)** One example is $\begin{bmatrix} 1 & 2 & 3 \\ 2 & 1 & 4 \\ 3 & 4 & 5 \end{bmatrix}$. **(b)** One example is $\begin{bmatrix} 0 & -1 & -1 \\ 1 & 0 & -1 \\ 1 & 1 & 0 \end{bmatrix}$. 22. Yes 23. $A^{-1} = \begin{bmatrix} \frac{1}{2} & \frac{1}{2} & -\frac{1}{2} \\ -\frac{1}{2} & \frac{1}{2} & \frac{1}{2} \\ \frac{1}{2} & -\frac{1}{2} & \frac{1}{2} \end{bmatrix}$

31. **(a)** For example, $A = \begin{bmatrix} 1 & 0 \\ 0 & 0 \end{bmatrix}$, $B = \begin{bmatrix} 0 & 1 \\ 0 & 0 \end{bmatrix}$ **(b)** $AB + BA$

32. **(a)** Same as 31(a) **(b)** $A^2 - AB + BA - B^2$ 33. $\begin{bmatrix} \pm 1 & 0 & 0 \\ 0 & \pm 1 & 0 \\ 0 & 0 & \pm 1 \end{bmatrix}$

34. **(a)** If A is not singular, then A^T is not singular. **(b)** True

35. **(a)** False **(b)** True **(c)** True **(d)** False

EXERCISE SET 1.5 [page 56]

1. (c), (d), (f)

2. **(a)** Add three times the first row to the second row.
   (b) Multiply the third row by $\frac{1}{3}$.
   (c) Interchange the first row and the fourth row.
   (d) Add $\frac{1}{7}$ times third row to the first row.

3. **(a)** $\begin{bmatrix} 0 & 0 & 1 \\ 0 & 1 & 0 \\ 1 & 0 & 0 \end{bmatrix}$ **(b)** $\begin{bmatrix} 0 & 0 & 1 \\ 0 & 1 & 0 \\ 1 & 0 & 0 \end{bmatrix}$ **(c)** $\begin{bmatrix} 1 & 0 & 0 \\ 0 & 1 & 0 \\ -2 & 0 & 1 \end{bmatrix}$ **(d)** $\begin{bmatrix} 1 & 0 & 0 \\ 0 & 1 & 0 \\ 2 & 0 & 1 \end{bmatrix}$

4. No, since C cannot be obtained by performing a single row operation on B.

5. **(a)** $\begin{bmatrix} -7 & 4 \\ 2 & -1 \end{bmatrix}$ **(b)** $\begin{bmatrix} -\frac{5}{39} & \frac{2}{13} \\ \frac{4}{39} & \frac{1}{13} \end{bmatrix}$ **(c)** Not invertible

6. **(a)** $\begin{bmatrix} \frac{3}{2} & -\frac{11}{10} & -\frac{6}{5} \\ -1 & 1 & 1 \\ -\frac{1}{2} & \frac{7}{10} & \frac{2}{5} \end{bmatrix}$ **(b)** Not invertible **(c)** $\begin{bmatrix} \frac{1}{2} & -\frac{1}{2} & \frac{1}{2} \\ -\frac{1}{2} & \frac{1}{2} & \frac{1}{2} \\ \frac{1}{2} & \frac{1}{2} & -\frac{1}{2} \end{bmatrix}$ **(d)** $\begin{bmatrix} \frac{7}{2} & 0 & -3 \\ -1 & 1 & 0 \\ 0 & -1 & 1 \end{bmatrix}$ **(e)** $\begin{bmatrix} \frac{1}{2} & -\frac{1}{2} & \frac{1}{2} \\ 0 & 0 & 1 \\ \frac{1}{2} & \frac{1}{2} & -\frac{1}{2} \end{bmatrix}$

7. **(a)** $\begin{bmatrix} 1 & 3 & 1 \\ 0 & 1 & -1 \\ -2 & 2 & 0 \end{bmatrix}$ **(b)** $\begin{bmatrix} \frac{\sqrt{2}}{26} & \frac{-3\sqrt{2}}{26} & 0 \\ \frac{4\sqrt{2}}{26} & \frac{\sqrt{2}}{26} & 0 \\ 0 & 0 & 1 \end{bmatrix}$ **(c)** $\begin{bmatrix} 1 & 0 & 0 & 0 \\ -\frac{1}{3} & \frac{1}{3} & 0 & 0 \\ 0 & -\frac{1}{5} & \frac{1}{5} & 0 \\ 0 & 0 & -\frac{1}{7} & \frac{1}{7} \end{bmatrix}$ **(d)** Not invertible **(e)** $\begin{bmatrix} -\frac{4}{5} & \frac{3}{5} & \frac{1}{5} & \frac{1}{5} \\ \frac{3}{2} & 0 & -1 & 0 \\ \frac{1}{2} & 0 & 0 & 0 \\ \frac{4}{5} & \frac{2}{5} & -\frac{1}{5} & -\frac{1}{5} \end{bmatrix}$

8. **(a)** $\begin{bmatrix} \frac{1}{k_1} & 0 & 0 & 0 \\ 0 & \frac{1}{k_2} & 0 & 0 \\ 0 & 0 & \frac{1}{k_3} & 0 \\ 0 & 0 & 0 & \frac{1}{k_4} \end{bmatrix}$ **(b)** $\begin{bmatrix} 0 & 0 & 0 & \frac{1}{k_4} \\ 0 & 0 & \frac{1}{k_3} & 0 \\ 0 & \frac{1}{k_2} & 0 & 0 \\ \frac{1}{k_1} & 0 & 0 & 0 \end{bmatrix}$ **(c)** $\begin{bmatrix} \frac{1}{k} & 0 & 0 & 0 \\ -\frac{1}{k^2} & \frac{1}{k} & 0 & 0 \\ \frac{1}{k^3} & -\frac{1}{k^2} & \frac{1}{k} & 0 \\ -\frac{1}{k^4} & \frac{1}{k^3} & -\frac{1}{k^2} & \frac{1}{k} \end{bmatrix}$

9. **(a)** $E_1 = \begin{bmatrix} 1 & 0 \\ 5 & 1 \end{bmatrix}$, $E_2 = \begin{bmatrix} 1 & 0 \\ 0 & \frac{1}{2} \end{bmatrix}$ **(b)** $A^{-1} = E_2E_1$ **(c)** $A = E_1^{-1}E_2^{-1}$

11. $\begin{bmatrix} 1 & 0 & 0 \\ 0 & 1 & 0 \\ 0 & -2 & 1 \end{bmatrix} \begin{bmatrix} 1 & 0 & 0 \\ 0 & 1 & 0 \\ 1 & 0 & 1 \end{bmatrix} \begin{bmatrix} 0 & 1 & 0 \\ 1 & 0 & 0 \\ 0 & 0 & 1 \end{bmatrix} \begin{bmatrix} 1 & 3 & 3 & 8 \\ 0 & 1 & 7 & 8 \\ 0 & 0 & 0 & 0 \end{bmatrix}$

16. **(b)** Add -1 times the first row to the second row.
Add -1 times the first row to the third row.
Add -1 times the second row to the first row.
Add the second row to the third row.

19. **(a)** False **(b)** False **(c)** True **(d)** True

20. **(a)** True **(b)** True **(c)** True **(d)** False

21. In general, no. Try $b = 1, a = c = d = 0$.

EXERCISE SET 1.6 [page 64]

1. $x_1 = 3, x_2 = -1$ 2. $x_1 = -3, x_2 = -3$ 3. $x_1 = -1, x_2 = 4, x_3 = -7$

4. $x_1 = 1, x_2 = -11, x_3 = 16$ 5. $x_1 = 1, x_2 = 5, x_3 = -1$

6. $w = -6, x = 1, y = 10, z = -7$ 7. $x_1 = 2b_1 - 5b_2, x_2 = -b_1 + 3b_2$

8. $x_1 = -\frac{15}{2}b_1 + \frac{1}{2}b_2 + \frac{5}{2}b_3, x_2 = \frac{1}{2}b_1 + \frac{1}{2}b_2 - \frac{1}{2}b_3, x_3 = \frac{5}{2}b_1 - \frac{1}{2}b_2 - \frac{1}{2}b_3$

9. **(a)** $x_1 = \frac{16}{3}, x_2 = -\frac{4}{3}, x_3 = -\frac{11}{3}$ **(b)** $x_1 = -\frac{5}{3}, x_2 = \frac{5}{3}, x_3 = \frac{10}{3}$
(c) $x_1 = 3, x_2 = 0, x_3 = -4$

11. **(a)** $x_1 = \frac{22}{17},\ x_2 = \frac{1}{17}$ **(b)** $x_1 = \frac{21}{17},\ x_2 = \frac{11}{17}$

12. **(a)** $x_1 = -18,\ x_2 = -1,\ x_3 = -14$ **(b)** $x_1 = -\frac{421}{2},\ x_2 = -\frac{25}{2},\ x_3 = -\frac{327}{2}$

13. **(a)** $x_1 = \frac{7}{15},\ x_2 = \frac{4}{15}$ **(b)** $x_1 = \frac{34}{15},\ x_2 = \frac{28}{15}$ **(c)** $x_1 = \frac{19}{15},\ x_2 = \frac{13}{15}$ **(d)** $x_1 = -\frac{1}{5},\ x_2 = \frac{3}{5}$

14. **(a)** $x_1 = 18,\ x_2 = -9,\ x_3 = 2$ **(b)** $x_1 = -23,\ x_2 = 11,\ x_3 = -2$
(c) $x_1 = 5,\ x_2 = -2,\ x_3 = 0$

15. **(a)** $x_1 = -12 - 3t,\ x_2 = -5 - t,\ x_3 = t$ **(b)** $x_1 = 7 - 3t,\ x_2 = 3 - t,\ x_3 = t$

16. $b_1 = 2b_2$ 17. $b_1 = b_2 + b_3$ 18. No restrictions 19. $b_1 = b_3 + b_4, b_2 = 2b_3 + b_4$

21. $X = \begin{bmatrix} 11 & 12 & -3 & 27 & 26 \\ -6 & -8 & 1 & -18 & -17 \\ -15 & -21 & 9 & -38 & -35 \end{bmatrix}$

22. **(a)** Only the trivial solution $x_1 = x_2 = x_3 = x_4 = 0$; invertible
(b) Infinitely many solutions; not invertible

27. **(a)** $I - A$ is invertible. **(b)** $\mathbf{x} = (I - A)^{-1}\mathbf{b}$

28. No. Try $A = I$. 29. Yes, for nonsquare matrices

EXERCISE SET 1.7 [page 71]

1. **(a)** $\begin{bmatrix} \frac{1}{2} & 0 \\ 0 & -\frac{1}{5} \end{bmatrix}$ **(b)** Not invertible **(c)** $\begin{bmatrix} -1 & 0 & 0 \\ 0 & \frac{1}{2} & 0 \\ 0 & 0 & 3 \end{bmatrix}$

2. **(a)** $\begin{bmatrix} 6 & 3 \\ 4 & -1 \\ 4 & 10 \end{bmatrix}$ **(b)** $\begin{bmatrix} -24 & -10 & 12 \\ 3 & -10 & 0 \\ 60 & 20 & -16 \end{bmatrix}$

3. **(a)** $A^2 = \begin{bmatrix} 1 & 0 \\ 0 & 4 \end{bmatrix}$, $A^{-2} = \begin{bmatrix} 1 & 0 \\ 0 & \frac{1}{4} \end{bmatrix}$, $A^{-k} = \begin{bmatrix} 1 & 0 \\ 0 & 1/(-2)^k \end{bmatrix}$

(b) $A^2 = \begin{bmatrix} \frac{1}{4} & 0 & 0 \\ 0 & \frac{1}{9} & 0 \\ 0 & 0 & \frac{1}{16} \end{bmatrix}$, $A^{-2} = \begin{bmatrix} 4 & 0 & 0 \\ 0 & 9 & 0 \\ 0 & 0 & 16 \end{bmatrix}$, $A^{-k} = \begin{bmatrix} 2^k & 0 & 0 \\ 0 & 3^k & 0 \\ 0 & 0 & 4^k \end{bmatrix}$

4. (b), (c) 5. (a) 6. $a = 11, b = -9, c = -13$ 7. $a = 2, b = -1$

8. **(a)** Does not commute **(b)** Commutes 10. **(a)** $\begin{bmatrix} 1 & 0 & 0 \\ 0 & -1 & 0 \\ 0 & 0 & -1 \end{bmatrix}$ **(b)** $\begin{bmatrix} \frac{1}{3} & 0 & 0 \\ 0 & \frac{1}{2} & 0 \\ 0 & 0 & 1 \end{bmatrix}$

11. **(a)** $\begin{bmatrix} a_{11} & a_{12} & a_{13} \\ a_{21} & a_{22} & a_{23} \\ a_{31} & a_{32} & a_{33} \end{bmatrix}\begin{bmatrix} 3 & 0 & 0 \\ 0 & 5 & 0 \\ 0 & 0 & 7 \end{bmatrix}$ **(b)** No

19. $\begin{bmatrix} 4 & 0 & 0 \\ 0 & 4 & 0 \\ 0 & 0 & 4 \end{bmatrix}, \begin{bmatrix} 4 & 0 & 0 \\ 0 & 4 & 0 \\ 0 & 0 & -1 \end{bmatrix}, \begin{bmatrix} 4 & 0 & 0 \\ 0 & -1 & 0 \\ 0 & 0 & 4 \end{bmatrix}, \begin{bmatrix} -1 & 0 & 0 \\ 0 & 4 & 0 \\ 0 & 0 & 4 \end{bmatrix},$

$\begin{bmatrix} -1 & 0 & 0 \\ 0 & -1 & 0 \\ 0 & 0 & 4 \end{bmatrix}, \begin{bmatrix} -1 & 0 & 0 \\ 0 & 4 & 0 \\ 0 & 0 & -1 \end{bmatrix}, \begin{bmatrix} 4 & 0 & 0 \\ 0 & -1 & 0 \\ 0 & 0 & -1 \end{bmatrix}, \begin{bmatrix} -1 & 0 & 0 \\ 0 & -1 & 0 \\ 0 & 0 & -1 \end{bmatrix}$

25. $A = \begin{bmatrix} 1 & 10 \\ 0 & -2 \end{bmatrix}$ 26. $\frac{n}{2}(1 + n)$ 27. Multiply corresponding diagonal entries.

28. A is diagonal. 30. **(a)** True **(b)** False **(c)** True **(d)** False

SUPPLEMENTARY EXERCISES [page 74]

1. $x' = \frac{3}{5}x + \frac{4}{5}y, y' = -\frac{4}{5}x + \frac{3}{5}y$ 2. $x' = x\cos\theta + y\sin\theta, y' = -x\sin\theta + y\cos\theta$

3. One possible answer is
$$\begin{aligned} x_1 - 2x_2 - x_3 - x_4 &= 0 \\ x_1 + 5x_2 + 2x_4 &= 0 \end{aligned}$$

4. 3 pennies, 4 nickels, 6 dimes

5. $x = 4, y = 2, z = 3$ 6. Infinitely many if $a = 2$ or $a = -\frac{3}{2}$; none otherwise

7. **(a)** $a \neq 0,\ b \neq 2$ **(b)** $a \neq 0,\ b = 2$ **(c)** $a = 0,\ b = 2$ **(d)** $a = 0,\ b \neq 2$

8. $x = \frac{5}{9}, y = 9, z = \frac{1}{3}$ 9. $K = \begin{bmatrix} 0 & 2 \\ 1 & 1 \end{bmatrix}$ 10. $a = 2, b = -1, c = 1$

11. **(a)** $X = \begin{bmatrix} -1 & 3 & -1 \\ 6 & 0 & 1 \end{bmatrix}$ **(b)** $X = \begin{bmatrix} 1 & -2 \\ 3 & 1 \end{bmatrix}$ **(c)** $X = \begin{bmatrix} -\frac{113}{37} & -\frac{160}{37} \\ -\frac{20}{37} & -\frac{46}{37} \end{bmatrix}$

12. **(a)** $Z = \begin{bmatrix} -1 & -7 & 11 \\ 14 & 10 & -26 \end{bmatrix} X$ **(b)** $z_1 = -x_1 - 7x_2 + 11x_3$
$z_2 = 14x_1 + 10x_2 - 26x_3$

13. mpn multiplications and $mp(n-1)$ additions 15. $a = 1, b = -2, c = 3$

16. $a = 1, b = -4, c = -5$ 26. $A = -\frac{7}{5}, B = \frac{4}{5}, C = \frac{3}{5}$

EXERCISE SET 2.1 [page 87]

1. **(a)** 5 **(b)** 9 **(c)** 6 **(d)** 10 **(e)** 0 **(f)** 2

2. **(a)** Odd **(b)** Odd **(c)** Even **(d)** Even **(e)** Even **(f)** Even

3. 22 4. 0 5. 52 6. $-3\sqrt{6}$ 7. $a^2 - 5a + 21$ 8. 0

9. -65 10. -4 11. -123 12. $-c^4 + c^3 - 16c^2 + 8c - 2$

13. **(a)** $\lambda = 1, \lambda = -3$ **(b)** $\lambda = -2, \lambda = 3, \lambda = 4$ 16. 275

17. **(a)** $= -120$ **(b)** $= -120$ 18. $x = \dfrac{3 \pm \sqrt{33}}{4}$ 22. Equals 0 if $n > 1$

24. The determinant is equal to the product of the diagonal entries.

25. The determinant is equal to the product of the diagonal entries.

EXERCISE SET 2.2 [page 94]

1. **(a)** -30 **(b)** -2 **(c)** 0 **(d)** 0 3. **(a)** -5 **(b)** -1 **(c)** 1

4. 30 5. 5 6. -17 7. 33 8. 39 9. 6 10. $-\frac{1}{6}$

11. -2 12. **(a)** -6 **(b)** 72 **(c)** -6 **(d)** 18

16. **(a)** $\det(A) = -1$ **(b)** $\det(A) = 1$ 17. $x = 0, -1, \frac{1}{2}$ 18. $x = 1, -3$

EXERCISE SET 2.3 [page 102]

1. **(a)** $\det(2A) = -40 = 2^2 \det(A)$ **(b)** $\det(-2A) = -448 = (-2)^3 \det(A)$

2. $\det AB = -170 = (\det A)(\det B)$

4. **(a)** Invertible **(b)** Not invertible **(c)** Not invertible **(d)** Not invertible

5. **(a)** -189 **(b)** $-\frac{1}{7}$ **(c)** $-\frac{8}{7}$ **(d)** $-\frac{1}{56}$ **(e)** 7

6. If $x = 0$, the first and third rows are proportional.
If $x = 2$, the first and second rows are proportional.

12. **(a)** $k = \dfrac{5 \pm \sqrt{17}}{2}$ **(b)** $k = -1$

14. **(a)** $\begin{bmatrix} \lambda - 1 & -2 \\ -2 & \lambda - 1 \end{bmatrix} \begin{bmatrix} x_1 \\ x_2 \end{bmatrix} = \begin{bmatrix} 0 \\ 0 \end{bmatrix}$ **(b)** $\begin{bmatrix} \lambda - 2 & -3 \\ -4 & \lambda - 3 \end{bmatrix} \begin{bmatrix} x_1 \\ x_2 \end{bmatrix} = \begin{bmatrix} 0 \\ 0 \end{bmatrix}$ **(c)** $\begin{bmatrix} \lambda - 3 & -1 \\ 5 & \lambda + 3 \end{bmatrix} \begin{bmatrix} x_1 \\ x_2 \end{bmatrix} = \begin{bmatrix} 0 \\ 0 \end{bmatrix}$

15. (i) $\lambda^2 - 2\lambda - 3 = 0$ (ii) $\lambda = -1, \lambda = 3$ (iii) $\begin{bmatrix} -t \\ t \end{bmatrix}, \begin{bmatrix} t \\ t \end{bmatrix}$

(i) $\lambda^2 - 5\lambda - 6 = 0$ (ii) $\lambda = -1, \lambda = 6$ (iii) $\begin{bmatrix} -t \\ t \end{bmatrix}, \begin{bmatrix} \frac{3}{4}t \\ t \end{bmatrix}$

(i) $\lambda^2 - 4 = 0$ (ii) $\lambda = -2, \lambda = 2$ (iii) $\begin{bmatrix} -\frac{t}{5} \\ t \end{bmatrix}, \begin{bmatrix} -t \\ t \end{bmatrix}$

20. No 21. AB is nonsingular if and only if both A and B are nonsingular.

22. **(a)** False **(b)** True **(c)** False **(d)** True

23. **(a)** True **(b)** True **(c)** False **(d)** True

EXERCISE SET 2.4 [page 112]

1. **(a)** $M_{11} = 29, M_{12} = 21, M_{13} = 27, M_{21} = -11, M_{22} = 13, M_{23} = -5, M_{31} = -19, M_{32} = -19, M_{33} = 19$
   (b) $C_{11} = 29, C_{12} = -21, C_{13} = 27, C_{21} = 11, C_{22} = 13, C_{23} = 5, C_{31} = -19, C_{32} = 19, C_{33} = 19$

2. **(a)** $M_{13} = 0,\ C_{13} = 0$ **(b)** $M_{23} = -96,\ C_{23} = 96$ 3. 152
   (c) $M_{22} = -48,\ C_{22} = -48$ **(d)** $M_{21} = 72,\ C_{21} = -72$

4. **(a)** $\text{adj}(A) = \begin{bmatrix} 29 & 11 & -19 \\ -21 & 13 & 19 \\ 27 & 5 & 19 \end{bmatrix}$ **(b)** $A^{-1} = \begin{bmatrix} \frac{29}{152} & \frac{11}{152} & -\frac{19}{152} \\ -\frac{21}{152} & \frac{13}{152} & \frac{19}{152} \\ \frac{27}{152} & \frac{5}{152} & \frac{19}{152} \end{bmatrix}$

5. -40 6. -66 7. 0 8. $k^3 - 8k^2 - 10k + 95$ 9. -240 10. 0

11. $A^{-1} = \begin{bmatrix} 3 & -5 & -5 \\ -3 & 4 & 5 \\ 2 & -2 & -3 \end{bmatrix}$ 12. $A^{-1} = \begin{bmatrix} 2 & 0 & \frac{3}{2} \\ \frac{2}{3} & \frac{1}{3} & \frac{2}{3} \\ -1 & 0 & -1 \end{bmatrix}$ 13. $A^{-1} = \begin{bmatrix} \frac{1}{2} & \frac{3}{2} & 1 \\ 0 & 1 & \frac{3}{2} \\ 0 & 0 & \frac{1}{2} \end{bmatrix}$

14. $A^{-1} = \begin{bmatrix} \frac{1}{2} & 0 & 0 \\ -4 & 1 & 0 \\ \frac{29}{12} & -\frac{1}{2} & \frac{1}{6} \end{bmatrix}$ 15. $A^{-1} = \begin{bmatrix} -4 & 3 & 0 & -1 \\ 2 & -1 & 0 & 0 \\ -7 & 0 & -1 & 8 \\ 6 & 0 & 1 & -7 \end{bmatrix}$

16. $x_1 = 1, x_2 = 2$ 17. $x = \frac{3}{11}, y = \frac{2}{11}, z = -\frac{1}{11}$

18. $x = -\frac{144}{55}, y = -\frac{61}{55}, z = \frac{46}{11}$ 19. $x_1 = -\frac{30}{11}, x_2 = -\frac{38}{11}, x_3 = -\frac{40}{11}$

20. $x_1 = 5, x_2 = 8, x_3 = 3, x_4 = -1$ 21. Cramer's rule does not apply.

22. $A^{-1} = \begin{bmatrix} \cos\theta & -\sin\theta & 0 \\ \sin\theta & \cos\theta & 0 \\ 0 & 0 & 1 \end{bmatrix}$ 23. $y = 0$ 24. $x = 1, y = 0, z = 2, w = 0$

31. $\det(A) = 10 \times (-108) = -1080$ 33. 12 34. One

35. **(a)** True **(b)** False **(c)** True **(d)** False

SUPPLEMENTARY EXERCISES [page 115]

1. $x' = \frac{3}{5}x + \frac{4}{5}y, y' = -\frac{4}{5}x + \frac{3}{5}y$ 2. $x' = x\cos\theta + y\sin\theta, y' = -x\sin\theta + y\cos\theta$

4. 2 5. $\cos\beta = \dfrac{c^2 + a^2 - b^2}{2ac}$, $\cos\gamma = \dfrac{a^2 + b^2 - c^2}{2ab}$ 10. **(b)** $\frac{19}{2}$

12. $\det(B) = (-1)^{n(n-1)/2}\det(A)$

13. **(a)** The ith and jth columns will be interchanged.
    (b) The ith column will be divided by c.
    (c) $-c$ times the jth column will be added to the ith column.

15. **(a)** $\lambda^3 + (-a_{11} - a_{22} - a_{33})\lambda^2 + (a_{11}a_{22} + a_{11}a_{33} + a_{22}a_{33} - a_{12}a_{21} - a_{13}a_{31} - a_{23}a_{32})\lambda +$
    $(a_{11}a_{23}a_{32} + a_{12}a_{21}a_{33} + a_{13}a_{22}a_{31} - a_{11}a_{22}a_{33} - a_{12}a_{23}a_{31} - a_{13}a_{21}a_{32})$

18. **(a)** $\lambda = -5, \lambda = 2, \lambda = 4;$ $\begin{bmatrix} -2t \\ t \\ t \end{bmatrix}, \begin{bmatrix} 5t \\ t \\ t \end{bmatrix}, \begin{bmatrix} 7t \\ 19t \\ t \end{bmatrix}$ **(b)** $\lambda = 1;$ $\begin{bmatrix} \frac{1}{2}t \\ -\frac{1}{2}t \\ t \end{bmatrix}$

EXERCISE SET 3.1 [page 125]

3. **(a)** $\overrightarrow{P_1P_2} = (-1, -1)$ **(b)** $\overrightarrow{P_1P_2} = (-7, -2)$ **(c)** $\overrightarrow{P_1P_2} = (2, 1)$ **(d)** $\overrightarrow{P_1P_2} = (a, b)$
(e) $\overrightarrow{P_1P_2} = (-5, 12, -6)$ **(f)** $\overrightarrow{P_1P_2} = (1, -1, -2)$ **(g)** $\overrightarrow{P_1P_2} = (-a, -b, -c)$ **(h)** $\overrightarrow{P_1P_2} = (a, b, c)$

4. **(a)** $Q(5, 10, -8)$ is one possible answer. **(b)** $Q(-7, -4, -2)$ is one possible answer.

5. **(a)** $P(-1, 2, -4)$ is one possible answer. **(b)** $P(7, -2, -6)$ is one possible answer.

6. **(a)** $(-2, 1, -4)$ **(b)** $(-10, 6, 4)$ **(c)** $(-7, 1, 10)$
(d) $(80, -20, -80)$ **(e)** $(132, -24, -72)$ **(f)** $(-77, 8, 94)$

7. $\mathbf{x} = (-\frac{8}{3}, \frac{1}{2}, \frac{8}{3})$ 8. $c_1 = 2, c_2 = -1, c_3 = 2$ 10. $c_1 = c_2 = c_3 = 0$

11. **(a)** $(\frac{9}{2}, -\frac{1}{2}, -\frac{1}{2})$ **(b)** $(\frac{23}{4}, -\frac{9}{4}, \frac{1}{4})$ 12. **(a)** $x' = 5,\ y' = 8$ **(b)** $x = -1,\ y = 3$

14. $\mathbf{u} = \left(\frac{\sqrt{3}}{2}, \frac{1}{2}\right), \mathbf{v} = \left(-\frac{1}{2}, -\frac{\sqrt{3}}{2}\right), \mathbf{u}+\mathbf{v} = \left(\frac{\sqrt{3}-1}{2}, \frac{1-\sqrt{3}}{2}\right), \mathbf{u}-\mathbf{v} = \left(\frac{\sqrt{3}+1}{2}, \frac{\sqrt{3}+1}{2}\right)$

EXERCISE SET 3.2 [page 128]

1. **(a)** 5 **(b)** $\sqrt{13}$ **(c)** 5 **(d)** $2\sqrt{3}$ **(e)** $3\sqrt{6}$ **(f)** 6

2. **(a)** $\sqrt{13}$ **(b)** $2\sqrt{26}$ **(c)** $\sqrt{209}$ **(d)** $3\sqrt{2}$

3. **(a)** $\sqrt{83}$ **(b)** $\sqrt{17}+\sqrt{26}$ **(c)** $4\sqrt{17}$ **(d)** $\sqrt{466}$ **(e)** $\left(\frac{3}{\sqrt{61}}, \frac{6}{\sqrt{61}}, -\frac{4}{\sqrt{61}}\right)$ **(f)** 1

4. $k = \pm\frac{4}{\sqrt{30}}$ 8. A sphere of radius 1 centered at (x_0, y_0, z_0) 12. Yes

13. **(a)** $a = c = 0$ **(b)** At least one of a or b is not zero, that is, $a^2 + b^2 > 0$

14. **(a)** The distance from x to the origin is less than 1. **(b)** $\|x - x_0\| > 1$

EXERCISE SET 3.3 [page 136]

1. **(a)** -11 **(b)** -24 **(c)** 0 **(d)** 0 2. **(a)** $-\frac{11}{\sqrt{13}\sqrt{74}}$ **(b)** $-\frac{3}{\sqrt{10}}$ **(c)** 0 **(d)** 0

3. **(a)** Orthogonal **(b)** Obtuse **(c)** Acute **(d)** Obtuse

4. **(a)** $(0, 0)$ **(b)** $(\frac{8}{13}, -\frac{12}{13})$ **(c)** $(-\frac{16}{13}, 0, -\frac{80}{13})$ **(d)** $(\frac{16}{89}, \frac{12}{89}, \frac{32}{89})$

5. **(a)** $(6, 2)$ **(b)** $(-\frac{21}{13}, -\frac{14}{13})$ **(c)** $(\frac{55}{13}, 1, -\frac{11}{13})$ **(d)** $(\frac{73}{89}, -\frac{12}{89}, -\frac{32}{89})$

6. **(a)** $\frac{2}{5}$ **(b)** $\frac{4\sqrt{5}}{5}$ **(c)** $\frac{18}{\sqrt{22}}$ **(d)** $\frac{43}{\sqrt{54}}$ 9. **(a)** 102 **(b)** $125\sqrt{2}$ **(c)** 170 **(d)** 170

10. For example, $(2, -5, 0), (-3, 0, 5), (0, 3, 2), (1, -5, -5), (-3, 3, 7)$

11. $\cos\theta_1 = \frac{\sqrt{10}}{10},\ \cos\theta_2 = \frac{3\sqrt{10}}{10},\ \cos\theta_3 = 0$ 12. The right angle is at B.

13. $\pm(1/\sqrt{3}, 1/\sqrt{3}, -1/\sqrt{3})$

15. **(a)** 1 **(b)** $\frac{1}{\sqrt{17}}$ **(c)** $\frac{6}{\sqrt{10}}$

19. **(b)** $\cos\beta = \frac{b}{\|\mathbf{v}\|},\ \cos\gamma = \frac{c}{\|\mathbf{v}\|}$ 20. $\theta_1 \approx 71°, \theta_2 \approx 61°, \theta_3 \approx 36°$

24. **(a)** The vector **u** is dotted with a scalar. **(b)** A scalar is added to the vector **w**.
(c) Scalars do not have norms. **(d)** The scalar k is dotted with a vector.

25. Yes; for example, if **a** and **u** are orthogonal 26. No; it merely says that **u** is orthogonal to $\mathbf{v} - \mathbf{w}$.

27. $\mathbf{r} = (\mathbf{u}\cdot\mathbf{r})\frac{\mathbf{u}}{\|\mathbf{u}\|} + (\mathbf{v}\cdot\mathbf{r})\frac{\mathbf{v}}{\|\mathbf{v}\|} + (\mathbf{w}\cdot\mathbf{r})\frac{\mathbf{w}}{\|\mathbf{w}\|}$ 28. Theorem of Pythagoras

EXERCISE SET 3.4 [page 147]

1. **(a)** $(32, -6, -4)$ **(b)** $(-14, -20, -82)$ **(c)** $(27, 40, -42)$
   (d) $(0, 176, -264)$ **(e)** $(-44, 55, -22)$ **(f)** $(-8, -3, -8)$

2. **(a)** $(18, 36, -18)$ **(b)** $(-3, 9, -3)$ 3. **(a)** $\sqrt{59}$ **(b)** $\sqrt{101}$ **(c)** 0

4. **(a)** $\frac{\sqrt{374}}{2}$ **(b)** $\sqrt{285}$ 7. For example, $(1, 1, 1) \times (2, -3, 5) = (8, -3, -5)$

8. **(a)** -10 **(b)** -110 9. **(a)** -3 **(b)** 3 **(c)** 3 **(d)** -3 **(e)** -3 **(f)** 0

10. **(a)** 16 **(b)** 45 11. **(a)** No **(b)** Yes **(c)** No 12. $\pm\left(0, \frac{2}{\sqrt{5}}, \frac{1}{\sqrt{5}}\right)$

13. $\left(\frac{6}{\sqrt{61}}, -\frac{3}{\sqrt{61}}, \frac{4}{\sqrt{61}}\right), \left(-\frac{6}{\sqrt{61}}, \frac{3}{\sqrt{61}}, -\frac{4}{\sqrt{61}}\right)$ 15. $2(\mathbf{v} \times \mathbf{u})$ 16. $\frac{12\sqrt{13}}{49}$

17. **(a)** $\frac{\sqrt{26}}{2}$ **(b)** $\frac{\sqrt{26}}{3}$ 19. **(a)** $\frac{2\sqrt{141}}{\sqrt{29}}$ **(b)** $\frac{\sqrt{137}}{3}$ 21. **(a)** $\sqrt{122}$ **(b)** $\theta \approx 40°19''$

23. **(a)** $\mathbf{m} = (0, 1, 0)$ and $\mathbf{n} = (1, 0, 0)$ **(b)** $(-1, 0, 0)$ **(c)** $(0, 0, -1)$

28. $(-8, 0, -8)$ 31. **(a)** $\frac{2}{3}$ **(b)** $\frac{1}{2}$ 35. **(b)** $\mathbf{u} \cdot \mathbf{w} \neq 0, \mathbf{v} \cdot \mathbf{w} = 0$

36. No, the equation is equivalent to $\mathbf{u} \times (\mathbf{v} - \mathbf{w}) = 0$ and hence to $\mathbf{v} - \mathbf{w} = k\mathbf{u}$ for some scalar k.

37. $\mathbf{u} \times (\mathbf{v} \times \mathbf{w}) \neq (\mathbf{u} \times \mathbf{v}) \times \mathbf{w}$, in general 38. They are collinear.

39. For example, $ab = ba$, $(ab)c = a(bc)$, and $ab = 0$ implies $a = 0$ or $b = 0$.

EXERCISE SET 3.5 [page 155]

1. **(a)** $-2(x+1) + (y-3) - (z+2) = 0$ **(b)** $(x-1) + 9(y-1) + 8(z-4) = 0$
   (c) $2z = 0$ **(d)** $x + 2y + 3z = 0$

2. **(a)** $-2x + y - z - 7 = 0$ **(b)** $x + 9y + 8z - 42 = 0$
   (c) $2z = 0$ **(d)** $x + 2y + 3y = 0$

3. **(a)** $(0, 0, 5)$ is a point in the plane and $\mathbf{n} = (-3, 7, 2)$ is a normal vector so that $-3(x-0) + 7(y-0) + 2(z-5) = 0$ is a point-normal form; other points and normals yield other correct answers.

4. **(a)** $2y - z + 1 = 0$ **(b)** $x + 9y - 5z - 26 = 0$

5. **(a)** Not parallel **(b)** Parallel **(c)** Parallel 6. **(a)** Parallel **(b)** Not parallel

7. **(a)** Not perpendicular **(b)** Perpendicular 8. **(a)** Perpendicular **(b)** Not perpendicular

9. **(a)** $x = 3 + 2t, y = -1 + t, z = 2 + 3t$ **(b)** $x = -2 + 6t, y = 3 - 6t, z = -3 - 2t$
   (c) $x = 2, y = 2 + t, z = 6$ **(d)** $x = t, y = -2t, z = 3t$

10. **(a)** $x = 5 + t, y = -2 + 2t, z = 4 - 4t$ **(b)** $x = 2t, y = -t, z = -3t$

11. **(a)** $x = -12 - 7t, y = -41 - 23t, z = t$ **(b)** $x = \frac{5}{2}t, y = 0, z = t$

12. **(a)** $(-2, 4, 1) \cdot (x+1, y-2, z-4) = 0$ **(b)** $(-1, 4, 3) \cdot (x-2, y, z+5) = 0$
    (c) $(-1, 0, 0) \cdot (x-5, y+2, z-1) = 0$ **(d)** $(a, b, c) \cdot (x, y, z) = 0$

13. **(a)** Parallel **(b)** Not parallel 14. **(a)** Perpendicular **(b)** Not perpendicular

15. **(a)** $(x, y, z) = (-1, 2, 3) + t(7, -1, 5)$ $(-\infty < t < +\infty)$
    (b) $(x, y, z) = (2, 0, -1) + t(1, 1, 1)$ $(-\infty < t < +\infty)$
    (c) $(x, y, z) = (2, -4, 1) + t(0, 0, -2)$ $(-\infty < t < +\infty)$
    (d) $(x, y, z) = (0, 0, 0) + t(a, b, c)$ $(-\infty < t < +\infty)$

17. $2x+3y-5z+36=0$ 18. **(a)** $z=0$ **(b)** $y=0$ **(c)** $x=0$

19. **(a)** $z-z_0=0$ **(b)** $x-x_0=0$ **(c)** $y-y_0=0$ 20. $7x+4y-2z=0$

21. $5x-2y+z-34=0$ 22. $\left(-\frac{173}{3},-\frac{43}{3},\frac{49}{3}\right)$ 23. $y+2z-9=0$

24. $x-y-4z-2=0$ 26. $x=\frac{11}{5}t-2, y=-\frac{2}{5}t+5, z=t$

27. $x+5y+3z-18=0$ 28. $(x-2)+(y+1)-3(z-4)=0$

29. $4x+13y-z-17=0$ 30. $3x+10y+4z-53=0$ 31. $3x-y-z-2=0$

32. $5x-3y+2z-5=0$ 33. $2x+4y+8z+13=0$ 36. $x-4y+4z+9=0$

37. **(a)** $x=\frac{11}{23}+\frac{7}{23}t, y=-\frac{41}{23}-\frac{1}{23}t, z=t$ **(b)** $x=-\frac{2}{5}t, y=0, z=t$

39. **(a)** $\frac{5}{3}$ **(b)** $\dfrac{1}{\sqrt{29}}$ **(c)** $\dfrac{4}{\sqrt{3}}$ 40. **(a)** $\dfrac{1}{2\sqrt{26}}$ **(b)** 0 **(c)** $\dfrac{2}{\sqrt{6}}$

42. **(a)** $\dfrac{x-3}{2}=y+1=\dfrac{z-2}{3}$ **(b)** $\dfrac{x+2}{6}=-\dfrac{y-3}{6}=-\dfrac{z+3}{2}$

43. **(a)** $x-2y-17=0$ and $x+4z-27=0$ is one possible answer.
(b) $x-2y=0$ and $-7y+2z=0$ is one possible answer.

44. **(a)** $\theta\approx 35°$ **(b)** $\theta\approx 79°$ 45. $\theta\approx 75°$ 46. They are identical.

47. They are perpendicular. 48. It is the line segment joining P_1 to P_2.

49. For example, $x=x_0+2t$, $y=y_0+3t$, $z=z_0+5t$ and $x=x_0+t$, $y=y_0-4t$, $z=z_0+2t$

EXERCISE SET 4.1 [page 170]

1. **(a)** $(-1,9,-11,1)$ **(b)** $(22,53,-19,14)$ **(c)** $(-13,13,-36,-2)$
(d) $(-90,-114,60,-36)$ **(e)** $(-9,-5,-5,-3)$ **(f)** $(27,29,-27,9)$

2. $\left(\frac{6}{5},\frac{2}{3},\frac{2}{3},\frac{2}{5}\right)$ 3. $c_1=1, c_2=1, c_3=-1, c_4=1$

5. **(a)** $\sqrt{29}$ **(b)** 3 **(c)** 13 **(d)** $\sqrt{31}$

6. **(a)** $\sqrt{133}$ **(b)** $\sqrt{30}+\sqrt{77}$ **(c)** $4\sqrt{30}$ **(d)** $\sqrt{1811}$ **(e)** $\left(\dfrac{1}{\sqrt{2}},\dfrac{1}{3\sqrt{2}},\dfrac{2}{3\sqrt{2}},\dfrac{2}{3\sqrt{2}}\right)$ **(f)** 1

8. $k=\pm\frac{5}{7}$ 9. **(a)** 7 **(b)** 14 **(c)** 7 **(d)** 11 10. **(a)** $\left(\dfrac{1}{\sqrt{10}},\dfrac{3}{\sqrt{10}}\right),\left(-\dfrac{1}{\sqrt{10}},-\dfrac{3}{\sqrt{10}}\right)$

11. **(a)** $\sqrt{10}$ **(b)** $\sqrt{56}$ **(c)** $\sqrt{59}$ **(d)** 10

14. **(a)** Yes **(b)** No **(c)** Yes **(d)** No **(e)** No **(f)** Yes

15. **(a)** $k=-3$ **(b)** $k=-2, k=-3$

16. $\pm\frac{1}{57}(-34,44,-6,11)$ 19. $x_1=1, x_2=-1, x_3=2$ 20. -6

22. The component in the **a** direction is $\text{proj}_{\mathbf{a}}\mathbf{u}=\frac{4}{15}(-1,1,2,3)$; the orthogonal component is $\frac{1}{15}(34,11,52,-27)$.

23. They do not intersect.

33. **(a)** Euclidean measure of "box" in R^n: $a_1a_2\cdots a_n$ **(b)** Length of diagonal: $\sqrt{a_1^2+a_2^2+\cdots+a_n^2}$

34. **(b)** The parallelogram law: The sum of the squares of the lengths of the four sides of a parallelogram is equal to the sum of the squares of the lengths of the two diagonals.

35. **(a)** $d(\mathbf{u},\mathbf{v})=\sqrt{2}$ 36. Yes, since any two vectors lie on a plane

37. **(a)** True **(b)** True **(c)** False **(d)** True **(e)** True, unless $\mathbf{u}=\mathbf{0}$

EXERCISE SET 4.2 [page 185]

1. **(a)** Linear; $R^3 \to R^2$ **(b)** Nonlinear; $R^2 \to R^3$ **(c)** Linear; $R^3 \to R^3$ **(d)** Nonlinear; $R^4 \to R^2$

2. **(a)** $\begin{bmatrix} 2 & -3 & 0 & 1 \\ 3 & 5 & 0 & -1 \end{bmatrix}$ **(b)** $\begin{bmatrix} 7 & 2 & -8 \\ 0 & -1 & 5 \\ 4 & 7 & -1 \end{bmatrix}$ **(c)** $\begin{bmatrix} -1 & 1 \\ 3 & -2 \\ 5 & -7 \end{bmatrix}$ **(d)** $\begin{bmatrix} 1 & 0 & 0 & 0 \\ 1 & 1 & 0 & 0 \\ 1 & 1 & 1 & 0 \\ 1 & 1 & 1 & 1 \end{bmatrix}$

3. $\begin{bmatrix} 3 & 5 & -1 \\ 4 & -1 & 1 \\ 3 & 2 & -1 \end{bmatrix}$; $T(-1, 2, 4) = (3, -2, -3)$

4. **(a)** $\begin{bmatrix} 2 & -1 \\ 1 & 1 \end{bmatrix}$ **(b)** $\begin{bmatrix} 1 & 0 \\ 0 & 1 \end{bmatrix}$ **(c)** $\begin{bmatrix} 1 & 2 & 1 \\ 1 & 5 & 0 \\ 0 & 0 & 1 \end{bmatrix}$ **(d)** $\begin{bmatrix} 4 & 0 & 0 \\ 0 & 7 & 0 \\ 0 & 0 & -8 \end{bmatrix}$

5. **(a)** $\begin{bmatrix} 0 & 1 \\ -1 & 0 \\ 1 & 3 \\ 1 & -1 \end{bmatrix}$ **(b)** $\begin{bmatrix} 7 & 2 & -1 & 1 \\ 0 & 1 & 1 & 0 \\ -1 & 0 & 0 & 0 \end{bmatrix}$ **(c)** $\begin{bmatrix} 0 & 0 & 0 \\ 0 & 0 & 0 \\ 0 & 0 & 0 \\ 0 & 0 & 0 \\ 0 & 0 & 0 \end{bmatrix}$ **(d)** $\begin{bmatrix} 0 & 0 & 0 & 1 \\ 1 & 0 & 0 & 0 \\ 0 & 0 & 1 & 0 \\ 0 & 1 & 0 & 0 \\ 1 & 0 & -1 & 0 \end{bmatrix}$

6. **(a)** $\begin{bmatrix} -1 \\ 1 \end{bmatrix}$ **(b)** $\begin{bmatrix} 3 \\ 13 \end{bmatrix}$ **(c)** $\begin{bmatrix} -2x_1 + x_2 + 4x_3 \\ 3x_1 + 5x_2 + 7x_3 \\ 6x_1 - x_3 \end{bmatrix}$ **(d)** $\begin{bmatrix} -x_1 + x_2 \\ 2x_1 + 4x_2 \\ 7x_1 + 8x_2 \end{bmatrix}$

7. **(a)** $T(-1, 4) = (5, 4)$ **(b)** $T(2, 1, -3) = (0, -2, 0)$

8. **(a)** $(-1, -2)$ **(b)** $(1, 2)$ **(c)** $(2, -1)$

9. **(a)** $(2, -5, -3)$ **(b)** $(2, 5, 3)$ **(c)** $(-2, -5, 3)$

10. **(a)** $(2, 0)$ **(b)** $(0, -5)$

11. **(a)** $(-2, 1, 0)$ **(b)** $(-2, 0, 3)$ **(c)** $(0, 1, 3)$

12. **(a)** $\left(\frac{3\sqrt{3}+4}{2}, \frac{3-4\sqrt{3}}{2}\right)$ **(b)** $\left(\frac{3-4\sqrt{3}}{2}, \frac{-3\sqrt{3}-4}{2}\right)$ **(c)** $\left(\frac{7\sqrt{2}}{2}, \frac{-\sqrt{2}}{2}\right)$ **(d)** $(4, 3)$

13. **(a)** $\left(-2, \frac{\sqrt{3}-2}{2}, \frac{1+2\sqrt{3}}{2}\right)$ **(b)** $(0, 1, 2\sqrt{2})$ **(c)** $(-1, -2, 2)$

14. **(a)** $\begin{bmatrix} 1 & 0 & 0 \\ 0 & 1/2 & \sqrt{3}/2 \\ 0 & -\sqrt{3}/2 & 1/2 \end{bmatrix}$ **(b)** $\begin{bmatrix} 1/2 & 0 & -\sqrt{3}/2 \\ 0 & 1 & 0 \\ \sqrt{3}/2 & 0 & 1/2 \end{bmatrix}$ **(c)** $\begin{bmatrix} 1/2 & \sqrt{3}/2 & 0 \\ -\sqrt{3}/2 & 1/2 & 0 \\ 0 & 0 & 1 \end{bmatrix}$

15. **(a)** $\left(-2, \frac{\sqrt{3}+2}{2}, \frac{-1+2\sqrt{3}}{2}\right)$ **(b)** $(-2\sqrt{2}, 1, 0)$ **(c)** $(1, 2, 2)$

16. **(a)** $\begin{bmatrix} 1 & 0 \\ 0 & -1 \end{bmatrix}$ **(b)** $\begin{bmatrix} 0 & 0 \\ 0 & \frac{1}{2} \end{bmatrix}$ **(c)** $\begin{bmatrix} 3 & 0 \\ 0 & -3 \end{bmatrix}$

17. **(a)** $\begin{bmatrix} 0 & 0 \\ 1/2 & -\sqrt{3}/2 \end{bmatrix}$ **(b)** $\begin{bmatrix} -\sqrt{2} & \sqrt{2} \\ \sqrt{2} & \sqrt{2} \end{bmatrix}$ **(c)** $\begin{bmatrix} -1 & 0 \\ 0 & -1 \end{bmatrix}$

18. **(a)** $\begin{bmatrix} -1 & 0 & 0 \\ 0 & 0 & 0 \\ 0 & 0 & 1 \end{bmatrix}$ **(b)** $\begin{bmatrix} 1 & 0 & 1 \\ 0 & \sqrt{2} & 0 \\ -1 & 0 & 1 \end{bmatrix}$ **(c)** $\begin{bmatrix} -1 & 0 & 0 \\ 0 & 1 & 0 \\ 0 & 0 & 0 \end{bmatrix}$

19. **(a)** $\begin{bmatrix} \sqrt{3}/8 & -\sqrt{3}/16 & 1/16 \\ 1/8 & 3/16 & -\sqrt{3}/16 \\ 0 & 1/8 & \sqrt{3}/8 \end{bmatrix}$ **(b)** $\begin{bmatrix} 0 & 0 & 0 \\ 0 & -1 & 0 \\ 0 & 0 & -1 \end{bmatrix}$ **(c)** $\begin{bmatrix} 0 & 1 & 0 \\ 0 & 0 & -1 \\ -1 & 0 & 0 \end{bmatrix}$

20. **(a)** Yes **(b)** Yes **(c)** No **21.** **(a)** Yes **(b)** No **22.** **(a)** $\begin{bmatrix} 1 & 0 & 0 \\ 0 & 0 & 0 \\ 0 & 0 & 0 \end{bmatrix} \begin{bmatrix} 0 & 0 & 0 \\ 0 & 1 & 0 \\ 0 & 0 & 0 \end{bmatrix} \begin{bmatrix} 0 & 0 & 0 \\ 0 & 0 & 0 \\ 0 & 0 & 1 \end{bmatrix}$

24. $$\begin{bmatrix} \frac{1}{3}(1-\cos\theta)+\cos\theta & \frac{1}{3}(1-\cos\theta)-\frac{1}{\sqrt{3}}\sin\theta & \frac{1}{3}(1-\cos\theta)-\frac{1}{\sqrt{3}}\sin\theta \\ \frac{1}{3}(1-\cos\theta)-\frac{1}{\sqrt{3}}\sin\theta & \frac{1}{3}(1-\cos\theta)+\cos\theta & \frac{1}{3}(1-\cos\theta)-\frac{1}{\sqrt{3}}\sin\theta \\ \frac{1}{3}(1-\cos\theta)-\frac{1}{\sqrt{3}}\sin\theta & \frac{1}{3}(1-\cos\theta)-\frac{1}{\sqrt{3}}\sin\theta & \frac{1}{3}(1-\cos\theta)+\cos\theta \end{bmatrix}$$

26. 135° **28.** **(c)** 90° **29.** **(a)** Twice the orthogonal projection on the x-axis
(b) Twice the reflection about the x-axis

30. **(a)** The x-coordinate is stretched by a factor of 2 and the y-coordinate is stretched by a factor of 3.
(b) Rotation through 30°

31. Rotation through the angle 2θ **32.** Rotation through the angle $-\theta$

EXERCISE SET 4.3 [page 198]

1. **(a)** Not one-to-one **(b)** One-to-one **(c)** One-to-one **(d)** One-to-one
(e) One-to-one **(f)** One-to-one **(g)** One-to-one

2. **(a)** $\begin{bmatrix} 8 & 4 \\ 2 & 1 \end{bmatrix}$; not one-to-one **(b)** $\begin{bmatrix} 2 & -3 \\ 5 & 1 \end{bmatrix}$; one-to-one **(c)** $\begin{bmatrix} -1 & 3 & 2 \\ 2 & 0 & 4 \\ 1 & 3 & 6 \end{bmatrix}$; not one-to-one **(d)** $\begin{bmatrix} 1 & 2 & 3 \\ 2 & 5 & 3 \\ 1 & 0 & 5 \end{bmatrix}$; one-to-one

3. For example, the vector (1, 3) is not in the range.

4. For example, the vector (1, 6, 2) is not in the range.

5. **(a)** One-to-one; $\begin{bmatrix} \frac{1}{3} & -\frac{2}{3} \\ \frac{1}{3} & \frac{1}{3} \end{bmatrix}$; $T^{-1}(w_1, w_2) = \left(\frac{1}{3}x_1 - \frac{2}{3}x_2, \frac{1}{3}x_1 + \frac{1}{3}x_2\right)$ **(b)** Not one-to-one

(c) One-to-one; $\begin{bmatrix} 0 & -1 \\ -1 & 0 \end{bmatrix}$; $T^{-1}(w_1, w_2) = (-x_2, -x_1)$ **(d)** Not one-to-one

6. **(a)** One-to-one; $\begin{bmatrix} 1 & -2 & 4 \\ -1 & 2 & -3 \\ -1 & 3 & -5 \end{bmatrix}$; $T^{-1}(w_1, w_2, w_3) = (x_1 - 2x_2 + 4x_3, -x_1 + 2x_2 - 3x_3, -x_1 + 3x_2 - 5x_3)$

(b) One-to-one; $\begin{bmatrix} \frac{1}{2} & \frac{1}{2} & -\frac{1}{2} \\ -\frac{5}{14} & \frac{5}{14} & \frac{3}{14} \\ -\frac{1}{7} & \frac{1}{7} & \frac{1}{7} \end{bmatrix}$; $T^{-1}(w_1, w_2, w_3) = \left(\dfrac{x_1 + x_2 - x_3}{2}, \dfrac{-5x_1 + 5x_2 + 3x_3}{14}, \dfrac{-x_1 + x_2 + x_3}{7}\right)$

(c) One-to-one; $\begin{bmatrix} -\frac{3}{2} & -\frac{3}{2} & \frac{11}{2} \\ \frac{1}{2} & \frac{1}{2} & -\frac{3}{2} \\ -\frac{1}{2} & \frac{1}{2} & -\frac{1}{2} \end{bmatrix}$; $T^{-1}(w_1, w_2, w_3) = \left(\dfrac{-3x_1 - 3x_2 + 11x_3}{2}, \dfrac{x_1 + x_2 - 3x_3}{2}, \dfrac{-x_1 + x_2 - x_3}{2}\right)$

(d) Not one-to-one

7. **(a)** Reflection about the x-axis **(b)** Rotation through the angle $-\pi/4$ **(c)** Contraction by a factor of $\frac{1}{3}$
(d) Reflection about the yz-plane **(e)** Dilation by a factor of 5

8. **(a)** Linear **(b)** Nonlinear **(c)** Linear **(d)** Linear

9. **(a)** Linear **(b)** Nonlinear **(c)** Linear **(d)** Nonlinear

10. **(a)** Linear **(b)** Nonlinear **11.** **(a)** Linear **(b)** Linear

12. **(a)** For a reflection about the y-axis, $T(\mathbf{e}_1) = \begin{bmatrix} -1 \\ 0 \end{bmatrix}$ and $T(\mathbf{e}_2) = \begin{bmatrix} 0 \\ 1 \end{bmatrix}$.

Thus, $T = \begin{bmatrix} -1 & 0 \\ 0 & 1 \end{bmatrix}$.

(b) For a reflection about the xz-plane, $T(\mathbf{e}_1) = \begin{bmatrix} 1 \\ 0 \\ 0 \end{bmatrix}$, $T(\mathbf{e}_2) = \begin{bmatrix} 0 \\ -1 \\ 0 \end{bmatrix}$, and $T(\mathbf{e}_3) = \begin{bmatrix} 0 \\ 0 \\ 1 \end{bmatrix}$.

Thus, $T = \begin{bmatrix} 1 & 0 & 0 \\ 0 & -1 & 0 \\ 0 & 0 & 1 \end{bmatrix}$.

(c) For an orthogonal projection on the x-axis, $T(\mathbf{e}_1) = \begin{bmatrix} 1 \\ 0 \end{bmatrix}$ and $T(\mathbf{e}_2) = \begin{bmatrix} 0 \\ 0 \end{bmatrix}$.

Thus, $T = \begin{bmatrix} 1 & 0 \\ 0 & 0 \end{bmatrix}$.

(d) For an orthogonal projection on the yz-plane, $T(\mathbf{e}_1) = \begin{bmatrix} 0 \\ 0 \\ 0 \end{bmatrix}$, $T(\mathbf{e}_2) = \begin{bmatrix} 0 \\ 1 \\ 0 \end{bmatrix}$, and $T(\mathbf{e}_3) = \begin{bmatrix} 0 \\ 0 \\ 1 \end{bmatrix}$.

Thus, $T = \begin{bmatrix} 0 & 0 & 0 \\ 0 & 1 & 0 \\ 0 & 0 & 1 \end{bmatrix}$.

(e) For a rotation through a positive angle θ, $T(\mathbf{e}_1) = \begin{bmatrix} \cos\theta \\ \sin\theta \end{bmatrix}$ and $T(\mathbf{e}_2) = \begin{bmatrix} -\sin\theta \\ \cos\theta \end{bmatrix}$.

Thus, $T = \begin{bmatrix} \cos\theta & -\sin\theta \\ \sin\theta & \cos\theta \end{bmatrix}$.

(f) For a dilation by a factor $k \geq 1$, $T(\mathbf{e}_1) = \begin{bmatrix} k \\ 0 \\ 0 \end{bmatrix}$, $T(\mathbf{e}_2) = \begin{bmatrix} 0 \\ k \\ 0 \end{bmatrix}$, $T(\mathbf{e}_3) = \begin{bmatrix} 0 \\ 0 \\ k \end{bmatrix}$.

Thus, $T = \begin{bmatrix} k & 0 & 0 \\ 0 & k & 0 \\ 0 & 0 & k \end{bmatrix}$.

13. **(a)** $T(\mathbf{e}_1) = \begin{bmatrix} -1 \\ 0 \end{bmatrix}$ and $T(\mathbf{e}_2) = \begin{bmatrix} 0 \\ 0 \end{bmatrix}$. Thus, $T = \begin{bmatrix} -1 & 0 \\ 0 & 0 \end{bmatrix}$.

(b) $T(\mathbf{e}_1) = \begin{bmatrix} 0 \\ -1 \end{bmatrix}$ and $T(\mathbf{e}_2) = \begin{bmatrix} 1 \\ 0 \end{bmatrix}$. Thus, $T = \begin{bmatrix} 0 & 1 \\ -1 & 0 \end{bmatrix}$.

(c) $T(\mathbf{e}_1) = \begin{bmatrix} 0 \\ 3 \end{bmatrix}$ and $T(\mathbf{e}_2) = \begin{bmatrix} 0 \\ 0 \end{bmatrix}$. Thus, $T = \begin{bmatrix} 0 & 0 \\ 3 & 0 \end{bmatrix}$.

14. **(a)** $T(\mathbf{e}_1) = \begin{bmatrix} \frac{1}{5} \\ 0 \\ 0 \end{bmatrix}$, $T(\mathbf{e}_2) = \begin{bmatrix} 0 \\ -\frac{1}{5} \\ 0 \end{bmatrix}$, and $T(\mathbf{e}_3) = \begin{bmatrix} 0 \\ 0 \\ \frac{1}{5} \end{bmatrix}$. Thus, $T = \begin{bmatrix} \frac{1}{5} & 0 & 0 \\ 0 & -\frac{1}{5} & 0 \\ 0 & 0 & \frac{1}{5} \end{bmatrix}$.

(b) $T(\mathbf{e}_1) = \begin{bmatrix} 1 \\ 0 \\ 0 \end{bmatrix}$, $T(\mathbf{e}_2) = \begin{bmatrix} 0 \\ 0 \\ 0 \end{bmatrix}$, and $T(\mathbf{e}_3) = \begin{bmatrix} 0 \\ 0 \\ 0 \end{bmatrix}$. Thus, $T = \begin{bmatrix} 1 & 0 & 0 \\ 0 & 0 & 0 \\ 0 & 0 & 0 \end{bmatrix}$.

(c) $T(\mathbf{e}_1) = \begin{bmatrix} -1 \\ 0 \\ 0 \end{bmatrix}$, $T(\mathbf{e}_2) = \begin{bmatrix} 0 \\ -1 \\ 0 \end{bmatrix}$, and $T(\mathbf{e}_3) = \begin{bmatrix} 0 \\ 0 \\ -1 \end{bmatrix}$. Thus, $T = \begin{bmatrix} -1 & 0 & 0 \\ 0 & -1 & 0 \\ 0 & 0 & -1 \end{bmatrix}$.

14. **(a)** $\begin{bmatrix} -\frac{2}{9} & \frac{7}{9} \\ \frac{1}{3} & -\frac{1}{6} \end{bmatrix}$ **(b)** $\begin{bmatrix} \frac{3}{4} & \frac{7}{2} \\ \frac{3}{2} & 1 \end{bmatrix}$ **(c)** $[\mathbf{p}]_B = \begin{bmatrix} 1 \\ -1 \end{bmatrix}$ **(d)** $[\mathbf{p}]_{B'} = \begin{bmatrix} -\frac{11}{4} \\ \frac{1}{2} \end{bmatrix}$

15. **(b)** $\begin{bmatrix} 2 & 0 \\ 1 & 3 \end{bmatrix}$ **(c)** $\begin{bmatrix} \frac{1}{2} & 0 \\ -\frac{1}{6} & \frac{1}{3} \end{bmatrix}$ **(d)** $[\mathbf{h}]_B = \begin{bmatrix} 2 \\ -5 \end{bmatrix}$, $[\mathbf{h}]_{B'} = \begin{bmatrix} 1 \\ -2 \end{bmatrix}$

16. **(a)** $(4\sqrt{2}, -2\sqrt{2})$ **(b)** $(-\frac{7}{2}\sqrt{2}, \frac{3}{2}\sqrt{2})$

17. **(a)** $(-1+3\sqrt{3}, 3+\sqrt{3})$ **(b)** $(\frac{5}{2}-\sqrt{3}, \frac{5}{2}\sqrt{3}+1)$

18. **(a)** $(\frac{1}{2}\sqrt{2}, \frac{3}{2}\sqrt{2}, 5)$ **(b)** $(-\frac{5}{2}\sqrt{2}, \frac{7}{2}\sqrt{2}, -3)$

19. **(a)** $(-\frac{1}{2}-\frac{5}{2}\sqrt{3}, 2, \frac{5}{2}-\frac{1}{2}\sqrt{3})$ **(b)** $(\frac{1}{2}-\frac{3}{2}\sqrt{3}, 6, -\frac{3}{2}-\frac{1}{2}\sqrt{3})$

20. **(a)** $(-1, \frac{3}{2}\sqrt{2}, -\frac{7}{2}\sqrt{2})$ **(b)** $(1, -\frac{3}{2}\sqrt{2}, \frac{9}{2}\sqrt{3})$

21. **(a)** $A = \begin{bmatrix} \cos\theta & 0 & -\sin\theta \\ 0 & 1 & 0 \\ \sin\theta & 0 & \cos\theta \end{bmatrix}$ **(b)** $A = \begin{bmatrix} 1 & 0 & 0 \\ 0 & \cos\theta & \sin\theta \\ 0 & -\sin\theta & \cos\theta \end{bmatrix}$

22. $\begin{bmatrix} \frac{\sqrt{2}}{4} & \frac{\sqrt{6}}{4} & -\frac{\sqrt{2}}{2} \\ -\frac{\sqrt{3}}{2} & \frac{1}{2} & 0 \\ \frac{\sqrt{2}}{4} & \frac{\sqrt{6}}{4} & \frac{\sqrt{2}}{2} \end{bmatrix}$ 23. $a^2 + b^2 = \frac{1}{2}$ 26. **(a)** Rotation **(b)** Rotation followed by a reflection

27. **(a)** Rotation followed by a reflection **(b)** Rotation

30. **(a)** Rotation and reflection **(b)** Rotation and dilation
(c) Any rigid operator is angle preserving. Any dilation or contraction with $k \neq 0, 1$ is angle preserving but not rigid.

31. $\det(A) = \pm 1$ 32. $a = 0, b = \sqrt{2/3}, c = -\sqrt{1/3}$ or $a = 0, b = -\sqrt{2/3}, c = \sqrt{1/3}$

SUPPLEMENTARY EXERCISES [page 334]

1. **(a)** $(0, a, a, 0)$ with $a \neq 0$ **(b)** $\pm\left(0, \frac{2}{\sqrt{5}}, \frac{1}{\sqrt{5}}, 0\right)$ 6. $\pm\left(\frac{1}{\sqrt{2}}, 0, \frac{1}{\sqrt{2}}\right)$

7. $w_k = \frac{1}{k}, k = 1, 2, \ldots, n$ 8. No 11. **(b)** θ approaches $\frac{\pi}{2}$

12. **(b)** The diagonals of a parallelogram are perpendicular if and only if its sides have the same length.

EXERCISE SET 7.1 [page 344]

1. **(a)** $\lambda^2 - 2\lambda - 3 = 0$ **(b)** $\lambda^2 - 8\lambda + 16 = 0$ **(c)** $\lambda^2 - 12 = 0$
(d) $\lambda^2 + 3 = 0$ **(e)** $\lambda^2 = 0$ **(f)** $\lambda^2 - 2\lambda + 1 = 0$

2. **(a)** $\lambda = 3, \lambda = -1$ **(b)** $\lambda = 4$ **(c)** $\lambda = \sqrt{12}, \lambda = -\sqrt{12}$
(d) No real eigenvalues **(e)** $\lambda = 0$ **(f)** $\lambda = 1$

3. **(a)** Basis for eigenspace corresponding to $\lambda = 3$: $\begin{bmatrix} \frac{1}{2} \\ 1 \end{bmatrix}$; basis for eigenspace corresponding to $\lambda = -1$: $\begin{bmatrix} 0 \\ 1 \end{bmatrix}$

   (b) Basis for eigenspace corresponding to $\lambda = 4$: $\begin{bmatrix} \frac{3}{2} \\ 1 \end{bmatrix}$

   (c) Basis for eigenspace corresponding to $\lambda = \sqrt{12}$: $\begin{bmatrix} \frac{3}{\sqrt{12}} \\ 1 \end{bmatrix}$; basis for eigenspace corresponding to $\lambda = -\sqrt{12}$: $\begin{bmatrix} -\frac{3}{\sqrt{12}} \\ 1 \end{bmatrix}$

   (d) There are no eigenspaces.

   (e) Basis for eigenspace corresponding to $\lambda = 0$: $\begin{bmatrix} 1 \\ 0 \end{bmatrix}, \begin{bmatrix} 0 \\ 1 \end{bmatrix}$

   (f) Basis for eigenspace corresponding to $\lambda = 1$: $\begin{bmatrix} 1 \\ 0 \end{bmatrix}, \begin{bmatrix} 0 \\ 1 \end{bmatrix}$

4. **(a)** $\lambda^3 - 6\lambda^2 + 11\lambda - 6 = 0$ **(b)** $\lambda^3 - 2\lambda = 0$ **(c)** $\lambda^3 + 8\lambda^2 + \lambda + 8 = 0$
   (d) $\lambda^3 - \lambda^2 - \lambda - 2 = 0$ **(e)** $\lambda^3 - 6\lambda^2 + 12\lambda - 8 = 0$ **(f)** $\lambda^3 - 2\lambda^2 - 15\lambda + 36 = 0$

5. **(a)** $\lambda = 1,\ \lambda = 2,\ \lambda = 3$ **(b)** $\lambda = 0,\ \lambda = \sqrt{2},\ \lambda = -\sqrt{2}$ **(c)** $\lambda = -8$
   (d) $\lambda = 2$ **(e)** $\lambda = 2$ **(f)** $\lambda = -4,\ \lambda = 3$

6. **(a)** $\lambda = 1$: basis $\begin{bmatrix} 0 \\ 1 \\ 0 \end{bmatrix}$; $\lambda = 2$: basis $\begin{bmatrix} -\frac{1}{2} \\ 1 \\ 1 \end{bmatrix}$; $\lambda = 3$: basis $\begin{bmatrix} -1 \\ 1 \\ 1 \end{bmatrix}$

   (b) $\lambda = 0$: basis $\begin{bmatrix} \frac{5}{3} \\ \frac{1}{3} \\ 1 \end{bmatrix}$; $\lambda = \sqrt{2}$: basis $\begin{bmatrix} \frac{1}{7}(15 + 5\sqrt{2}) \\ \frac{1}{7}(-1 + 2\sqrt{2}) \\ 1 \end{bmatrix}$; $\lambda = -\sqrt{2}$: basis $\begin{bmatrix} \frac{1}{7}(15 - 5\sqrt{2}) \\ \frac{1}{7}(-1 - 2\sqrt{2}) \\ 1 \end{bmatrix}$

   (c) $\lambda = -8$: basis $\begin{bmatrix} -\frac{1}{6} \\ -\frac{1}{6} \\ 1 \end{bmatrix}$ **(d)** $\lambda = 2$: basis $\begin{bmatrix} \frac{1}{3} \\ \frac{1}{3} \\ 1 \end{bmatrix}$ **(e)** $\lambda = 2$: basis $\begin{bmatrix} -\frac{1}{3} \\ -\frac{1}{3} \\ 1 \end{bmatrix}$

   (f) $\lambda = -4$: basis $\begin{bmatrix} -2 \\ \frac{8}{3} \\ 1 \end{bmatrix}$; $\lambda = 3$: basis $\begin{bmatrix} 5 \\ -2 \\ 1 \end{bmatrix}$

7. **(a)** $(\lambda - 1)^2(\lambda + 2)(\lambda + 1) = 0$ **(b)** $(\lambda - 4)^2(\lambda^2 + 3) = 0$

8. **(a)** $\lambda = 1,\ \lambda = -2,\ \lambda = -1$ **(b)** $\lambda = 4$

9. **(a)** $\lambda = 1$: basis $\begin{bmatrix} 0 \\ 0 \\ 0 \\ 1 \end{bmatrix}$ and $\begin{bmatrix} 2 \\ 3 \\ 1 \\ 0 \end{bmatrix}$; $\lambda = -2$; basis $\begin{bmatrix} -1 \\ 0 \\ 1 \\ 0 \end{bmatrix}$; $\lambda = -1$: basis $\begin{bmatrix} -2 \\ 1 \\ 1 \\ 0 \end{bmatrix}$ **(b)** $\lambda = 4$: basis $\begin{bmatrix} \frac{3}{2} \\ 1 \\ 0 \\ 0 \end{bmatrix}$

10. **(a)** $\lambda = -1,\ \lambda = 5$ **(b)** $\lambda = 3,\ \lambda = 7,\ \lambda = 1$ **(c)** $\lambda = -\frac{1}{3},\ \lambda = 1,\ \lambda = \frac{1}{2}$

11. $\lambda = 1,\ \lambda = \frac{1}{512},\ \lambda = 512,\ \lambda = 0$

12. For A^{25}, $\lambda = 1, -1$; basis for $\lambda = 1$: $\begin{bmatrix} -1 \\ 1 \\ 0 \end{bmatrix}, \begin{bmatrix} -1 \\ 0 \\ 1 \end{bmatrix}$; basis for $\lambda = -1$: $\begin{bmatrix} 2 \\ -1 \\ 1 \end{bmatrix}$

13. **(a)** $y = x$ and $y = 2x$ **(b)** No lines **(c)** $y = 0$ 14. **(a)** -5 **(b)** 7

22. **(a)** $\lambda_1 = 1$: $\begin{bmatrix}1\\0\\1\end{bmatrix}$; $\lambda_2 = \frac{1}{2}$: $\begin{bmatrix}\frac{1}{2}\\1\\0\end{bmatrix}$; $\lambda_3 = \frac{1}{3}$: $\begin{bmatrix}1\\1\\1\end{bmatrix}$ **(b)** $\lambda_1 = -2$: $\begin{bmatrix}1\\0\\1\end{bmatrix}$; $\lambda_2 = -1$: $\begin{bmatrix}\frac{1}{2}\\1\\0\end{bmatrix}$; $\lambda_3 = 0$: $\begin{bmatrix}1\\1\\1\end{bmatrix}$

(c) $\lambda_1 = 3$: $\begin{bmatrix}1\\0\\1\end{bmatrix}$; $\lambda_2 = 4$: $\begin{bmatrix}\frac{1}{2}\\1\\0\end{bmatrix}$; $\lambda_3 = 5$: $\begin{bmatrix}1\\1\\1\end{bmatrix}$

24. **(a)** False **(b)** True **(c)** True **(d)** True

25. **(a)** A is 6×6. **(b)** A is invertible. **(c)** A has three eigenspaces.

EXERCISE SET 7.2 (page 354)

1. 1, 2, or 3

2. **(a)** $\lambda = 3,\ \lambda = 5$
(b) For $\lambda = 3$, the rank of $3I - A$ is 1 and the nullity is 2. For $\lambda = 5$, the rank of $5I - A$ is 2 and the nullity is 1.
(c) A is diagonalizable since the eigenspaces produce a total of three basis vectors.

3. Not diagonalizable 4. Not diagonalizable 5. Not diagonalizable

6. Not diagonalizable 7. Not diagonalizable

8. $P = \begin{bmatrix}\frac{4}{5} & \frac{3}{4}\\1 & 1\end{bmatrix}$; $P^{-1}AP = \begin{bmatrix}1 & 0\\0 & 2\end{bmatrix}$ 9. $P = \begin{bmatrix}\frac{1}{3} & 0\\1 & 1\end{bmatrix}$; $P^{-1}AP = \begin{bmatrix}1 & 0\\0 & -1\end{bmatrix}$

10. $P = \begin{bmatrix}0 & 1 & 0\\1 & 0 & 1\\-1 & 0 & 1\end{bmatrix}$; $P^{-1}AP = \begin{bmatrix}0 & 0 & 0\\0 & 1 & 0\\0 & 0 & 2\end{bmatrix}$ 11. $P = \begin{bmatrix}-2 & 0 & 1\\0 & 1 & 0\\1 & 0 & 0\end{bmatrix}$; $P^{-1}AP = \begin{bmatrix}3 & 0 & 0\\0 & 3 & 0\\0 & 0 & 2\end{bmatrix}$

12. Not diagonalizable 13. $P = \begin{bmatrix}1 & 2 & 1\\1 & 3 & 3\\1 & 3 & 4\end{bmatrix}$; $P^{-1}AP = \begin{bmatrix}1 & 0 & 0\\0 & 2 & 0\\0 & 0 & 3\end{bmatrix}$

14. Not diagonalizable 15. $P = \begin{bmatrix}-\frac{1}{3} & 0 & 0\\0 & 1 & 0\\1 & 0 & 1\end{bmatrix}$; $P^{-1}AP = \begin{bmatrix}0 & 0 & 0\\0 & 0 & 0\\0 & 0 & 1\end{bmatrix}$ 16. Not diagonalizable

17. $P = \begin{bmatrix}1 & 1 & 0 & 0\\0 & 1 & 1 & 0\\0 & 0 & 1 & 1\\0 & 0 & 0 & 1\end{bmatrix}$; $P^{-1}AP = \begin{bmatrix}-2 & 0 & 0 & 0\\0 & -2 & 0 & 0\\0 & 0 & 3 & 0\\0 & 0 & 0 & 3\end{bmatrix}$

18. $\begin{bmatrix}1 & 0\\-1023 & 1024\end{bmatrix}$ 19. $\begin{bmatrix}-1 & 10237 & -2047\\0 & 1 & 0\\0 & 10245 & -2048\end{bmatrix}$

20. **(a)** $\begin{bmatrix}1 & 0 & 0\\0 & 1 & 0\\0 & 0 & 1\end{bmatrix}$ **(b)** $\begin{bmatrix}1 & 0 & 0\\0 & 1 & 0\\0 & 0 & 1\end{bmatrix}$ **(c)** $\begin{bmatrix}1 & -2 & 8\\0 & -1 & 0\\0 & 0 & -1\end{bmatrix}$ **(d)** $\begin{bmatrix}1 & -2 & 8\\0 & -1 & 0\\0 & 0 & -1\end{bmatrix}$

21. $A^n = PD^nP^{-1} = \begin{bmatrix}1 & 1 & 1\\2 & 0 & -1\\1 & -1 & 1\end{bmatrix}\begin{bmatrix}1^n & 0 & 0\\0 & 3^n & 0\\0 & 0 & 4^n\end{bmatrix}\begin{bmatrix}\frac{1}{6} & \frac{1}{3} & \frac{1}{6}\\\frac{1}{2} & 0 & -\frac{1}{2}\\\frac{1}{3} & -\frac{1}{3} & \frac{1}{3}\end{bmatrix}$

25. **(a)** False **(b)** False **(c)** True **(d)** True

20. **(a)** $\begin{bmatrix} 2 \\ 6 \\ 12 \end{bmatrix}$ **(d)** $-3x^2+3$ **(e)**

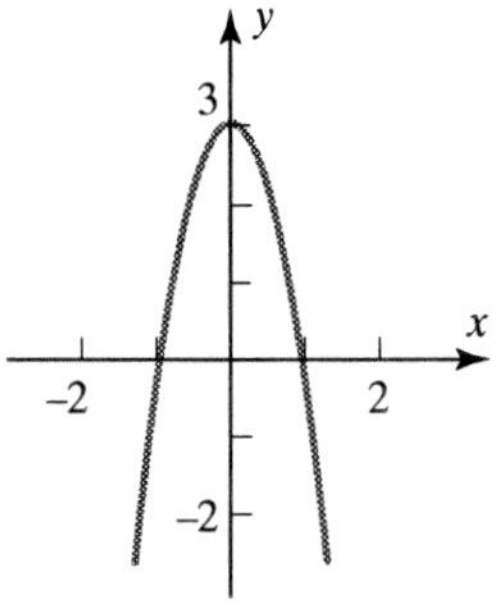

24. $\begin{bmatrix} 0 & 1 & 0 & 0 & \cdots & 0 \\ 0 & 0 & 1 & 0 & \cdots & 0 \\ 0 & 0 & 0 & 1 & \cdots & 0 \\ \vdots & \vdots & \vdots & \vdots & & \vdots \\ 0 & 0 & 0 & 0 & \cdots & 1 \\ 0 & 0 & 0 & 0 & \cdots & 0 \end{bmatrix}$

25. $\begin{bmatrix} 0 & 0 & 0 & \cdots & 0 \\ 1 & 0 & 0 & \cdots & 0 \\ 0 & \frac{1}{2} & 0 & \cdots & 0 \\ 0 & 0 & \frac{1}{3} & \cdots & 0 \\ \vdots & \vdots & \vdots & & \vdots \\ 0 & 0 & 0 & \cdots & \frac{1}{n+1} \end{bmatrix}$

EXERCISE SET 9.1 [page 424]

1. **(a)** $y_1 = c_1e^{5x} - 2c_2e^{-x}$, $y_2 = c_1e^{5x} + c_2e^{-x}$ **(b)** $y_1 = 0$, $y_2 = 0$

2. **(a)** $y_1 = c_1e^{7x} - 3c_2e^{-x}$, $y_2 = 2c_1e^{7x} + 2c_2e^{-x}$ **(b)** $y_1 = -\frac{1}{40}e^{7x} + \frac{81}{40}e^{-x}$, $y_2 = -\frac{1}{20}e^{7x} - \frac{27}{20}e^{-x}$

3. **(a)** $y_1 = -c_2e^{2x} + c_3e^{3x}$, $y_2 = c_1e^x + 2c_2e^{2x} - c_3e^{3x}$, $y_3 = 2c_2e^{2x} - c_3e^{3x}$ **(b)** $y_1 = e^{2x} - 2e^{3x}$, $y_2 = e^x - 2e^{2x} + 2e^{3x}$, $y_3 = -2e^{2x} + 2e^{3x}$

4. $y_1 = (c_1 + c_2)e^{2x} + c_3e^{8x}$, $y_2 = -c_2e^{2x} + c_3e^{8x}$, $y_3 = -c_1e^{2x} + c_3e^{8x}$

7. $y = c_1e^{3x} + c_2e^{-2x}$

8. $y = c_1e^x + c_2e^{2x} + c_3e^{3x}$

EXERCISE SET 9.2 [page 434]

1. **(a)** $\begin{bmatrix} 0 & -1 \\ -1 & 0 \end{bmatrix}$ **(b)** $\begin{bmatrix} -1 & 0 \\ 0 & -1 \end{bmatrix}$ **(c)** $\begin{bmatrix} 1 & 0 \\ 0 & 0 \end{bmatrix}$ **(d)** $\begin{bmatrix} 0 & 0 \\ 0 & 1 \end{bmatrix}$

2. **(a)** $(-1, -2)$ **(b)** $(-2, -1)$ **(c)** $(2, 0)$ **(d)** $(0, 1)$

3. **(a)** $\begin{bmatrix} 1 & 0 & 0 \\ 0 & 1 & 0 \\ 0 & 0 & -1 \end{bmatrix}$ **(b)** $\begin{bmatrix} 1 & 0 & 0 \\ 0 & -1 & 0 \\ 0 & 0 & 1 \end{bmatrix}$ **(c)** $\begin{bmatrix} -1 & 0 & 0 \\ 0 & 1 & 0 \\ 0 & 0 & 1 \end{bmatrix}$

4. **(a)** $(1, 1, -1)$ **(b)** $(1, -1, 1)$ **(c)** $(-1, 1, 1)$

5. **(a)** $\begin{bmatrix} 0 & -1 & 0 \\ 1 & 0 & 0 \\ 0 & 0 & 1 \end{bmatrix}$ **(b)** $\begin{bmatrix} 1 & 0 & 0 \\ 0 & 0 & -1 \\ 0 & 1 & 0 \end{bmatrix}$ **(c)** $\begin{bmatrix} 0 & 0 & 1 \\ 0 & 1 & 0 \\ -1 & 0 & 0 \end{bmatrix}$

6. **(a)** Rectangle with vertices at $(0, 0), (1, 0), (1, -2), (0, -2)$
   (b) Rectangle with vertices at $(0, 0), (-1, 0), (-1, 2), (0, 2)$
   (c) Rectangle with vertices at $(0, 0), (1, 0), (1, \frac{1}{2}), (0, \frac{1}{2})$
   (d) Square with vertices at $(0, 0), (2, 0), (2, 2), (0, 2)$
   (e) Parallelogram with vertices at $(0, 0), (1, 0), (7, 2), (6, 2)$
   (f) Parallelogram with vertices at $(0, 0), (1, -2), (1, 0), (0, 2)$

7. Rectangle with vertices at $(0, 0), (-3, 0), (0, 1), (-3, 1)$

8. **(a)** $\begin{bmatrix} \frac{1}{\sqrt{2}} & -\frac{1}{\sqrt{2}} \\ \frac{1}{\sqrt{2}} & \frac{1}{\sqrt{2}} \end{bmatrix}$ **(b)** $\begin{bmatrix} 0 & -1 \\ 1 & 0 \end{bmatrix}$ **(c)** $\begin{bmatrix} -1 & 0 \\ 0 & -1 \end{bmatrix}$ **(d)** $\begin{bmatrix} 0 & 1 \\ -1 & 0 \end{bmatrix}$ **(e)** $\begin{bmatrix} \frac{\sqrt{3}}{2} & \frac{1}{2} \\ -\frac{1}{2} & \frac{\sqrt{3}}{2} \end{bmatrix}$

9. **(a)** $\begin{bmatrix} 1 & 0 \\ 4 & 1 \end{bmatrix}$ **(b)** $\begin{bmatrix} 1 & -2 \\ 0 & 1 \end{bmatrix}$

10. **(a)** $\begin{bmatrix} 1 & 0 \\ 0 & \frac{1}{3} \end{bmatrix}$ **(b)** $\begin{bmatrix} 6 & 0 \\ 0 & 1 \end{bmatrix}$

11. **(a)** Expansion by a factor of 3 in the x-direction
    (b) Expansion by a factor of -5 in the y-direction
    (c) Shear by a factor of 4 in the x-direction

12. **(a)** $\begin{bmatrix} 2 & 0 \\ 0 & 1 \end{bmatrix}\begin{bmatrix} 3 & 0 \\ 0 & 1 \end{bmatrix}$; expansion in the y-direction by a factor of 3, then expansion in the x-direction by a factor of 2

(b) $\begin{bmatrix} 1 & 0 \\ 2 & 1 \end{bmatrix}\begin{bmatrix} 1 & 4 \\ 0 & 1 \end{bmatrix}$; shear in the x-direction by a factor of 4, then shear in the y- direction by a factor of 2

(c) $\begin{bmatrix} 0 & 1 \\ 1 & 0 \end{bmatrix}\begin{bmatrix} 4 & 0 \\ 0 & 1 \end{bmatrix}\begin{bmatrix} 1 & 0 \\ 0 & -2 \end{bmatrix}$; expansion in the y-direction by a factor of -2, then expansion in the x-direction by a factor of 4, then reflection about $y = x$

(d) $\begin{bmatrix} 1 & 0 \\ 4 & 1 \end{bmatrix}\begin{bmatrix} 1 & 0 \\ 1 & 18 \end{bmatrix}\begin{bmatrix} 1 & -3 \\ 0 & 1 \end{bmatrix}$; shear in the x-direction by a factor of -3, then expansion in the y-direction by a factor of 18, then shear in the y-direction by a factor of 4

13. **(a)** $\begin{bmatrix} \frac{1}{2} & 0 \\ 0 & 5 \end{bmatrix}$ **(b)** $\begin{bmatrix} 1 & 0 \\ 2 & 5 \end{bmatrix}$ **(c)** $\begin{bmatrix} 0 & -1 \\ -1 & 0 \end{bmatrix}$ 14. **(a)** $\begin{bmatrix} 0 & 1 \\ -5 & 0 \end{bmatrix}$ **(b)** $\frac{1}{2}\begin{bmatrix} \sqrt{3} & -1 \\ -6\sqrt{3}+3 & 6+3\sqrt{3} \end{bmatrix}$

16. $16y - 11x - 3 = 0$ 17. **(a)** $y = \frac{2}{7}x$ **(b)** $y = x$ **(c)** $y = \frac{1}{2}x$ **(d)** $y = -2x$

18. $\begin{bmatrix} 1 & -2 \\ 0 & 1 \end{bmatrix}$ 19. **(b)** No. A is not invertible.

22. **(a)** $\begin{bmatrix} 1 & 0 & 0 \\ 0 & 0 & 1 \\ 0 & 1 & 0 \end{bmatrix}$ **(b)** $\begin{bmatrix} 0 & 0 & 1 \\ 0 & 1 & 0 \\ 1 & 0 & 0 \end{bmatrix}$ **(c)** $\begin{bmatrix} 0 & 1 & 0 \\ 1 & 0 & 0 \\ 0 & 0 & 1 \end{bmatrix}$

23. **(a)** $\begin{bmatrix} 1 & 0 & k \\ 0 & 1 & k \\ 0 & 0 & 1 \end{bmatrix}$ **(b)** xz-direction: $\begin{bmatrix} 1 & k & 0 \\ 0 & 1 & 0 \\ 0 & k & 1 \end{bmatrix}$; yz-direction: $\begin{bmatrix} 1 & 0 & 0 \\ k & 1 & 0 \\ k & 0 & 1 \end{bmatrix}$

24. **(a)** $\lambda_1 = 1$: $\begin{bmatrix} 1 \\ 0 \end{bmatrix}$; $\lambda_2 = -1$: $\begin{bmatrix} 0 \\ 1 \end{bmatrix}$ **(b)** $\lambda_1 = 1$: $\begin{bmatrix} 0 \\ 1 \end{bmatrix}$; $\lambda_2 = -1$: $\begin{bmatrix} 1 \\ 0 \end{bmatrix}$

(c) $\lambda_1 = 1$: $\begin{bmatrix} 1 \\ 1 \end{bmatrix}$; $\lambda_2 = -1$: $\begin{bmatrix} -1 \\ 1 \end{bmatrix}$ **(d)** $\lambda = 1$: $\begin{bmatrix} 1 \\ 0 \end{bmatrix}$

(e) $\lambda = 1$: $\begin{bmatrix} 0 \\ 1 \end{bmatrix}$ **(f)** (θ an odd integer multiple of π) $\lambda = -1$: $\begin{bmatrix} 1 \\ 0 \end{bmatrix}$

(θ an even integer with multiple of π) $\lambda = 1$: $\begin{bmatrix} 1 \\ 0 \end{bmatrix}, \begin{bmatrix} 0 \\ 1 \end{bmatrix}$

(θ not an integer multiple of π) no real eigenvalues

EXERCISE SET 9.3 (page 441)

1. $y = \frac{1}{2} + \frac{7}{2}x$ 2. $y = \frac{2}{3} + \frac{1}{6}x$ 3. $y = 2 + 5x - 3x^2$ 4. $y = -5 + 3x - 4x^2 + 2x^3$

8. $y = 4 - .2x + .2x^2$; if $x = 12$, then $y = 30.4$ ($30.4 thousand)

EXERCISE SET 9.4 (page 447)

1. **(a)** $(1+\pi) - 2\sin x - \sin 2x$ **(b)** $(1+\pi) - 2\left[\sin x + \frac{\sin 2x}{2} + \frac{\sin 3x}{3} + \cdots + \frac{\sin nx}{n}\right]$

2. **(a)** $\frac{4}{3}\pi^2 + 4\cos x + \cos 2x + \frac{4}{9}\cos 3x - 4\pi \sin x - 2\pi \sin 2x - \frac{4\pi}{3}\sin 3x$

(b) $\frac{4}{3}\pi^2 + 4\sum_{k=1}^{n} \frac{\cos kx}{k^2} - 4\pi \sum_{k=1}^{n} \frac{\sin kx}{k}$

EXERCISE SET 9.9 [page 484]

1. $x_1 = 2, x_2 = 1$ 2. $x_1 = -2, x_2 = 1, x_3 = -3$ 3. $x_1 = 3, x_2 = -1$

4. $x_1 = 4, x_2 = -1$ 5. $x_1 = -1, x_2 = 1, x_3 = 0$ 6. $x_1 = 1, x_2 = -2, x_3 = 1$

7. $x_1 = -1, x_2 = 1, x_3 = 0$ 8. $x_1 = -1, x_2 = 1, x_3 = 1$

9. $x_1 = -3, x_2 = 1, x_3 = 2, x_4 = 1$ 10. $x_1 = 2, x_2 = -1, x_3 = 0, x_4 = 0$

11. **(a)** $A = LU = \begin{bmatrix} 2 & 0 & 0 \\ -2 & 1 & 0 \\ 2 & 1 & 1 \end{bmatrix}\begin{bmatrix} 1 & \frac{1}{2} & -\frac{1}{2} \\ 0 & 0 & 1 \\ 0 & 0 & 0 \end{bmatrix}$

(b) $A = L_1DU = \begin{bmatrix} 1 & 0 & 0 \\ -1 & 1 & 0 \\ 1 & 1 & 1 \end{bmatrix}\begin{bmatrix} 2 & 0 & 0 \\ 0 & 1 & 0 \\ 0 & 0 & 1 \end{bmatrix}\begin{bmatrix} 1 & \frac{1}{2} & -\frac{1}{2} \\ 0 & 0 & 1 \\ 0 & 0 & 0 \end{bmatrix}$

(c) $A = L_2U_2 = \begin{bmatrix} 1 & 0 & 0 \\ -1 & 1 & 0 \\ 1 & 1 & 1 \end{bmatrix}\begin{bmatrix} 2 & 1 & -1 \\ 0 & 0 & 1 \\ 0 & 0 & 0 \end{bmatrix}$

13. **(b)** $\begin{bmatrix} a & b \\ b & d \end{bmatrix} = \begin{bmatrix} 1 & 0 \\ \frac{c}{a} & 1 \end{bmatrix}\begin{bmatrix} a & b \\ 0 & \frac{ad - bc}{a} \end{bmatrix}$

14. Additions: $\frac{n^3}{3} + \frac{n^2}{2} - \frac{5n}{6}$; multiplications: $\frac{n^3}{3} + n^2 - \frac{n}{3}$

18. $A = PLU = \begin{bmatrix} 1 & 0 & 0 \\ 0 & 0 & 1 \\ 0 & 1 & 0 \end{bmatrix}\begin{bmatrix} 3 & 0 & 0 \\ 0 & 2 & 0 \\ 3 & 0 & 1 \end{bmatrix}\begin{bmatrix} 1 & -\frac{1}{3} & 0 \\ 0 & 1 & \frac{1}{2} \\ 0 & 0 & 1 \end{bmatrix}$

EXERCISE SET 10.1 [page 492]

1. (a–d)

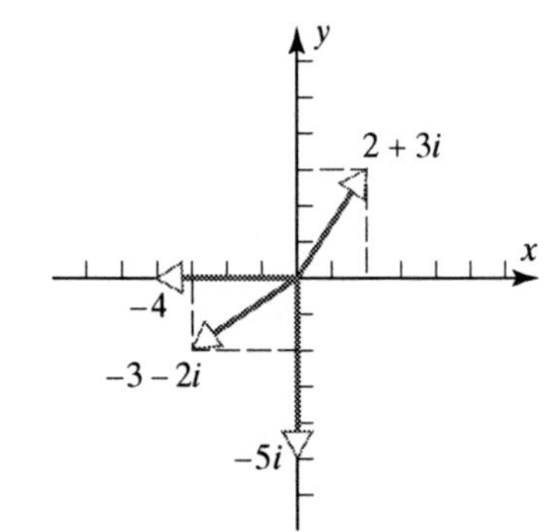

2. **(a)** (2, 3) **(b)** (−4, 0) **(c)** (−3, −2) **(d)** (0, −5)

3. **(a)** $x = -2, y = -3$ **(b)** $x = 2, y = 1$

4. **(a)** $5 + 3i$ **(b)** $-3 - 7i$ **(c)** $4 - 8i$ **(d)** $-4 - 5i$ **(e)** $19 + 14i$ **(f)** $-\frac{11}{2} - \frac{17}{2}i$

5. **(a)** $2 + 3i$ **(b)** $-1 - 2i$ **(c)** $-2 + 9i$

6.

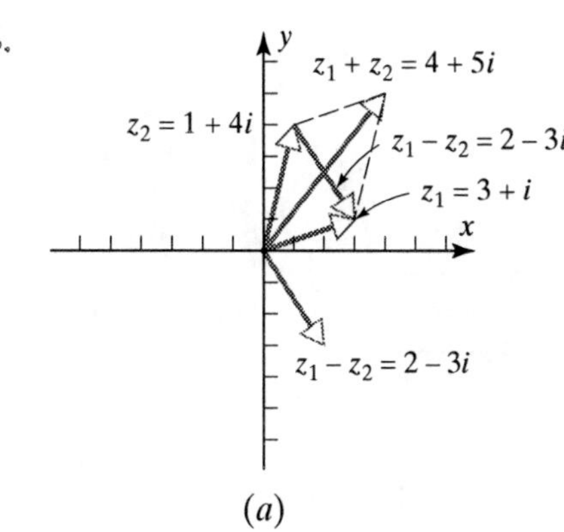

(a)

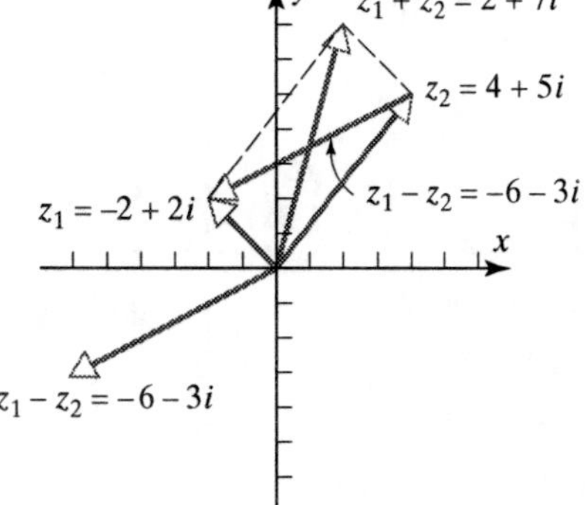

(b)

7.

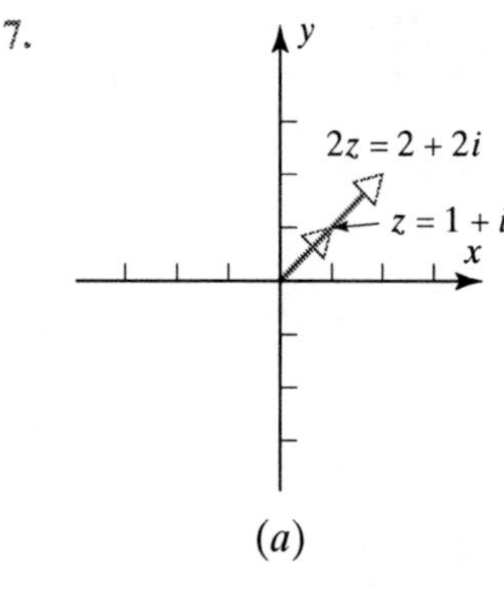

(a)

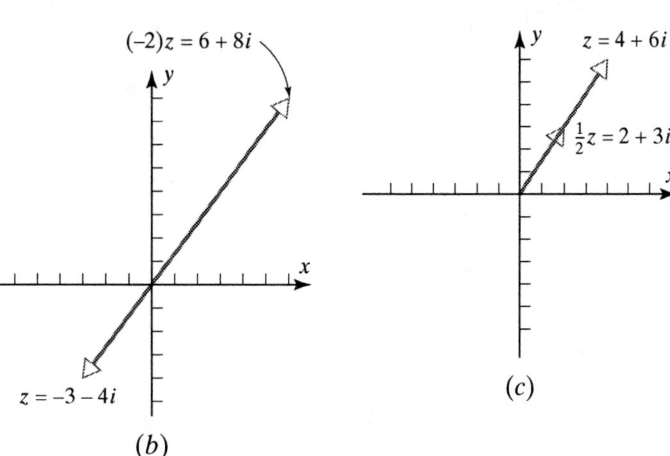

(b) (c)

8. **(a)** $k_1 = -5, k_2 = 3$ **(b)** $k_1 = 3, k_2 = 1$

9. **(a)** $z_1z_2 = 3 + 3i, z_1^2 = -9, z_2^2 = -2i$
(b) $z_1z_2 = 26, z_1^2 = -20 + 48i, z_2^2 = -5 - 12i$
(c) $z_1z_2 = \frac{11}{3} - i, z_1^2 = \frac{4}{9}(-3 + 4i), z_2^2 = -6 - \frac{5}{2}i$

10. **(a)** $9 - 8i$ **(b)** $-63 + 16i$ **(c)** $-32 - 24i$ **(d)** $22 + 19i$ 11. $76 - 88i$ 12. $26 - 18i$

13. $-26 + 18i$ 14. $-1 - 11i$ 15. $-\frac{63}{16} + i$ 16. $(2 + \sqrt{2}) + i(1 - \sqrt{2})$ 17. 0 18. $-24i$

19. **(a)** $\begin{bmatrix} 1+6i & -3+7i \\ 3+8i & 3+12i \end{bmatrix}$ **(b)** $\begin{bmatrix} 3-2i & 6+5i \\ 3-5i & 13+3i \end{bmatrix}$ **(c)** $\begin{bmatrix} 3+3i & 2+5i \\ 9-5i & 13-2i \end{bmatrix}$ **(d)** $\begin{bmatrix} 9+i & 12+2i \\ 18-2i & 13+i \end{bmatrix}$

20. **(a)** $\begin{bmatrix} 13+13i & -8+12i & -33-22i \\ 1+i & 0 & i \\ 7+9i & -6+6i & -16-16i \end{bmatrix}$ **(b)** $\begin{bmatrix} 6+2i & -11+19i \\ -1+6i & -9-5i \end{bmatrix}$ **(c)** $\begin{bmatrix} 6i & 1+i \\ -6-i & 5-9i \end{bmatrix}$ **(d)** $\begin{bmatrix} 22-7i & 2+10i \\ -5-4i & 6-8i \\ 9-i & -1-i \end{bmatrix}$

22. **(a)** $z = -1 \pm i$ **(b)** $z = \dfrac{1}{2} \pm \dfrac{\sqrt{3}}{2}i$ 23. **(b)** i

EXERCISE SET 10.2 [page 498]

1. **(a)** $2 - 7i$ **(b)** $-3 + 5i$ **(c)** $-5i$ **(d)** i **(e)** -9 **(f)** 0

2. **(a)** 1 **(b)** 7 **(c)** 5 **(d)** $\sqrt{2}$ **(e)** 8 **(f)** 0

4. **(a)** $-\frac{17}{25} - \frac{19}{25}i$ **(b)** $\frac{23}{25} + \frac{11}{25}i$ **(c)** $\frac{23}{25} - \frac{11}{25}i$ **(d)** $-\frac{17}{25} + \frac{19}{25}i$ **(e)** $\frac{1}{5} - i$ **(f)** $\dfrac{\sqrt{26}}{5}$

5. **(a)** $-i$ **(b)** $\frac{1}{26} + \frac{5}{26}i$ **(c)** $7i$ 6. **(a)** $\frac{6}{5} + \frac{2}{5}i$ **(b)** $-\frac{2}{5} + \frac{1}{5}i$ **(c)** $\frac{3}{5} + \frac{11}{5}i$ **(d)** $\frac{3}{5} + \frac{1}{5}i$

7. $\frac{1}{2} + \frac{1}{2}i$ 8. $\frac{2}{5} + \frac{1}{5}i$ 9. $-\frac{7}{625} - \frac{24}{625}i$ 10. $-\frac{11}{25} + \frac{2}{25}i$ 11. $\dfrac{1-\sqrt{3}}{4} + \dfrac{1+\sqrt{3}}{4}i$

12. $-\frac{1}{26}-\frac{5}{26}i$ **13.** $-\frac{1}{10}+\frac{1}{10}i$ **14.** $-\frac{2}{5}$ **15.** **(a)** $-1-2i$ **(b)** $-\frac{3}{25}-\frac{4}{25}i$

17. **(a)**

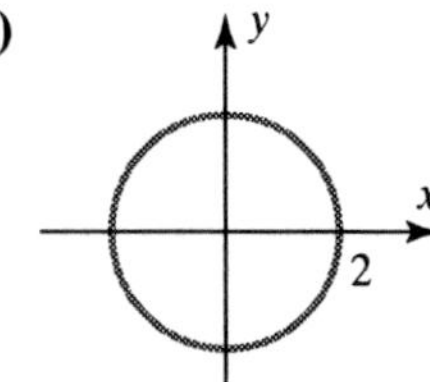

(b)

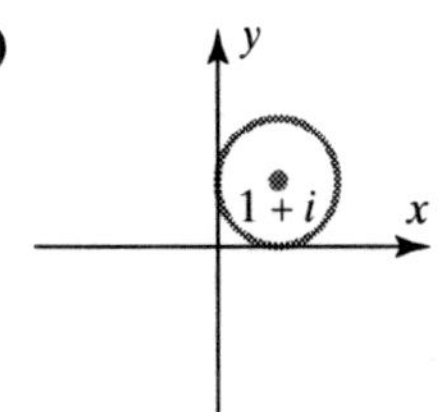

(c)

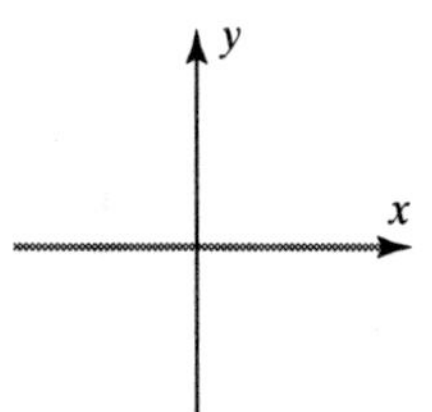

(d)

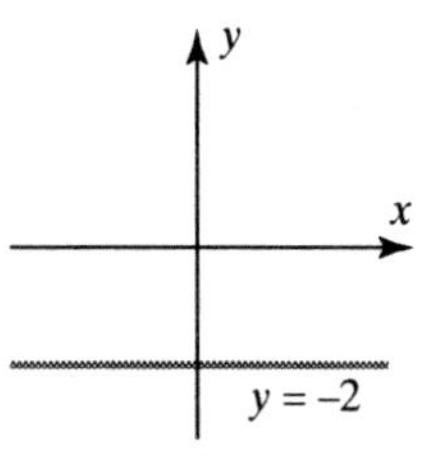

18. **(a)**

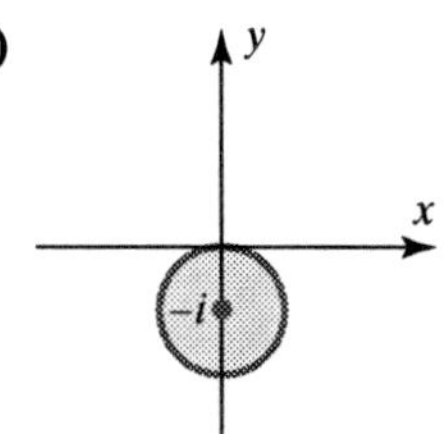

(b)

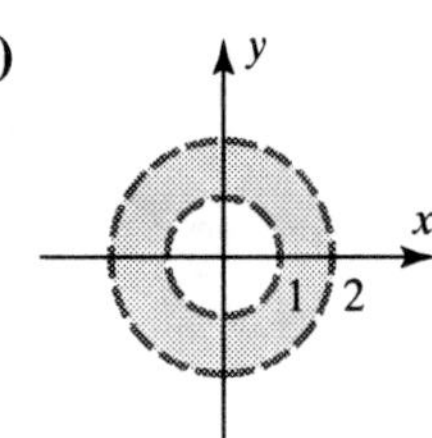

(c)

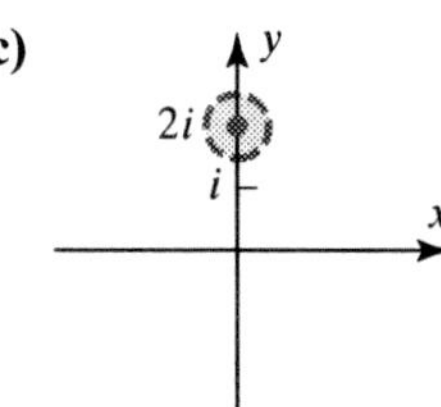

(d)

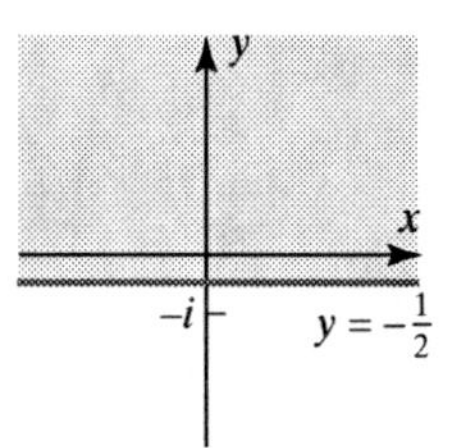

19. **(a)** $-y$ **(b)** $-x$ **(c)** y **(d)** x **20.** **(b)** $-i$ **23.** **(a)** $\dfrac{x_1x_2+y_1y_2}{x_2^2+y_2^2}$ **(b)** $\dfrac{x_2y_1-x_1y_2}{x_2^2+y_2^2}$

27. **(c)** Yes, if $z \neq 0$. **28.** $x_1=i, x_2=-i$ **29.** $x_1=1+i, x_2=1-i$

30. $x_1=\frac{1}{2}+i, x_2=2, x_3=\frac{1}{2}-i$ **31.** $x_1=i, x_2=0, x_3=-i$

32. $x_1=-(1+i)t, x_2=t$ **33.** $x_1=(1+i)t, x_2=2t$

34. $x_1=-(1-i)t, x_2=-it, x_3=t$ **35.** **(a)** $\begin{bmatrix} i & 2 \\ -1 & i \end{bmatrix}$ **(b)** $\begin{bmatrix} 0 & 1 \\ -i & 2i \end{bmatrix}$

39. **(a)** $\begin{bmatrix} -i & -2-2i & -1+i \\ 1 & 2 & -i \\ i & i & 1 \end{bmatrix}$ **(b)** $\begin{bmatrix} 1+i & -i & 1 \\ -7+6i & 5-i & 1+4i \\ 1+2i & -i & 1 \end{bmatrix}$

41. **(a)** $|z_1-z_2|=\sqrt{(a_1-a_2)^2+(b_1-b_2)^2}$
(b) $|(8+8i)-12|^2=80=40+40=|12-(6+2i)|^2+|(6+2i)-(8+8i)|^2$

EXERCISE SET 10.3 [page 505]

1. **(a)** 0 **(b)** $\pi/2$ **(c)** $-\pi/2$ **(d)** $\pi/4$ **(e)** $2\pi/3$ **(f)** $-\pi/4$

2. **(a)** $5\pi/3$ **(b)** $-\pi/3$ **(c)** $5\pi/3$

3. **(a)** $2\left[\cos\left(\frac{\pi}{2}\right)+i\sin\left(\frac{\pi}{2}\right)\right]$ **(b)** $4[\cos\pi+i\sin\pi]$ **(c)** $5\sqrt{2}\left[\cos\left(\frac{\pi}{4}\right)+i\sin\left(\frac{\pi}{4}\right)\right]$
(d) $12\left[\cos\left(\frac{2\pi}{3}\right)+i\sin\left(\frac{2\pi}{3}\right)\right]$ **(e)** $3\sqrt{2}\left[\cos\left(-\frac{3\pi}{4}\right)+i\sin\left(-\frac{3\pi}{4}\right)\right]$ **(f)** $4\left[\cos\left(-\frac{\pi}{6}\right)+i\sin\left(-\frac{\pi}{6}\right)\right]$

4. **(a)** $6\left[\cos\left(\frac{5\pi}{12}\right)+i\sin\left(\frac{5\pi}{12}\right)\right]$ **(b)** $\frac{2}{3}\left[\cos\left(\frac{\pi}{12}\right)+i\sin\left(\frac{\pi}{12}\right)\right]$
(c) $\frac{3}{2}\left[\cos\left(-\frac{\pi}{12}\right)+i\sin\left(-\frac{\pi}{12}\right)\right]$ **(d)** $\frac{32}{9}\left[\cos\left(\frac{11\pi}{12}\right)+i\sin\left(\frac{11\pi}{12}\right)\right]$

5. 1 **6.** **(a)** -64 **(b)** $-i$ **(c)** $-64\sqrt{3}-64i$ **(d)** $-\dfrac{1+\sqrt{3}i}{2048}$

7. **(a)**

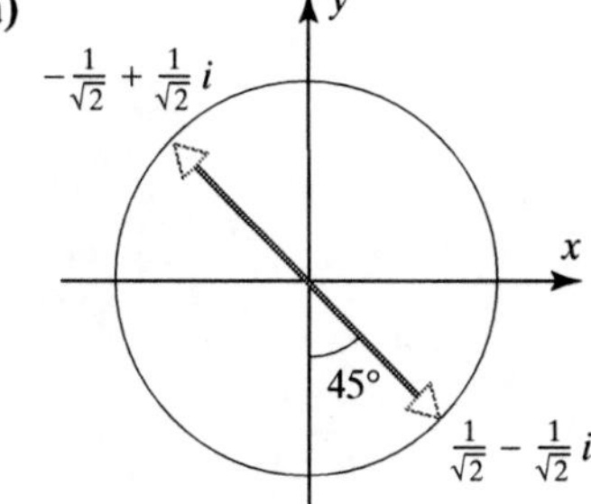

(b)

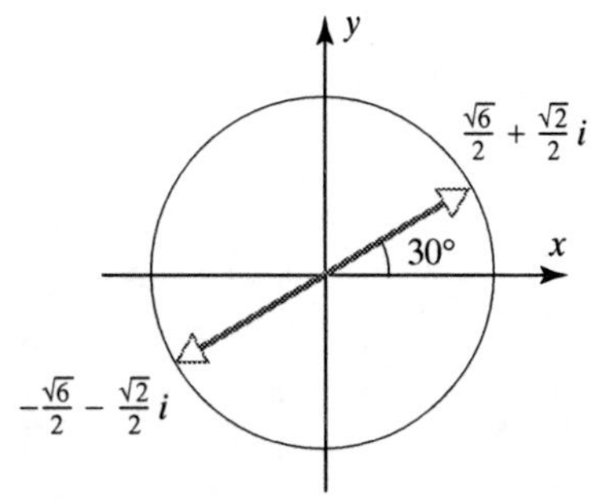

(c)

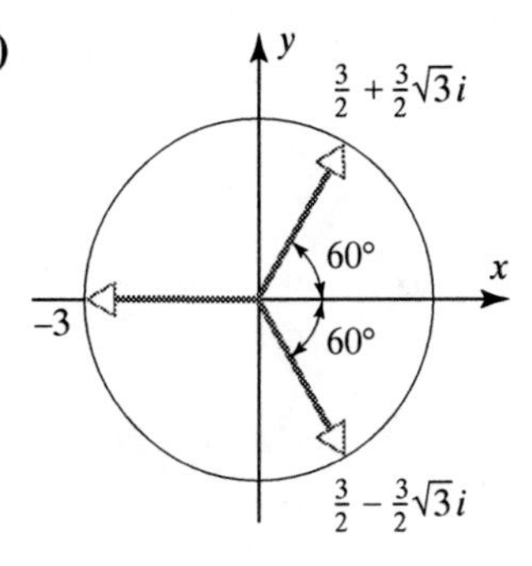

(d)

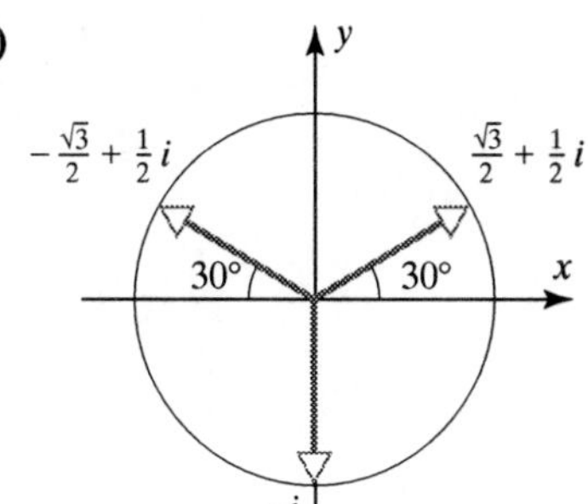

(e)

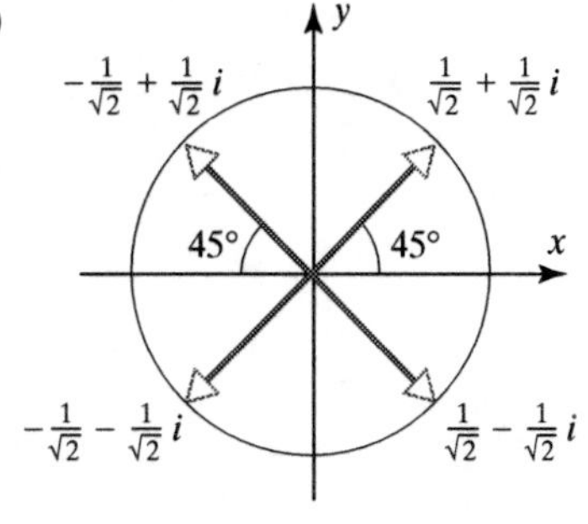

(f)

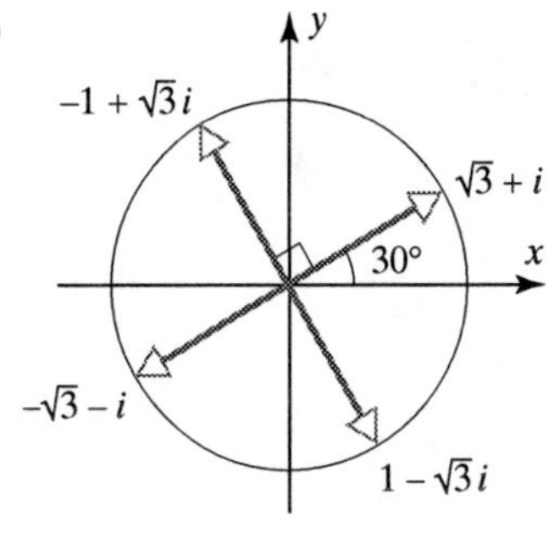

8.

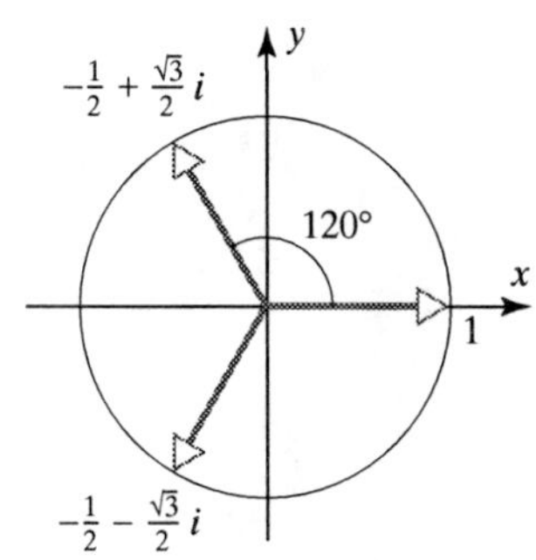

9.

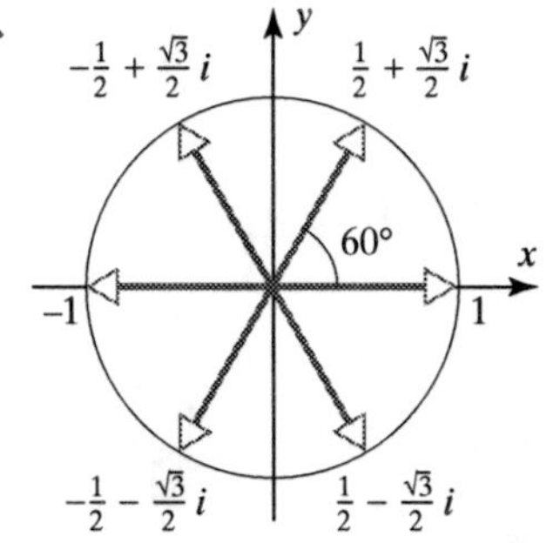

10. $\sqrt[4]{2}\left[\cos\left(\frac{\pi}{8}\right)+i\sin\left(\frac{\pi}{8}\right)\right]$, $\sqrt[4]{2}\left[\cos\left(\frac{9\pi}{8}\right)+i\sin\left(\frac{9\pi}{8}\right)\right]$ 11. $\pm 2, \pm 2i$

12. The roots are $\pm(2^{1/4}+2^{1/4}i)$, $\pm(2^{1/4}-2^{1/4}i)$ and the factorization is $z^4+8=(z^2-2^{5/4}z+2^{3/2})\cdot(z^2+2^{5/4}z+2^{3/2})$.

13. Rotates z clockwise by 90° 14. **(a)** 16 **(b)** $\frac{i}{4^9}$

15. **(a)** $\text{Re}(z)=-3$, $\text{Im}(z)=0$ **(b)** $\text{Re}(z)=-3$, $\text{Im}(z)=0$
(c) $\text{Re}(z)=0$, $\text{Im}(z)=-\sqrt{2}$ **(d)** $\text{Re}(z)=-3$, $\text{Im}(z)=0$

20. $\cos 2\theta=\cos^2\theta-\sin^2\theta$, $\sin 2\theta = 2\sin\theta\cos\theta$
$\cos 3\theta=\cos^3\theta-3\sin^2\theta\cos\theta$, $\sin 3\theta=3\sin\theta\cos^2\theta-\sin^3\theta$

EXERCISE SET 10.4 [page 510]

1. **(a)** $(3i, -i, -2-i, 4)$ **(b)** $(3+2i, -1-2i, -3+5i, -i)$
(c) $(-1-2i, 2i, 2-i, -1)$ **(d)** $(-3+9i, 3-3i, -3-6i, 12+3i)$
(e) $(-3+2i, 3, -3-3i, i)$ **(f)** $(-1-5i, 3i, 4, -5)$

2. $(2+i, 0, -3+i, -4i)$ 3. $c_1=-2-i, c_2=0, c_3=2-i$

5. **(a)** $\sqrt{2}$ **(b)** $2\sqrt{3}$ **(c)** $\sqrt{10}$ **(d)** $\sqrt{37}$

6. **(a)** $\sqrt{43}$ **(b)** $\sqrt{10}+\sqrt{29}$ **(c)** $\sqrt{10}+\sqrt{10}i$ **(d)** $\sqrt{699}$ **(e)** $\left(\frac{1+i}{\sqrt{6}}, \frac{2i}{\sqrt{6}}, 0\right)$ **(f)** 1

8. All k such that $|k|=\frac{1}{5}$ 9. **(a)** 3 **(b)** $2-27i$ **(c)** $-5-10i$

Index

D

E

F

G

H

I